FUNDAMENTAL SYMMETRIES IN NUCLEI AND PARTICLES

FUNDAMENTAL SYMMETRIES IN NUCLEI AND PARTICLES

Symposium honoring Felix Boehm's 65th Birthday

CALTECH, PASADENA, September 7-9, 1989

edited by

H. Henrikson & P. Vogel

California Institute of Technology

USA

World Scientific

Singapore • New Jersey • London • Hong Kong

Published by

World Scientific Publishing Co. Pte. Ltd.
P O Box 128, Farrer Road, Singapore 9128
USA office: 687 Hartwell Street, Teaneck, NJ 07666
UK office: 73 Lynton Mead, Totteridge, London N20 8DH

Cover photograph: On the cover the bronze sculpture of Tenjin, the tenth century philosopher, sitting atop his water buffalo, casts a contemplative eye on the doors of Dabney Hall, the meeting room of the Caltech Symposium. "Tenjin" is the posthumous title bestowed on Sugawara Michizane, the Japanese scholar of Chinese literature. A major festival in his honor is held annually in Osaka. In local shrines throughout Japan students buy "Tenjin" amulets for luck during the period of school entrance examinations.

FUNDAMENTAL SYMMETRIES IN NUCLEI & PARTICLES

ISBN 981-02-0060-9

Printed in Singapore by JBW Printers & Binders Pte. Ltd.

DEDICATION

It is with pleasure that we dedicate these proceedings to Felix Boehm on his 65th birthday. We cannot hope to adequately chronicle his distinguished career in this space, we can only try to outline a few of the highlights.

Felix Hans Boehm was born on June 9, 1924 in Basel, Switzerland, the second of five sons of Hans Boehm, publisher and printer in his hometown. He graduated from the Humanistic Gymnasium in Basel in 1943; his early scientific interests centered on electro physiology (study of the hypothalamus, which entailed the building of sensitive amplifiers and recorders). But he turned to physics later that year when he enrolled at the ETH Zürich. Among his teachers were Paul Scherrer (his thesis advisor) and Wolfgang Pauli; some of his fellow students at ETH were K. Alder, J. P. Blaser, H. Frauenfelder, P. Marmier, R. Steffen, and V. Telegdi, with all of whom he has established an enduring friendship. His thesis on the "Excitation Functions and Cross Sections of the (p,n) Reactions" was submitted in 1951. Even though ETH was then a leading center of nuclear physics (and physics in general), the European laboratories had not recovered from the ravages of WWII, and many young physicists from Europe went to the US for their "Wanderjahre", some making their home here. Felix was among them; he sailed aboard the SS Liberte in March 1952 for a postdoctoral appointment with Professor C. S. Wu at Columbia University. The following year he headed further west, to Caltech in Pasadena, to join the laboratory of Jesse W. M. DuMond. He became an assistant professor in 1958, a professor in 1961, and now is the Valentine Professor. It is his scientific work during the 36 years at Caltech that inspired many of the talks in these proceedings.

The first few years at Caltech Felix participated in measurements of nuclear gamma ray energies using the recently completed DuMond bent crystal spectrometer. These data helped to provide the experimental basis of the Bohr-Mottelson model. However, Felix's interests soon shifted to "fundamental" problems with the advent of the golden era of beta decay: the discovery of parity violation. Felix (with A. Wapstra) have pioneered the $\beta-\gamma$ circular polarization correlation method which has helped to establish the now accepted V$-$A model.

Soon thereafter, the current $\times$ current model of weak interaction has been proposed (in particular, by Feynman and Gell-Mann here at Caltech). To verify its validity (by demonstrating that the hadron current $\times$ hadron current term also violates parity conservation) Felix started a series of experiments on parity nonconservation in nuclear γ transitions, and was among the first (with E. Kankeleit, E. Lipson, and H. Vanderleeden) to observe these small effects. Several talks in these proceedings are direct descendants of those early efforts.

In the late sixties and the early seventies Felix utilized the bent crystal γ spectrometer which, as so many instruments in his laboratory was designed and built by Herb Henrikson, to measure (with E. Hatch, P. Lee, R. Chesler, A. Hahn, and K. C. Wang) the isotope and chemical shifts of atomic K x−rays, a technique that permitted the determination of the variation of the nuclear mean square charge radius in a model independent way. He also became involved in the then emerging meson factories, in particular at LAMPF, and succeeded in using the crystal spectrometer in precision measurements of mesic x−ray energies. Here, at Caltech, Felix (with N. K. Cheung and J. Gimlett) used the nuclear orientation techniques in the search for the time−reversal noninvariance in nuclear γ transitions. They achieved a remarkable sensitivity and demonstrated that the final state effects can be measured in some cases. An echo of these experiments can be felt in several talks in these proceedings.

In the last decade Felix's main activity was devoted to the study of neutrino properties. Following frequent discussions with M. Gell-Mann as well as with H. Fritzsch and P. Minkowski (who were then at Caltech) Felix decided that it would be worthwhile to begin a search for neutrino oscillations using a nuclear reactor as a source. He found an enthusiastic collaborator in R. Mössbauer; the first detector was built at Caltech (A. Hahn, H. Kwon, and J.-L. Vuilleumier were the other members of the Caltech contingent) and an experiment (in fact, a first dedicated neutrino oscillation experiment anywhere) was performed at ILL in Grenoble during 1979−80. The neutrino detector was then moved to the power reactor at Gösgen, and a series of three measurements at varying distances was carried out. These experiments, perhaps unfortunately, did not lead to the discovery of neutrino oscillations; they have established, however, limits on the oscillation parameters that withstood many challenges, and could serve as a "standard" for this kind of study. More recently, emphasis in the work of Felix and his team has shifted to double beta decay, a process uniquely suited to search for the neutrino Majorana mass. In collaboration with Neuchatel (J.-L. Vuilleumier) and PSI (K. Gabathuler) a large Ge detector system was installed in the St. Gotthard Tunnel (attended by P. Fisher and H. Wong), and a large Xe TPC is about to be launched. Once more, many groups are trying to discover neutrinoless decay. It is not clear that anybody will succeed, but it is clear that Felix and his group will perform a thorough and definitive experiment.

These few remarks and the accompanying volume, we hope, gives a hint of the quality that characterizes the accomplishments in physics of Felix Boehm. All that remains is to transmit the best wishes of all the participants of the symposium: "Happy birthday, Felix".

FOREWORD

The Symposium on Fundamental Symmetries in Nuclei and Particles was held September 7–9 on the Caltech campus in Pasadena, California. The symposium was organized with two goals in mind. We wished first to bring together physicists whose interests in symmetries span the boundaries of nuclear, atomic, and particle physics and who, consequently, do not meet often. Secondly, we desired an occasion to celebrate the 65th birthday of Felix Boehm, who is one of the leading practitioners in this field. Among the participants were many friends, colleagues, and students of Felix; for them it was a reunion.

The program of invited talks was selected with the help of an advisory committee consisting of E. Adelberger, J. deBoer, B.Kayser, P. Rosen, B. Stech, J.-L. Vuilleumier, and A. Winther. An effort was made to have a very balanced meeting. Thus we had nine experimentalists and nine theorists among the speakers. Eight came from overseas and ten from the United States. In addition, the seven orally presented contributed talks, the summaries of which are also included in these proceedings, are status reports on some of the more exciting experiments that might open windows to new physics. The invited talks cover a wide range of topics and taken together offer an overview of state-of-the-art efforts to test the "standard model" and to find new and unexpected phenomena leading to deeper understanding, generalization, or reformulation of physical laws. The experimental techniques were not neglected, because no matter how compelling theoretical ideas are, their final proof comes when an ingenious experiment is performed and verified. Moreover, good experiments are distinguished by their originality and elegance.

It is, perhaps, worthwhile to briefly summarize the topics covered in these proceedings, and to emphasize the connection of their seemingly disconnected subjects, particularly since the symposium did not have a summary talk.

This was not a meeting on the "big physics" where the highest priority is finding the Higgs boson and top quark, the particles that must exist if the "standard model" is valid at any scale. Instead, what concerned us was nonaccelerator physics, or the accelerator physics on a more modest scale. Among the outstanding issues there are the mass, mixing, and charge conjugation properties of the neutrino, and the role and proper description of CP non-invariance. These two issues were the central ones at the symposium and were treated in several interconnected talks.

The status of searches for the neutrino mass and mixing was reviewed by Robertson, Rosen, Vuilleumier, Wu, and in the contribution of Wong. Probably the most important experimental result reported by Robertson was an upper limit of 13.4 eV (95% C.L.) for the electron antineutrino mass obtained in the tritium decay experiment at Los Alamos. This is in a clear contradiction to the longstanding claim of the ITEP Moscow group to have seen a mass larger than 20 eV in the same tritium beta decay. This finding, combined with the mass limit extracted from the neutrino signal of supernova SN1987A means that all indications of a nonvanishing neutrino mass are contradicted by other experiments of similar sensitivity.

The solar neutrino problem may be an indication of nonstandard neutrino behavior. Among the various possible explanations, the elegant Mikheyev-Smirnov-Wolfenstein mechanism of neutrino oscillations, discussed by Rosen, looks particularly promising. However, it remains to be seen whether the apparent anticorrelation of the neutrino flux with sunspot number is real (or whether it is just a statistical fluctuation), and whether the gallium experiments, which are beginning to take data, will help in deciding among the various possible explanations of the present situation. Reactor neutrino oscillation experiments, covered by Vuilleumier, are sensitive to small values of δm^2, the neutrino mass−squared difference. By scaling up the size of the detectors and moving them further away from the reactors, one can begin to explore the region of δm^2 between 10^{-2} and 10^{-4} eV^2, totally unexplored so far. Nuclear double beta decay is the most sensitive test of the charge conjugation properties of neutrinos. (Are they Dirac or Majorana particles?) To put that problem in perspective, Wu talked about the early experiments, where all the present problems (elimination of the background from cosmic rays and radioactive contamination, etc.) were first encountered. Wong's contribution dealt with a novel double beta decay search using a ^{136}Xe filled Time Projection Chamber.

The other large complex of problems, addressed in the talks by Bowman, Weidenmüller, Kayser, Mampe, and Wolfenstein was the search for CP and T noninvariance. So far, the effects of CP noninvariance has been seen only in K-meson decays, the theoretical description of which remains largely uncertain (see the general discussion of Wolfenstein). It is quite likely that similar and maybe larger effects will be present in the decay of B-mesons; Kayser showed how CP violation might manifest itself in these decays. However, at some level CP or T noninvariance will be present also in very different systems, for example in nuclear moments and transitions. This is a totally open problem and thus very different avenues are being explored. The classical test is the search for the neutron dipole moment, which has now reached a superb sensitivity of $\approx 10^{-25}$ ecm (see the talk of Mampe). A very different approach to T nonconservation tests was introduced by Weidenmüller. It has been realized recently that compound-nucleus reactions are potentially sensitive to violation of time-reversal invariance. Weidenmüller described the elegant theoretical basis of this approach, and possible strategies in looking for an enhancement of symmetry violation. A realization of some of these ideas was put forward by Bowman, who reported on an experiment in progress on neutron–nucleus scattering. (Experiments of a similar kind found very large enhancements in parity noninvariance effects.) In his talk, Herczeg showed how studies of nuclear beta decay can help in pinpointing the origin (or find another manifestations of) CP or T noninvariance.

The history of parity nonconservation in weak interactions goes back more than 30 years. However, there are still many subtle effects worth pursuing, as discussed by Khriplovich, Telegdi, and in the contribution of Ramsay. Nuclear anapole moments, described by Kriplovich, are electromagnetic moments that arise in a nucleus when parity is not conserved. These moments could be observed in atomic parity testing experiments of the kind discussed by Telegdi. Parity violation in atomic transitions has been clearly demonstrated in these elegant experiments, which are now bringing quantitative information about the electron–quark interaction. Ultimately, the atomic parity violation experiments should give us an accurate value of the Weinberg angle $\sin^2\theta_W$. (We heard from Zacek how $\sin^2\theta_W$ is determined from muon neutrino–electron scattering in the CHARM II experiment.)

Parity violation in nuclear systems, where weak and strong interactions compete, was also observed a long time ago, in particular in the pioneering experiments by Felix Boehm here at Caltech. The new generation of experiments on the two nucleon system was the topic of the contribution by Ramsay. At 220 MeV the parity violation in p–p scattering would be dominated by ρ meson exchange and thus its effect could be isolated. That example illustrates that in processes involving hadrons one has to worry about the strong interactions, the most difficult (so far) to treat theoretically. Various aspects of this problem were discussed by Henley and Stech. Henley talked about an approximate symmetry, the charge symmetry of nuclear physics. This symmetry is broken by (primarily) the mass difference between u and d quarks. Stech described another aspect of QCD: the interplay of the weak interaction and QCD in studies of heavy meson decays. He stressed that such decays can help us to understand the mysteries of nonperturbative QCD. A different aspect of strong interactions was discussed by Yamazaki who showed how one can, at least in principle, study systems with real pions in deeply bound orbits.

An important new feature of present day physics is the close connection between "laboratory physics" and problems that traditionally belong to astrophysics and cosmology. To stress this feature we heard the talk of Steigman on primordial nucleosynthesis and limits that it provides on the number of neutrino families (a number somewhat less than four). One of the important factors for the treatment of primordial nucleosynthesis is the accuracy with which the neutron lifetime is known; an improved result was reported by Mampe. Another astrophysical connection is the observation of high-energy particles emitted by point astronomical sources. Nagle in his contribution described one of the new detectors (the CYGNUS array at Los Alamos) used for that purpose.

Of course, as stressed already, sophisticated experimental techniques are necessary if progress in physics is to be achieved. Low temperature detectors, discussed by Zehnder, signal a possible opening of new avenues of research. The talk by Adelberger and the contribution of Kankeleit dealt with experimental techniques and unexpected phenomena on very different scales. Adelberger reported on his search for a "fifth" force of range >1 meter, which would couple to baryon or lepton number (or to their combination). Despite initial claims (by others), Adelberger and his collaborators could not find any evidence for such an interaction in a series of exceedingly accurate experiments. Kankeleit summarized the work on positron creation in collision of very heavy ions. Several groups observed positron "lines" in these experiments, sometimes in coincidence with similar electron lines. There is no satisfactory theoretical explanation of this phenomenon. Kankeleit stressed that the very existence of these lines has not been firmly established.

Thus we see in all of these talk that despite the truly impressive progress in experimental techniques we have not seen signs of "new" physics yet. The standard model seem to work surprisingly well and we have to try harder or be more clever to see deviations which will point us toward a more general and more satisfactory all encompassing theory. I hope that the symposium helped to clarify some of the problems encountered along this difficult route.

The symposium was not just a serious discourse covering important physics experiments and theories. There was also a time for social events, in particular the banquet on Friday evening at the Athenaeum. There we all enjoyed the witty after dinner speech of Jorrit deBoer, and toasted Felix's health (with California wines only, of course) encouraged by the remarks of Caltech provost Barclay Kamb and of Felix's old friends Chien Shiung Wu and Valentine Telegdi.

Finally, there are many thanks to be given: To the Division of Physics, Mathematics, and Astronomy of Caltech for welcome financial assistance, to Caltech President Thomas Everhart for his opening remarks, to the founding member of the Physics 34 group and coeditor of these proceedings Herb Henrikson for his help at all stages of the organization of the symposium, and last but far from least to the conference secretary Elsa Garcia, who devoted considerable effort so that all participants were housed and fed and could enjoy their stay in Pasadena.

Petr Vogel
Pasadena, November 1989.

Table of Contents

FUNDAMENTAL SYMMETRIES IN NUCLEI AND PARTICLES

SEARCH FOR PARITY AND TIME REVERSAL SYMMETRY VIOLATION USING POLARIZED NEUTRON-NUCLEUS INTERACTIONS

J. D. Bowman, C. D. Bowman, J. Knudson, S. Penttila, S. J. Seestrom, J. J. Szymanski, V. W. Yuan
Los Alamos National Laboratory, Los Alamos, NM

C. R. Gould, D. G. Haase, G. E. Mitchell
North Carolina State University, Raleigh, NC and
Triangle Universities Nuclear Laboratory, Durham, NC

N. R. Roberson
Duke University, Durham, NC and
Triangle Universities Nuclear Laboratory, Durham, NC

P. P. J. Delheij
TRIUMF, Vancouver, British Columbia

H. Postma
University of Technology, Delft, The Netherlands

E. D. Davis
Max Planck Institute, Heidelberg, FRG and
University of Arizona, Tucson, AZ

ABSTRACT:

I describe a program of experiments to search for parity (P) violation and time-reversal (T) violation using the scattering of 1 to 10,000 eV neutrons from nuclei. Transmission measurements will be used to determine polarization observables in the forward elastic scattering amplitude. The large fluxes of pulsed neutrons available at the Los Alamos Proton Storage Ring make possible the detection of $\sim 10^{12}$ interactions in a few months of running. Properties of the compound nucleus, e.g. the small spacings of levels, provide large enhancement factors for T-odd as well as P-odd observables. The T-violation experiments are detailed-balance tests in which the cross section for a process and its time reverse are compared. I characterize and give explicit strategies for dealing with various sources of systematic error. The first round of measurements will study parity violation in the compound nucleus at the 10^{-7} level. These results will test statistical models that relate symmetry violations in the nucleon-nucleon force to symmetry violations in the compound nucleus. The secound round of measurements will test time-reversal violation in the P-even nucleon-nucleon (N-N) force at the few times 10^{-6} level.

INTRODUCTION

The pioneering work of Alfimenkov [Alf 83] et al. demonstrated that for suitable observables symmetry violations in the effective N-N interaction are enhanced in the many -body nuclear system. The existence of P-violation in the nucleon-nucleon interaction is well established and is of order 10^{-7}. Considerable experimental and theoretical effort has been devoted to determining a set of P-violating nucleon-meson couplings. The size of the compound nucleus (CN) effects can be related to the close spacing of the compound nuclear levels, and the small value of the ratio of p-wave to s-wave neutron penetrabilities. The quantitative connection between the experimentally measured parity violation and the effective nucleon-nucleon interaction is made via a stochastic description of the CN system. In a first round of experiments we will carry out a series of transmission measurements of parity violation in the neutron-nucleus interaction in heavy nuclei, to extract the variance of the CN parity-violating matrix elements, and to connect the measured variance with the P-odd N-N interaction. We will greatly extend both the range of neutron energies and the sensitivities of neutron transmission P tests. Success at this stage, the ability to quantitatively relate parity violation at the CN and N-N levels, will justify a parallel analysis procedure for T-violation experiments.

Both P-odd, T-odd and P-even, T-odd Cartesian invariants can be constructed from neutron spin, target spin, and neutron momentum. Large enhancements are also expected for tests of T-violation via neutron-nucleus transmission measurements based on these invariants. A P-even T-test with a polarized beam and aligned target is technically feasible at present, while the P-odd T-test will not be possible to design until more experience with systematic errors, material properties, and nuclear structure has been accumulated. It is possible to improve the sensitivity of our knowledge of the P-even, T-odd part of the N-N interaction by two to three orders of magnitude. In a second round of experiments we will perform a P-even, T-test. We will evaluate the prospects for a P-odd T-test during the course of this work.

Systematic errors are crucial for all symmetry test experiments, and hence special attention must be paid to possible instrumental asymmetries, as well as more fundamental sources of systematic error. An issue particular to T-tests is that of simulation of a false T-odd effect by the sequential action of two T-even interactions. These false effects are smaller in the P-even test than in the P-odd test.

SCIENTIFIC MOTIVATION

The neutron-nucleus scattering process provides excellent opportunities for testing parity and time reversal with unprecedented sensitivity. The high density of CN levels is one obvious reason. Figure 1 shows the total cross section for n + ^{165}Ho scattering as a function of neutron energy [McL 88]. The large peaks correspond to compound nuclear states in ^{166}Ho. They are formed at an excitation energy of ~ 6 MeV and their spacing is a few eV. By contrast, level spacings close to the ground state are typically ~ 10 - 10^2 keV. Consequently, admixtures in CN states due to a symmetry-violating interaction in the nuclear Hamiltonian are much larger than for levels near the ground state.

For the present transmission experiments with polarized neutrons, the relevant dynamical variables are neutron spin **s**, neutron momentum **k**, and target spin **I**, and the correlations of interest are:

P-odd, T-even	$\mathbf{s}\cdot\mathbf{k}$	Helicity
P-odd, T-odd	$\mathbf{s}\cdot\mathbf{I}\times\mathbf{k}$	Three fold correlation (TC)
P-even, T-odd	$(\mathbf{s}\cdot\mathbf{I}\times\mathbf{k})(\mathbf{I}.\mathbf{k})$	Five fold correlation (FC)

The helicity parity-violation test is carried out with a polarized beam and an unpolarized target. Figure 2 shows schematically how such measurements are performed. A polarizing filter and spin flipper prepare a neutron beam in either of two helicity states. The neutron beam is transmitted through a thick unpolarized sample and the asymmetry,

asymmetry = $(\sigma(+)-\sigma(-)) / (\sigma(+)+\sigma(-))$ where
$\sigma(+/-)$ = Resonance cross section for $\mathbf{s}\cdot\mathbf{k}$ = +/-,

is determined from the count rates in the two helicity states. Remarkably large parity-violating asymmetries have been observed for a number of nuclei, the most spectacular case being the 0.734 eV p-wave resonance in ^{139}La for which the effect is between 7 and 10% [Alf 83, Mas 87].

The existence of P-violation in the nucleon-nucleon (N-N) interaction is well established. The characteristic size is ~10^{-7}; considerable effort has been devoted to extracting the strengths of the P-violating meson-exchange couplings from p-p scattering data [Kis 87] and from γ decays in light nuclei [Ade 85]. The interest in CN resonance studies lies not so much in showing that P-violation occurs, but rather in clarifying how the underlying symmetry violation in the effective N-N interaction manifests itself in the complicated many-body nuclear system. This general issue is of considerable current interest in quantum chaos [Boh 88, Mit 88] and has been studied in recent years by the techniques of Random Matrix Theory (RMT) [Fre 88a,b]. A symmetry-violation is characterized by the variance of the CN matrix elements of the symmetry-violating

Hamiltonian. The goals of the P-violation studies are to determine the corresponding variance, either via on-resonance measurements or measurements of energy averages, and then to validate the theoretical basis for relating it to the parameter of fundamental interest, namely α_P, the strength of the P-odd effective N-N interaction relative to the P-even part. The quantity α_P is expected to be of order 10^{-7}. Successful testing of this conversion from the neutron-nucleus regime to the N-N regime would support the use of a similar procedure for T-violation.

The P-odd T-test must be performed with a polarized target and the P-even T-test with an aligned target. In both cases the neutron beam is polarized perpendicular to the (**I**,**k**) plane. Time-reversal tests are in principle more complicated than the parity measurements because of problems relating to precession of the neutron spin in the target caused by the real **s**·**I** interaction (pseudomagnetic field) and the **s**·**B** interaction (magnetic field) followed by an absorptive **s**·**k** interaction; a false T-violating asymmetry results. Consider an experiment to search for the threefold or fivefold correlation patterned after the approach used to study parity violation. Such an experiment is illustrated in Figure 3. A polarized or aligned target is substituted for the unpolarized target in Figure 2. When **s** is reversed, the T-violating correlations change sign and the spin dependence of the transmission is measured. However, a transmission difference does not necessarily imply T-violation, because changing the state of the polarizer does not correspond to time-reversing the apparatus.

Stodolsky [Sto 86] showed how to deal with this by adding an analyzer between the polarized target and the detector. The dependence of the transmission on the neutron spin is measured by simultaneously reversing the states of the polarizer and the analyzer. The two states of the apparatus are now time reversal partners of one another and detailed balance is tested. Later Kabir [Kab 88] showed that a comparison of a polarizer-target experiment with a target-analyzer experiment done with a single polarizer/analyzer was fully equivalent to the above polarizer-target-analyzer detailed balance test.

Although both the P-odd TC and the P-even FC experiments are conceptually feasible, the TC suffers from a particularly severe practical problem, namely the precession of the neutron spin in the magnetic and pseudomagnetic fields of the polarized target. Polarized targets with sufficiently small fields, such that the neutron will not precess many times (thereby averaging out the T-violating signal), are not available at present, although the KEK group is exploring the development of a frozen-spin Lanthanum target. By contrast, since this problem is absent for an aligned target (no external magnetic field and a small pseudomagnetic field arising only from the weak interaction), the FC experiment is far simpler to perform. In fact it is possible to identify T-violation not only from the polarizer-target-analyzer or the polarizer/analyzer comparison methods mentioned earlier, but also from measurements which determine the target thickness dependence of the asymmetry signal. Simulation of T-odd effects by the sequential action of two T-even interactions depends quadratically on the target thickness, while true T-violation depends linearly on target thickness. This is the approach we propose to use. As part of the P-violation study we will search for good candidate resonances for the P-odd, T-odd test. The experience gained with systematic errors in the P study and P-even T-test will also be applicable to any future P-odd T-test experiments.

There is no experimental evidence for P-even T-violation in the strong force. We follow French [Fre 88b] and adopt the strength α_T of the P-even T-odd part of the effective N-N interaction appropriate to the shell model relative to the P-even T-even part as the relevant parameter. Analysis of the fluctuations in level spacings in heavy nuclei [Fre 88a] implies $\alpha_T \leq (1\text{-}2) \times 10^{-3}$ (99% confidence) [Fre 88b]. An analysis [Fre 87] of the most precise detailed-balance measurement to date [Bla 83] yields $\alpha_T \leq (2\text{-}3) \times 10^{-3}$ (99% confidence). The planned FC measurements represent a marked improvement: they will achieve a sensitivity of $\alpha_T \sim 10^{-6}$. Detailed constraints on α_T implied by experimental limits on the neutron electric dipole moment and on P-even T-odd correlations associated with nuclear γ-transitions are unknown. It has been customary to relate these measurements to parameters of an effective one-body interaction whereas α_T is defined in terms of an effective two-body interaction. It is, however, unlikely that these limits are more restrictive.

THEORETICAL BACKGROUND

The remarkably large effects observed in the helicity dependence of neutron total cross sections can be explained plausibly with simple models and conventional nuclear physics. These enhancement mechanisms make possible detailed study of parity violation in the nucleon-nucleus interaction and hold the promise of providing similar enhancements for time-reversal invariance violation tests. The parity-violating N-N force is reasonably well known. We will measure the size of parity-violating matrix elements in the CN regime. These data will test statistical theories which relate the two regimes. Having validated the theoretical procedures we will be able to relate upper limits, or a non-zero effect, for time-reversal violation in the CN to the corresponding N-N force.

We take as the quantities to be determined the spreading widths of the symmetry-violating Hamiltonians. For a P-even, T-odd interaction H^T, this is

$$\Gamma^{TV} = 2\pi \langle (H^{T}_{\mu\mu'})^2 \rangle / D^J$$

where $\langle\ \rangle$ represents a local energy-averaging operation, the matrix element $H^{T}_{\mu\mu'}$ is between CN levels μ and μ' of the same spin J and parity, and D^J is the average spacing of levels of this spin and parity. Similarly, the spreading width for a P-odd interaction H^P is

$$\Gamma^{PV} = 2\pi \langle (H^{P}_{\mu\mu'})^2 \rangle / D^J ,$$

where now μ and μ' are of the same spin J but opposite parity. Unlike the more basic variances $\langle (H^{T}_{\mu\mu'})^2 \rangle$ and $\langle (H^{P}_{\mu\mu'})^2 \rangle$, spreading widths have the important advantage of being essentially independent of the mass number and excitation energy of the CN. Furthermore, the relation of Γ^{TV} to α_T has been established in a variety of studies - viz. according to [Fre 88b],

$$\Gamma^{TV} \sim 2\pi\,(10^5\ \text{eV})\,(\alpha_T)^2 .$$

Assuming the generality of the reduction developed in [Fre 88b], Γ^{PV} is likewise related to α_P by

$$\Gamma^{PV} \sim 2\pi\,(10^5\ \text{eV})\,(\alpha_P)^2 .$$

Given its relation to the weak N-N potential, α_P is assumed to be of order 10^{-7}; α_T is only known to be less than ~10^{-3}.

Below, we estimate the sensitivity of the planned experiments to α_T and α_P. We consider both conventional on-resonance measurements of the type performed by the Dubna group, as well as measurements of energy averages (the rationale for which is discussed in [Dav 89a]). We now summarize the theory essential to these estimates.

On-resonance tests of P and T

The parity-violation experiments consist of measurements of the helicity dependence of the neutron total cross section at epithermal p-wave neutron resonances in heavy nuclei. Following [Alf 83], we consider the analyzing power Q defined by $\sigma^{\pm} = \sigma_p(1 \pm Q)$ where $\sigma^{+(-)}$ is the cross section for +(-) helicity neutrons and σ_p is the unpolarized p-wave neutron resonance cross section. For a neutron of energy E, wave number $k = 1/\lambda\!\!\!^{-}$:

$$\sigma_p = \pi \lambda\!\!\!^{-2} \; g_J \frac{\Gamma_p \Gamma_p^n}{(E_p - E)^2 + (\Gamma_p/2)^2},$$

where g_J is the spin statistical factor and E_p, Γ_p, and Γ_p^n are, respectively, the resonance energy, total width and neutron width.

In the simplest case, only interference between the p-wave resonance and the nearest s-wave resonance is relevant [Bun 83]. A detailed description is given by Vanhoy et al. [Van 88]. The small, (real) parity-violating piece H^P of the Hamiltonian acts in the internal region of the compound nuclear system, and first-order perturbation theory is used to obtain the perturbed wavefunctions and the perturbed reduced width amplitudes γ'. S-matrix elements are calculated with standard R-matrix theory, utilizing the "level" representation. The net result is that one has parity non-conserving T-matrix elements of the form

$$T_{c'c} = i\, e^{i(\phi_{c'} + \phi_c)} \frac{g_{sc'}\, V\, g_{pc}}{(E_s - E - i\Gamma_s/2)(E_p - E - i\Gamma_p/2)} \quad .$$

Here, all quantities refer to the unperturbed basis: $V = H^P_{\mu\mu'}$ is the matrix element of H^P between the two compound-nuclear states, and $g^2_{\mu c}$, E_μ and Γ_μ are the neutron width in channel c, the resonance energy, and total width of the resonance μ, respectively.

The expression for the observable $\delta\sigma = \sigma^+ - \sigma^-$ is for the example of target spin I = 1/2 and compound nucleus spin J = 1,

$$\delta\sigma(E) \approx \sigma_p \left(\frac{V}{\Delta E_s}\right) x_j \sqrt{\Gamma_s^n / \Gamma_p^n}$$

where $\Delta E_s = E_s - E$, and x_j is the entrance-channel mixing parameter. Note that, quite generally, $\delta\sigma(E)$ is linear in V. For order of magnitude estimates, the analyzing power at the peak of the p-wave resonance is thus of the form

$$Q \sim \frac{V}{|E_s - E_p|} \sqrt{\Gamma_s^n / \Gamma_p^n} \; .$$

The ratio $V / |E_s - E_p|$ experiences what is usually termed a dynamical enhancement [Sha 68, Sus 82], referring to an enhancement in its value over what is expected on the basis of single-particle estimates. Clearly CN energy spacings are much smaller than typical single-particle energy spacings. On the other hand, the CN matrix element V is also smaller than the single-particle matrix elements of H^P. To estimate these effects, it is customary to resort to the "black-nucleus" model for the CN, in which it is assumed that a CN wavefunction is a superposition of "simple" components of random relative phase. If the number of "simple" components is N, then V is estimated to be reduced by a factor ~ $1/\sqrt{N}$, while the CN energy spacings are reduced by a factor ~ 1/N. This leads to a dynamical enhancement of order $\sqrt{N}$. The specific value depends on the level density -- typical estimates lead to an enhancement of ~10^3. This enhancement manifests itself in the earlier expressions for Γ^{TV} and Γ^{PV} in terms of α_T and α_P as the factor of 10^5 eV.

Additional enhancement is associated with the factor of $\sqrt{\Gamma^n_s/\Gamma^n_p}$. The penetrabilities present in Γ^n_s and Γ^n_p imply that, at low neutron energies, $\sqrt{\Gamma^n_s/\Gamma^n_p}$ is proportional to 1/kR, where R is the nuclear radius. Thus, for epithermal neutrons in heavy nuclei, $\sqrt{\Gamma^n_s/\Gamma^n_p}$ is of order 10^2. Various aspects of enhancement mechanisms have been discussed in detail by Bunakov [Bun 87, Bun 88a,b].

Parity violation can also be studied via the complementary method of using a polarized target rather than a polarized beam. The method is described by Haase et al. [Haa 88] and the parallel between **s**· **k** and **I**· **k** experiments has been explored by Vanhoy et al. [Van 89].

Treatment of a T-violating Hamiltonian H^T with matrix elements $W = H^T_{\mu\mu'}$ parallels the parity analysis. The essential difference is that W is purely imaginary. For the P-even, T-odd case, enhancements are expected at a single p-wave resonance or for two interfering p-wave resonances [Bun 88b]. Detailed results are summarized by Gould et al. [Gou 89]. The cross section difference for two interfering J = 3 p-wave resonances (1 and 2) for an I = 7/2 target (e.g. ^{165}Ho) is, at the center of resonance 2,

$$\delta\sigma \approx \hat{t}_{20}(\tfrac{7}{2})\,(x_j^{(1)} y_j^{(2)} - x_j^{(2)} y_j^{(1)})\, 4\pi \lambda\!\!\!^{-2}\, \frac{W}{E_1 - E_2}\, \frac{\sqrt{\Gamma^n_{p1}\Gamma^n_{p2}}}{\Gamma_{p2}}$$

where $\hat{t}_{20}(7/2) \sim 0.7$ is the statistical tensor describing the alignment of the target, and x_j and y_j are entrance-channel mixing parameters. Again neglecting non-essential factors of order unity, the corresponding T-violating analyzing power is

$$Q \sim \frac{W}{|E_{p1} - E_{p2}|}\,.$$

The argument employed with parity violation indicates that this test of T-invariance also benefits from dynamical enhancement [Bun 88b]. Significantly lower limits on T-violation in the parity-even nucleon-nucleon interaction are thus attainable.

Tests of P and T with Energy Averages

Use of energy averages in the resolved resonance regime has recently been considered in connection with both the FC [Dav 88] and parity violation [Dav 89]. The analysis applies to energy averages constructed from an arbitrary set of measurements of $\delta\sigma$ and not just on-resonance values, which is an advantage in the event p-wave resonances are not identified. The sensitivity is not as great as for optimal pairs of resonances. Nevertheless, such observables do benefit from enhancements. More importantly, the interpretation of a result is, by design, simpler than for the case of a single on-resonance measurement. No detailed spectroscopic information must be known and the energy averages of interest are automatically proportional to the symmetry-violating spreading widths.

The relevant energy average is the variance of the cross-section asymmetry:

$$\langle (\delta\sigma)^2 \rangle \equiv \langle (\sigma^+ - \sigma^-)^2 \rangle,$$

where for parity, $\sigma^{+(-)}$ denote total cross sections for +(-) helicity neutrons, and for time-reversal, $\sigma^{+(-)}$ denote total cross sections for neutrons polarized parallel (anti-parallel) to $\mathbf{k} \times \mathbf{I}$. $\langle (\delta\sigma)^2 \rangle$ is simply related to the variance of the symmetry-violating part of the S-matrix.

Exact evaluation of $\langle |S_J^{(PV)}|^2 \rangle$ and $\langle |S_{J\pi}^{(TV)}|^2 \rangle$ using Random Matrix Theory is feasible. Introducing the sensitivity parameters β_J and $\gamma_{J\pi}$ such that

$$\lambda\!\!\!^{-2} \langle |S_J^{(PV)}|^2 \rangle^{1/2} = \beta_J \, \alpha_P$$

$$\lambda\!\!\!^{-2} \langle |S_{J\pi}^{(TV)}|^2 \rangle^{1/2} = \gamma_{J\pi} \, \alpha_T ,$$

one finds that, on specializing to s- and p-wave neutrons ($E_n < 10$ keV),

$$\beta_J \sim 78 \, (s_0 \, s_1)^{1/2} \frac{1}{(t^{J\pi} \, t^{J(-\pi)})^{1/2}} \frac{A^{1/3}}{(D^J)^{1/2}} \text{ barns}$$

and

$$\gamma_{J\pi} \sim 9.3 \times 10^{-3} \, s_1 \left(\frac{2.1 - \ln t^{J\pi}}{t^{J\pi}} \right)^{1/2} \left(\frac{E}{D^J} \right)^{1/2} A^{2/3} \text{ barns},$$

where

$$t^{J\pi} = 2\pi \frac{\Gamma^{J\pi}}{D^J} ,$$

and A is the target mass number, E is the neutron energy (in eV), D^J ($\Gamma^{J\pi}$) is the average spacing (width) in eV of J^π CN resonances and $s_l = 2 \times 10^4 S_l$, where S_l is the standard *l*th wave neutron strength function tabulated in [Mug 81].

Fluctuations are suppressed if the experimental resolution ΔE becomes greater than $\Gamma^{J\pi}$. Closed-form expressions are obtained if the experimental resolution is assumed Lorentzian with FWHM = ΔE. Defining

$$\bar{t}^{J\pi} \equiv t^{J\pi} \left(1 + \frac{\Delta E}{\Gamma^{J\pi}}\right) ,$$

the new sensitivity parameters, $\bar{\gamma}_{J\pi}$ and $\bar{\beta}_J$, are obtained from $\gamma_{J\pi}$ and β_J, respectively, by the simple step of replacing $t^{J\pi}$ with $\bar{t}^{J\pi}$.

The quantities $\bar{\gamma}_{J\pi}$ and $\bar{\beta}_J$ are plotted in Figure 4 for the J = 3 states in holmium ($D^{J\pi} \sim D_0 \sim 4.6$ eV, $\Gamma^{J\pi} \sim 0.08$ eV, $s_0 \sim 1.8$, $s_1 \sim 0.6$). The same values are obtained for J = 4 states. The $\bar{\gamma}_{J\pi}$ from s- and d-wave interference is of similar magnitude. The experimental resolution ΔE is that of the new ^{10}B counter (see Section 8) at a flight path of 60 m.

SENSITIVITY OF P AND T TESTS AT LANSCE

The LANSCE facility for testing P and T violation is described by Bowman et al. [Bow 88]. It consists of an LMN polarized proton target, a magnetic spin-flipper system, and neutron detector stations at 11, 25 and 60 m. To estimate the sensitivity we need the neutron count rate as a function of energy. We consider an 18" diameter by 4-cm thick ^{10}B detector with efficiency ~ 0.65, situated at 60 m. The effective solid angle is 2.7×10^{-6} sr, less than the geometric solid angle because we view the 13 x 13 cm^2 neutron moderator through the 10 cm^2 aperture of the LMN polarizer. At present, the polarizer transmission is 0.14 and the neutron polarization $P_n \sim 0.5$. We choose target parameters appropriate to the aligned holmium target presently under construction at TUNL - i.e. we take a sample of thickness n = 0.06 atoms/barn and transmission to be 0.55, corresponding to the measurements for which $\sigma_{tot} \approx \sigma_{pot} \sim 10$ barns. For a 50 μA beam, the neutron count rate is

$$N \sim \frac{2.0 \times 10^{12}}{E} (0.65)(0.14)(0.55)(2.7 \times 10^{-6})$$

$$\sim \frac{2.7 \times 10^{5}}{E} \quad \text{neutrons/eV-sec.}$$

The new butanol polarizer will have a 50 cm^2 aperture and a higher transmission (0.30 compared to 0.14). Hence the count rate compared to LMN would be a factor of 10 larger. The polarization of the neutrons will also be higher, typically 0.80.

On-Resonance Tests

Both P- and T-violation measurements are performed on p-wave resonances. We restrict our analysis to neutron energies for which $\Gamma_p^n << \Gamma_p$ and for which the experimental resolution is $< \Gamma_p$. Typically this means $E \leq 100$ eV, and $\sigma_p << \sigma_{pot}$. The optimum target thickness is ~ 2 interaction lengths, which, with $\sigma_{tot} \sim \sigma_{pot}$, is readily attainable for P-experiments but not necessarily for T-experiments, which require cryogenic targets. The peak cross-section difference determines the analyzing power Q:

$$\delta\sigma = 2 P_n \sigma_p (\max) Q.$$

We recall that, for order of magnitude estimates,

$$Q \sim \frac{W}{|E_{p1} - E_{p2}|} \quad \text{for T-violation,}$$

$$Q \sim \frac{V}{|E_s - E_p|} \sqrt{\frac{\Gamma_s^n}{\Gamma_p^n}} \quad \text{for P-violation,}$$

and

$$\sigma_p\,(\text{max}) \sim \frac{4\pi}{k^2}\,\frac{\Gamma_p^n}{\Gamma_p}$$

neglecting Doppler broadening.

The statistical error in the analyzing power Q in t seconds is

$$\delta Q \sim 1 / (P_n\, n\, \sigma_p\,(\text{max}) \sqrt{N\, \Delta E_B\, t}\,)$$

where ΔE_B is the width of the energy bin used. We determine Q from counts under the p-wave resonance, meaning we take $\Delta E_B \sim \Gamma_p$, the width of the p-wave resonance ($\Gamma_p \sim \Gamma_\gamma$ for non-fissile nuclei).

From the spreading-width relation given earlier, we have

$$W \sim x\ 10^3 \left(\frac{D_1}{10}\right)^{1/2} \alpha_T$$

where D_1 is the average p wave spacing in eV and x is a Gaussian-distributed variable of unit variance. The uncertainty in α_T (corresponding to x = 1 above) is therefore

$$\delta\,\alpha_T \sim \frac{k^2}{2\,P_n n}\ \frac{\left(|E_{p1} - E_{p2}|\right)^{1/2}}{D_1^{1/2}} \cdot \frac{\Gamma_p^{1/2}}{\Gamma_p^n} \cdot \frac{1}{(N\,t)^{1/2}}\ .$$

Similarly,

$$\delta\,\alpha_p \sim \frac{k^2}{2\,P_n n} \cdot \frac{(|E_s - E_p|)^{1/2}}{(D_0 D_1)^{1/4}} \cdot \frac{\Gamma_p^{1/2}}{(\Gamma_p^n\,\Gamma_s^n)^{1/2}} \cdot \frac{1}{(N\,t)^{1/2}}$$

where D_0 is the average s-wave spacing.

These spectroscopic parameters above ($|E_s - E_p|$, Γ_s^n , Γ_p^n, etc.) will vary randomly with choice of resonance (or pair of resonances), and much of the potential of our experiments rests on the fact that there may be fortuitous combinations of these parameters giving rise to unusually large enhancements. Nevertheless, for the purpose of discussion, we shall estimate "reasonable" values for bounds on $\delta\alpha_T$ and $\delta\alpha_P$ by assuming that these parameters have their average values. Thus the energy differences between resonances are set equal to the average s-wave spacing: $|E_s - E_p| \sim |E_{p1} - E_{p2}| \sim D_0 \sim D_1$. Neutron widths are evaluated from the strength functions S_l and average spacings D_l (taken from [Mug 81]) by

$$\Gamma_s^n \sim S_0 D_0 E^{1/2} \text{ and } \Gamma_p^n \sim S_1 D_1 E^{3/2} \ (0.088 A^{2/3} \times 10^{-6}) .$$

For convenience we tabulate the values used:

	D_0 (eV)	$\overline{\Gamma}_p$ (eV)	S_0 (x10^4)	S_1 (x10^4)
^{139}La	208	0.055	0.78	0.4
^{165}Ho	4.6	0.077	1.8	0.63
^{235}U	0.44	0.3	1.0	1.8
^{238}U	20.9	0.023	1.2	1.7

This procedure gives the following statistical uncertainties in α_T and α_P (E in eV, t in seconds and n = 0.06 atoms/b except for ^{235}U where n = 0.006 atoms/b):

	$\delta\,\alpha_P$	$\delta\,\alpha_T$
^{139}La	$6 \times 10^{-6}\sqrt{\frac{E}{t}}$	$5 \times 10^{-3}\sqrt{\frac{1}{t}}$
^{165}Ho	$2 \times 10^{-5}\sqrt{\frac{E}{t}}$	$2 \times 10^{-2}\sqrt{\frac{1}{t}}$
^{235}U	$1 \times 10^{-4}\sqrt{\frac{E}{t}}$	$4 \times 10^{-1}\sqrt{\frac{1}{t}}$
^{238}U	$4 \times 10^{-6}\sqrt{\frac{E}{t}}$	--

Parity-violation measurements are most sensitive at lowest energies, while T-violation measurements have the same sensitivity at all energies. P-violation studies must reach $\alpha_P < 10^{-7}$ to see effects. This is readily attainable in 10^6 seconds up to a few hundred eV in ^{238}U. Holmium is less satisfactory, and ^{235}U appears unsuitable for on-resonance tests -- its small level spacing means the p-wave strength is fragmented over many levels and any one resonance is only weakly excited. T-violation experiments in aligned ^{165}Ho will set bounds on α_T which approach a few x 10^{-5} in 10^6 seconds. ^{235}U is not competitive unless a thick target can be prepared. ^{139}La shows good sensitivity, but there is no evidence that two suitable p-wave resonances exist.

With the new butanol polarizer, the sensitivities are a factor of four better. A long run of 10^7 seconds will then set bounds approaching 10^{-6} on α_T for holmium, a factor of 10^3 better than the best current limit.

Tests with Energy Averages

These tests exploit fluctuations in the cross-section difference both on and off resonance, averaged over the whole spectrum. As discussed earlier, the P-violation sensitivity is greatest when the energy resolution $\Delta E \leq \Gamma$, i.e. at neutron energies above below ~ 100 eV. From the plot of $\overline{\gamma}_{J\pi}$ in Figure 4, the T-violation sensitivity is a maximum in the region around 100 eV.

The statistical uncertainty $\Delta(\delta\sigma)$ in any single measurement of the cross-section difference $\delta\sigma$ for a bin width equal to the energy resolution ΔE is

$$\Delta(\delta\sigma) \sim 1 / \left(P_n\, n \sqrt{N\, \Delta E\, t}\,\right) .$$

Averaging over M bins of width ΔE in an interval $I = M\,\Delta E$ ($I \sim 20$ eV) yields the statistical uncertainty in $\langle (\delta\sigma)^2 \rangle_I$

$$\sim \left(\frac{1}{M^2} \sum_{i=1}^{M} \left[\Delta(\delta\sigma_i) \right]^4 \right)^{1/2} .$$

Weighting each such local uncertainty by the average value of $\overline{\gamma_{J\pi}}^{-2}$ or $\overline{\beta_J}^{-2}$ at that energy yields the following uncertainties in α_P and α_T in t seconds (same input parameters as for the isolated resonance tests):

	$\delta\alpha_P$	$\delta\alpha_T$
^{139}La	$\frac{1 \times 10^{-5}}{\sqrt{t}}$	$\frac{5 \times 10^{-2}}{\sqrt{t}}$
^{165}Ho	$\frac{6 \times 10^{-5}}{\sqrt{t}}$	$\frac{5 \times 10^{-2}}{\sqrt{t}}$
^{235}U	$\frac{4 \times 10^{-4}}{\sqrt{t}}$	$\frac{4 \times 10^{-1}}{\sqrt{t}}$
^{238}U	$\frac{1 \times 10^{-5}}{\sqrt{t}}$	--

In 10^6 seconds the uncertainty in α_T for holmium is ~ 7 x 10^{-5}, comparable to the average on-resonance value. A 10^7-second run with the new polarizer will approach ~ 5 x 10^{-6}, a factor of 200 better than the best previous limit.

^{238}U shows promise for energy-averaged parity studies. A 10^6 second run will approach an accuracy of ~ 10^{-8} in α_P.

SYSTEMATIC ERRORS

In this section I discuss the systematic errors associated with transmission measurements of P- and T-violation. We consider instrumental effects, as well as physical effects arising from interactions of the neutron spin which are important in T-tests. In addition we sketch some techniques which will be used to reduce noise introduced by drifts. For T-violation, we discuss only the five-fold correlation (FC). The systematic errors associated with the three-fold correlation (TC), although tractable, are more formidable [Sto 86, Bow 87].

The experiments will be performed using a double modulation technique. The neutron spin will be reversed every few seconds by the use of a magnetic "spin flipper" and once or twice a day by changing the pumping frequency of the dynamically polarized proton spin filter. (See section 7.) The spin flipper reversals are not precise since the magnetic field which guides the neutron spin is changed, but this method has the advantage that the spin can be reversed in milliseconds. Changing the polarizer state precisely reverses the spin direction, which is determined by the direction of the spin filter's field. The magnitude of the neutron polarizations may differ by one or two percent between the two polarizer states and will be continuously monitored using nuclear magnetic resonance (NMR) techniques. This reversal method requires about one hour.

Instrumental Effects

Instrumental systematic errors arise from the fact that the spin-flipper fringe field, although small, might affect the detector or the target, and from the fact that the spin-flipper reversals are not exactly 180°. In this section we discuss how both these effects are cancelled by an eight-step reversal sequence and the double modulation technique.

The spin flipper's operation depends on using adiabatic and diabatic magnetic field transitions. If a neutron spin is in an adiabatically (slowly) varying magnetic field,

$$\frac{dB}{dt} << \omega_L B,$$

the projection of the spin on the field direction is a constant. If on the other hand the field changes diabatically (suddenly),

$$\frac{dB}{dt} >> \omega_L B,$$

the spin direction does not change and the projection of the spin on the field direction is reversed. The neutron spin is initially along the homogeneous field of the polarized proton filter and is parallel or anti-parallel to the field depending on the radio-frequency pumping field. The longitudinal (z) fringing field adiabatically guides the longitudinal neutron spin to the spin flipper. The spin flipper has a time-independent longitudinal magnetic field which reverses sign at the midpoint. A transverse (y) field is turned on or off to reverse or not reverse the spin. When the transverse field is off the neutron spin diabatically passes through the field-reversal region and emerges without reversing its direction. When the transverse field is on (up or down) the spin adiabatically follows the magnetic field direction and rotates by 180 degrees in a distance of one meter. Figure 5 illustrates the operation of the spin flipper.

The neutron spin is reversed equally well by an up or down transverse field, and the eight-step sequence will consist of the transverse field off, up, down, off, up, off, off, down. To the extent that the detector (target) response is a linear function of the magnetic field, this reversal sequence eliminates the influence of the field on the detector (target). Furthermore, the use of this reversal sequence eliminates the effects of slow drifts up to terms of order $(dt / \tau)^2$ where dt is the measurement time in one spin state and τ is a time characteristic of the drifts.

The spin reversal caused by the spin flipper is not exact because the magnetic field configurations in the reversal and non-reversal states are different. By using a double modulation technique with the spin flipper and polarizing filter a fast exact spin reversal can be achieved. The transmission through the target depends approximately linearly on each component of the neutron average spin vector, and data will be measured in eight step sequences for both polarizer states. Let **S**(P,F) be the average spin vector at a particular neutron time of flight. The quantity **S**(P,F) is a function of the polarizer state P (pumping frequency), either A or B, and of the flipper state F (transverse field), either off, up or down. The four spin vectors for which the transmissions are measured are:

Pumping Frequency	Flipper Field	Spin Components
A	off	**S1** = **S**(A,off)
A	(up+down)/2	**S2** = 1/2(**S**(A,up) + **S**(A,down))
B	off	**S3** = **S**(B,off)
B	(up+down)/2	**S4** = 1/2(**S**(B,up)) + **S**(B,down))

The spin flipper approximately reverses the spin direction and therefore: **S1** ~ -**S2** and **S3** ~ -**S4**. Since changing the polarizer state exactly reverses the spin direction, **S1** = -**S3** and **S2** = -**S4**.

For small changes in the average spin vector the transmission T(P,F) will consist of a spin-independent transmission T0 plus a linear function of the average spin vector **S**,

$$T(P,F)=T0 + \mathbf{E}\cdot\mathbf{S},$$

where **E** is a vector. There will also be possible contributions D(F) which represent the (linear) influence of the flipper stray field in state F on the detector and target. The four measured transmissions are T1 ....T4

Pumping Frequency	Flipper Field	Transmission (**E** a vector)
A	off	T1 = T0 + **E**·**S**1 + D(off)
A	(up + down)/2	T2 = T0 + **E**·**S**2 + 1/2[D(up) + D(down)]
B	off	T3 = T0 + **E**·**S**3 + D(off)
B	(up + down)/2	T4 = T0 + **E**·**S**4 + 1/2[D(up) + D(down)]

The desired combination of transmission differences ΔT that eliminates the direct influence terms as well as the spin independent term is:

$$\Delta T = 1/2[(T1 - T2) - (T3 + T4)] =$$

$$\mathbf{E}\cdot[(\mathbf{S}1-\mathbf{S}2)/2-(\mathbf{S}3-\mathbf{S}4)/2] =$$

$$\mathbf{E}\cdot[\mathbf{S}(A)-\mathbf{S}(B)].$$

Furthermore ΔT is itself the difference of transmissions for two exactly reversed spin vectors **S**(A) and **S**(B). To summarize: 1) The double modulation procedure eliminates the effects of the spin flipper field acting directly on the detector and target. 2) The quantity ΔT is the difference of transmissions for two reversed spin states.

Physical Effects

In order to control systematic errors which depend on the neutron spin direction, we also must develop a procedure which establishes that ΔT is non-zero only as a result of T-violation. We consider two sources of systematic error: 1) sequential T-even effects and 2) misalignments of **S**(A/B). We find that the first can be detected by the quadratic dependence of ΔT on target thickness, and that the second can be identified by subsidiary measurements with the neutron spin aligned along different axes.

Polarized neutrons propagate through a medium according to a spin-dependent index of refraction. If χ_i is the initial spin state and χ_f is the final spin state then

$$\chi_f = \exp(i\,(n-1)kz)\,\chi_i$$

where

$$k\,(n-1) = \frac{2\pi}{k}\,\rho\, f(0^\circ) = b\sigma_x + c\sigma_z + d\sigma_y$$

Here n is the index of refraction, k is the neutron momentum, ρ is the number of scattering centers per unit volume, σ_x, σ_y and σ_z are Pauli operators, b, c and d are complex coefficients which depend on the polarization or alignment of the target and f(0°) is the forward scattering amplitude. The neutron momentum k is taken to be in the z direction. For the five-fold correlation (Figure 3) the alignment axis is in the (z, x) plane at an angle of 45 degrees to the z axis and the polarizer prepares the spin to be either up or down (±y). For a purely aligned target the average value of **I** is zero, and the coefficients b and c are small, arising only from the weak interaction. It is the absence of a large strong interaction allowed spin-spin interaction which makes the FC less subject to systematic error than the TC.

In order to understand how b, c and d arise, it is informative to write down a complete set of Cartesian invariants made from the operators **s, I, k.** These are given as follows [Gou 89]:

Operator	Parity	Time Reversal	Components for Pure Alignment in the x-z plane
s·k	-	+	z
s·I	+	+	none
s·I × k	-	-	none
(s·k)(I·k)	+	+	none
(s·I)(I·k)	-	+	x,z
(s·I × k)(I·k)	+	-	y
(s·k)(I·k)^2	-	+	z
(s·I)(I·k)^2	+	+	none
(s·I × k)(I·k)^2	-	-	none

(Terminates when power of I is 2 I+1)

An invariant which involves **I** to an odd power averages to zero in a purely aligned target and induces no spin dependence in the neutron index of refraction. Note also that terms in $\mathbf{k}^3$ and higher will not contribute if only s-wave and p-wave neutrons are considered.

The coefficients b, c and d which arise from the spin dependence of the forward elastic scattering amplitude are:

$$b = \frac{1}{\sqrt{2}} \sum_{N}^{\text{odd}} (\mathbf{s \cdot k})(\mathbf{I \cdot k})^N \equiv \frac{b_1}{\sqrt{2}}$$

$$c = \sum_{N}^{\text{even}} (\mathbf{s \cdot k})(\mathbf{I \cdot k})^N + \frac{1}{\sqrt{2}} \sum_{N}^{\text{odd}} (\mathbf{s \cdot k})(\mathbf{I \cdot k})^N = c_1 + \frac{b_1}{\sqrt{2}}$$

$$d = \sum_{N}^{\text{odd}} (\mathbf{s \cdot I \times k})\ (\mathbf{I \cdot k})^N .$$

The FC is the leading term in d. The problem is to measure d in the presence of b and c.

First consider the transmission through the polarizer and target. The operator which represents the combined action is

$$U = [(1+gs_y)/2] \exp [iz(b\sigma_x + c\sigma_z + d\sigma_y)],$$

where g = +1 for neutron polarization up and g = -1 for polarization down. Let

$$m^2 = b^2 + c^2 + d^2,$$
$$P = \cos(mz),$$
$$\text{and } Q = \sin(mz),$$

and define T(g) as the transmission through the apparatus when the polarizer is in state g. Then

$$\Delta T = T(+) - T(-) =$$

$$\mathrm{Im}(Pd^*Q^*/m^*) + (QQ^*/mm^*)\mathrm{Im}(bc^*) =$$

$$\text{Im}(Pd^*Q^*/m^*) + \frac{1}{\sqrt{2}}(QQ^*/mm^*)\text{Im}\,(b_1c_1^*).$$

The three interactions represented by b, c and d act coherently. The real parts precess the neutron spin and the imaginary parts attenuate the neutron flux. The first term depends linearly on the imaginary part of d. The flux is attenuated differently when the imaginary part of the five-fold correlation is positive or negative. The second term represents a precession of the y-component of the spin about the x axis by the real (**s·I**)(**I·k**) interaction followed by an attenuation due to the imaginary **s·k** interaction acting on the resulting z spin component, and vice versa. This term represents the sequential action of two T-even interactions resulting in a false T-odd interaction. In neutron transmission these terms can be removed by the addition of an analyzer after the target, or by comparing induced polarization with transmission asymmetry [Kab 88]. These effects can also be removed by less expensive techniques. For mz small

$$\text{Im}(Pd^*Q^*/m^*) \sim -z\text{Im}(d)$$

and

$$\frac{1}{\sqrt{2}}(QQ^*/mm^*)\text{Im}(b_1c_1^*) \sim \frac{1}{\sqrt{2}}z^2\,\text{Im}\,(b_1c_1^*).$$

The effect of the five-fold correlation increases linearly with target thickness, while the sequential interaction increases quadratically with target thickness. This difference provides a direct means of testing whether or not a transmission asymmetry is the result of T-violation or of this sequential process. Since the coefficient c_1 may depend on whether or not the target is aligned, establishing that the longitudinal asymmetry is zero when the target is warm does not ensure the absence of this false simulation of T-odd effects.

In treating misalignments of **s**, **k** and **I**, it may assumed without loss of generality that **k** is in the z direction and **I** is in the (z, x) plane. The effects of misalignments of the polarizer and the target are calculated by inserting a rotation operator between the polarizer operator and the target operator. Since only asymmetries which falsify time reversal are of interest, d may be set to zero.

$$U = [(1+gs_y)/2]\,(1+(1/2)(\theta\sigma_x + \phi\sigma_y + \eta\sigma_z))\,\exp[iz(b\sigma_x + c\sigma_z)]\,.$$

The change in the transmission asymmetry is given by

$$\Delta T'-\Delta T = \theta\, \mathrm{Im}(P^*Qc/m) + \eta\, \mathrm{Im}(P^*Qb/m)\,.$$

Where $\Delta T'$ is the transmission difference with misalignments present. This expression has a straightforward interpretation. The polarizer produces neutrons polarized in the y direction. A misalignment of the polarizer relative to the target corresponds to a rotation about the x axis (θ) and gives a small polarization in the z direction. The imaginary part of c then produces a transmission asymmetry. The same argument applies to misalignment about the z-axis which rotates y into x and induces a term proportional to Im(b). Misalignment with respect to the polarization axis y has no effect on the transmission asymmetry. The terms $\mathrm{Im}(P^*Qc/m)$ and $\mathrm{Im}(P^*Qb/m)$ can be measured directly by polarizing the beam in the x and z directions, respectively ($\theta = \pi/2$ and $\eta = \pi/2$). Any small asymmetries which arise will be due to P-odd effects. The reduction of these false asymmetries for the situation where the neutron spin is along the y axis and will be suppressed two or three orders of magnitude. The procedures for spin alignment can be studied using the ^{139}La resonance.

CONCLUSION

I have argued that many-body enhancements of symmetry-violating nuclear forces make the compound nucleus a good laboratory in which to search for time-reversal violation. The large fluxes of epithermal neutrons available at the Los Alamos Neutron Scattering Center provide good statistical precision. The study of parity violation in the compound nucleus will test the theoretical procedure used to relate average symmetry-violating matrix elements between states of the compound nucleus to the size of symmetry-violating components of the nucleon-nucleon interaction. Tests involving both isolated resonances and energy averages have been defined and their sensitivities have been estimated. Systematic errors of both instrumental and physical origin have been characterized, and strategies for controlling them have been presented. The prospects are excellent to test-time reversal symmetry in the nucleon-nucleon force to a part in 10^6.

REFERENCES

Ade 85 E. G. Adelberger and W. C. Haxton, Ann. Rev. Nucl. Part. Sci. **35**, 501 (1988)

Alf 83 V. P. Alfimenkov et al., Nucl. Phys. A**398**, 93 (1983)

Bla 83 E. Blanke et al., Phys. Rev. Letters **51**, 355 (1983)

Boh 88 O. Bohigas and H. A. Weidenmüller, Ann. Rev. Nucl. Part. Sci. **38**, 421 (1988)

Boo 86 D. Boose, H. L. Harney and H. A. Weidenmüller, Phys. Rev. Letters **56**, 2012 (1986)

Bow 87 J. David Bowman, in Tests of Time Reversal Invariance in Neutron Physics (World Scientific, 1987) p. 121

Bow 88 C. D. Bowman et al., Hyperfine Int. **43**, 119 (1988)

Bow 89 C. D. Bowman, J. D. Bowman and V. W. Yuan, Phys. Rev. C**39**, 1721 (1989)

Bun 83 V. E. Bunakov and V. P. Gudkov, Nucl. Phys. **A401**, 93 (1983)

Bun 87 V. E. Bunakov and V. P. Gudkov, in Tests of Time Reversal Invariance in Neutron Physics (World Scientific, 1987) p. 175

Bun 88a V. E. Bunakov, Phys. Lett. **207B**, 233 (1988)

Bun 88b V. E. Bunakov, Phys. Rev. Lett. **60**, 2250 (1988)

Dav 89a E. D. Davis, Phys. Lett. **226B**, 197 (1989)

Dav 89b E. D. Davis, Symposium/Workshop on Spin/Symmetries, TRIUMF, June/July 1989

Dav 89c E. D. Davis (to be published)

Fre 87 J. B. French, A. Pandey and J. Smith, in Tests of Time Reversal Invariance in Neutron Physics (World Scientific, 1987) p. 80

Fre 88a J. B. French et al., Ann. Phys. (N.Y.) **181**, 198 (1988)

Fre 88b J. B. French et al., Ann. Phys. (N.Y.) **181**, 235 (1988)

Gou 89 C. R. Gould, D. G. Haase, N. R. Roberson, H. Postma and J. D. Bowman, Int. J. Mod. Phys. (to be published)

Haa 88 D. G. Haase, C. R. Gould, G. E. Mitchell and N. R. Roberson, Hyperfine Interactions **43**, 127 (1988)

Kab 88 P. K. Kabir, Phys. Rev. Lett. **60**, 686 (1988)

Key 73 G. A. Keyworth et al., Phys. Rev. C**8**, 2352 (1973)

Kis 87 S. Kistryn et al., Phys. Rev. Letters **58**, 1616 (1987)

Mas 87 Y. Masuda, T. Adachi, S. Ishimoto, E. Kikutani, H. Koiso, K. Morimoto and A. Masaike, Hyperfine Interactions **34**, 143 (1987)

McL 88 V. McLane, C. L. Dunford and P. F. Rose, Neutron Cross Sections, Vol. 2 (Academic Press 1988)

Mit 88 G. E. Mitchell et al., Phys. Rev. Lett. **61**, 1473 (1988)

Mug 81 S. F. Mughabghab, M. Divadeenam and N. E. Holden, Neutron Cross Sections, Vol. 1 (NNDC and Academic Press 1981)

Rob 87 Tests of Time Reversal Invariance in Neutron Physics, ed. N. R. Roberson, C. R. Gould and J. D. Bowman (World Scientific, 1987)

Sha 68 I. S. Shapiro, Sov. Phys. Usp. **11**, 582 (1968)

Sto 86 L. Stodolsky, Phys. Lett. **172B**, 5 (1986)

Sus 82 O. P. Sushkov and V. V. Flambaum, Sov. Phys. Usp. **25**, 1 (1982)

Van 88 J. R. Vanhoy, E. G. Bilpuch, J. F. Shriner, Jr. and G. E. Mitchell, Z. Phys. A - Atoms and Nuclei **331**, 1 (1988)

Van 89 J. R. Vanhoy, E. G. Bilpuch, J. F. Shriner, Jr. and G. E. Mitchell, Z. Phys. A - Atoms and Nuclei **333**, 229 (1989)

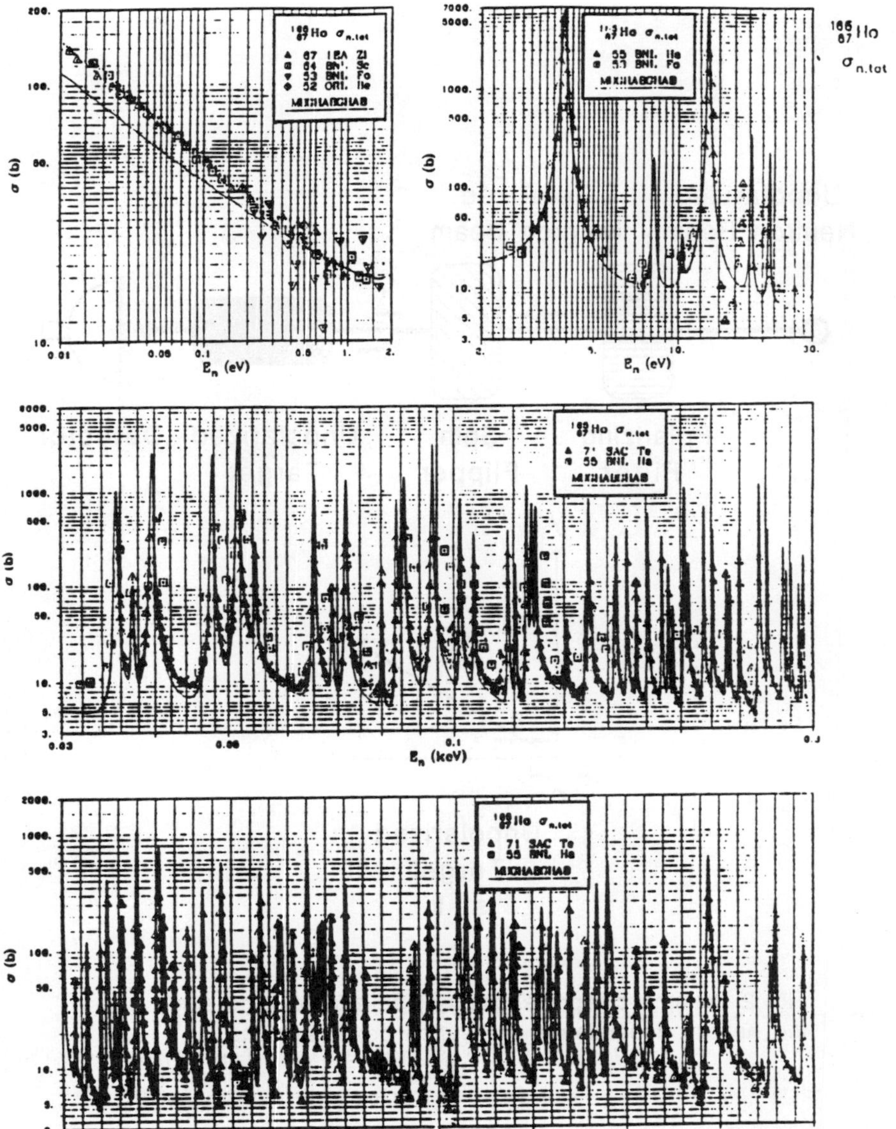

Fig. 1 Neutron total cross section for ^{165}Ho.

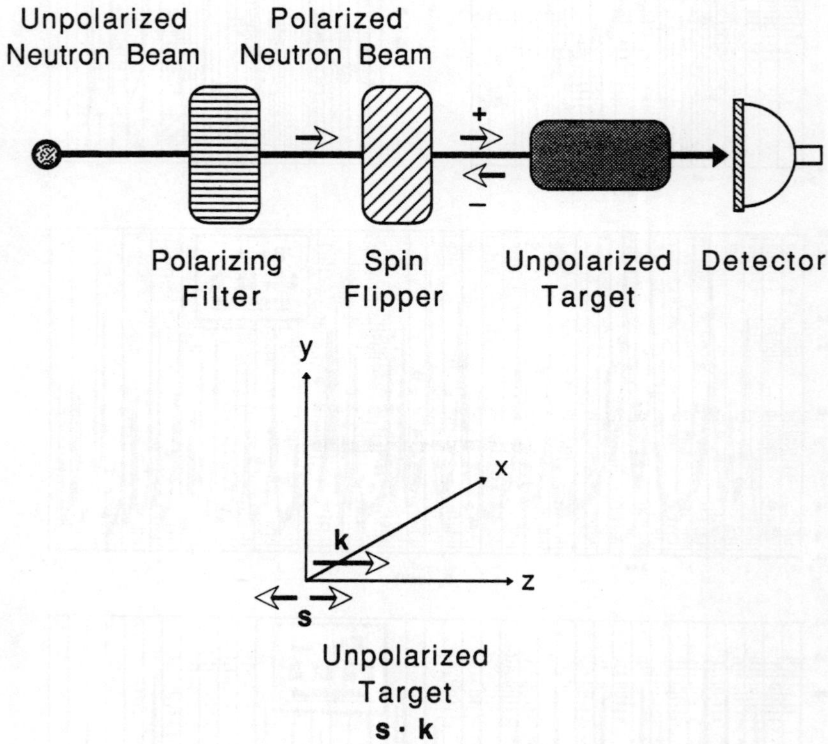

Fig. 2 Schematic representation of an apparatus to study the helicity dependence of neutron cross sections.

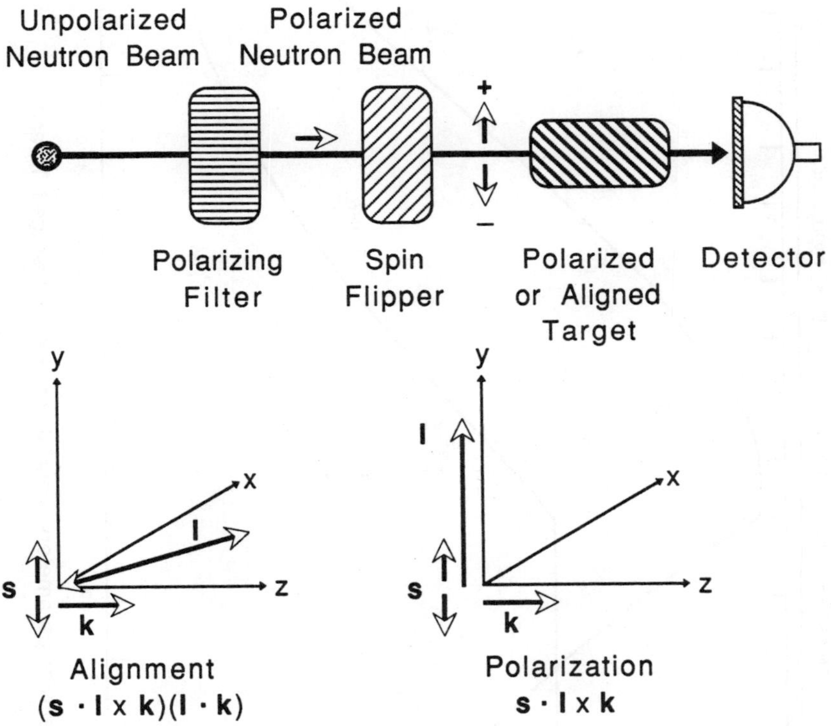

Fig. 3 Schematic representation of an apparatus to search for time-reversal violation.

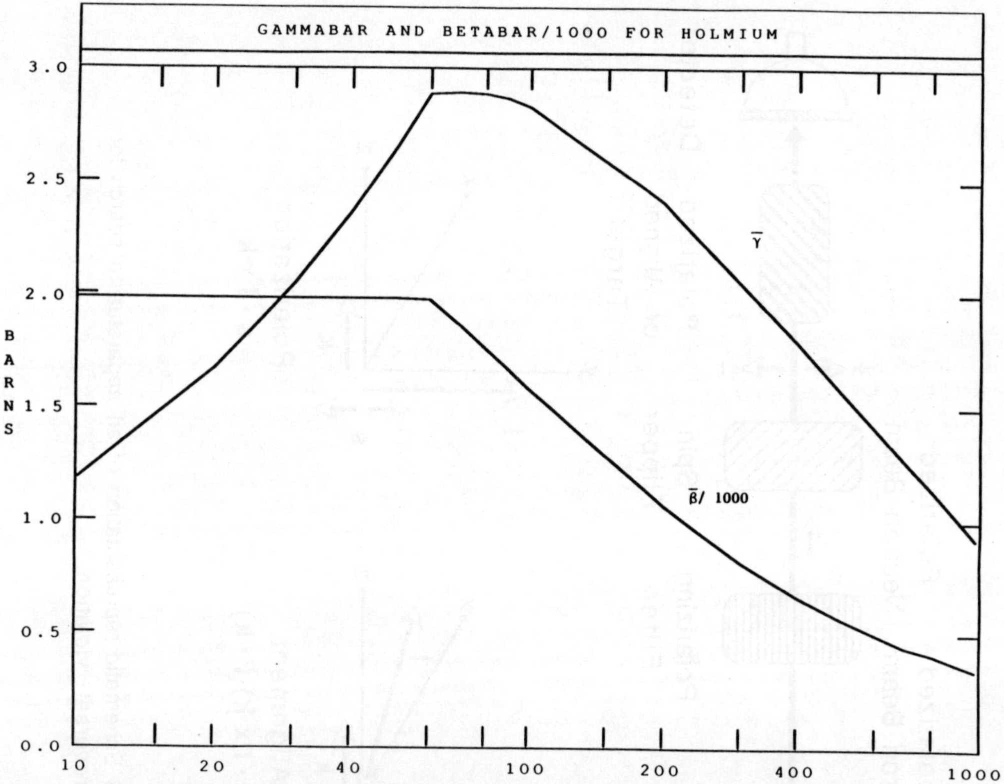

Fig. 4 Sensitivity parameters $\bar{\gamma}$ and β for energy-averaged T- and P-violation for J=3 states in ^{165}Ho.

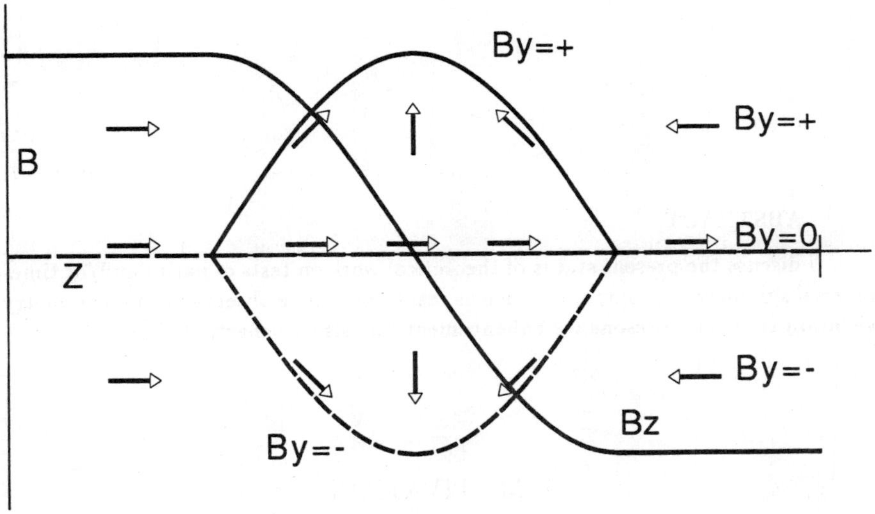

Fig. 5 Longitudinal and transverse magnetic fields as functions of the longitudinal coordinate z in the spin flipper. The neutrons travel in the z direction. The transverse field is either up, off or down. The direction of the spin vector as a function of z is shown for each of these states. For the Transverse field up or down, the spin rotates counterclockwise or clockwise respectively. For the transverse field off, the spin direction does not change.

What can we learn from tests of fundamental symmetries in compound-nucleus Reactions?

Hans A. Weidenmüller

Max-Planck-Institut für Kernphysik, (69) Heidelberg, F.R. of Germany

ABSTRACT

I discuss the present status of theoretical work on tests of parity and/or time-reversal symmetry in compound-nucleus reactions, where observables for symmetry violation show large resonance enhancement but also stochastic behaviour.

1. MOTIVATION

Three years ago I had the privilege to return to Caltech and to visit for six months. When we discussed the status of experimental and theoretical work on symmetry tests in the compound-nucleus regime, Felix Boehm in his friendly but incisive manner would keep asking nagging questions which then I could not answer. Progress has meanwhile been made, mostly through work done at Heidelberg, by E. David Davis and others.[1-7] When Peter Vogel asked me to speak at this meeting in honour of Felix, I was happy to accept. I felt it would be appropriate to discuss recent developments in a wider context.

Tests of fundamental symmetries in strong - interaction physics are always timely and of general interest either if an existing upper bound on symmetry vio-

⋆ Dedicated to Felix Boehm on the occasion of his 65th birthday.

lation can be significantly lowered, or if more detailed information on a symmetry-breaking interaction can be obtained. During the last two decades, the interest in using nuclear reactions in the compound-nucleus (CN) regime for tests of parity violation and time-reversal symmetry has sharply risen, particularly after the discovery at Dubna[8], of a big parity violation effect in ^{139}La, amounting to a signal of the order of 10 per cent. This discovery demonstrated the existence of huge enhancement factors for symmetry breaking forces in the CN regime, in the present case caused mainly by the closeness of two CN resonances of opposite parity, and led to a flurry of activity, including plans for extensive further work[9].

In attempting to analyse data on symmetry violatation in CN reactions, theorists face three problems: (i) To understand quantitatively the large enhancement factors, and to work out proposals for an optimal choice of an observable for each kind of symmetry test - chosen optimally with regard to enhancement factors and attainable experimental resolution and accuracy; (ii) to relate upper bounds or actual data on symmetry violation with the stochastic Hamiltonian governing CN processes, and to deduce an upper bound on or the strength of the symmetry-violating piece in this Hamiltonian. This must be done by an averaging process, and the question is: Does enhancement survive the averaging?; (iii) to reduce said bound or value to information on the strength of the symmetry-violating piece in the effective nucleon-nucleon interaction in the shell model. In the present talk, I review the status of theory with regard to these questions.

2. THE DATA ON ^{139}La. FURTHER EXPERIMENTS

In the CN regime, tests of parity violation have used beams of polarized neutrons, and looked for neutron spin rotation in transmission[10], for an asymmetry in gammas or fission fragments following neutron capture[8,11,12], and for the spin-dependence of neutron transmission[10,12,9]. Sizeable effects have been seen in all these experiments, the biggest observation having been made for neutron transmission in a ^{139}La target. With $N_+(N_-)$ the number of neutrons transmitted with initial spin in (opposite to) their momentum, the original measurement in Dubna, carried out at the 0.734 eV p-wave resonance in ^{140}La, gave for $R = (N_+ - N_-)\ /\ (N_+ + N_-)$ the value $R\ =\ 7.3 \pm 0.4\%$; a later measurement at KEK yielded $9.5 \pm 0.3\%$; recent Los Alamos data seem to agree with the latter value. These numbers must be compared with the relative strength (relative to the strong force) of the weak parity non-conserving nuclear force of 10^{-6} or 10^{-7}. To account for the data, the following enhancement factors have been considered[13]. (i) Dynamic or resonance enhancement: Near neutron threshold and in heavy nuclei, resonances of the same spin and parity have a mean spacing of typically 10 eV. Therefore, it is very likely that one finds an s-wave resonance in the immediate vicinity (a few eV or less) of a given p-wave resonance. The parity-violating interaction V mixes the two states; in first order the mixing amplitude is $V^{+-}/\Delta E$ with V^{+-} the matrix element and ΔE the resonance spacing. The spreading width $\Gamma^{\downarrow}\ =\ 2\pi\overline{(V^{+-})^2}/d$ [with d the mean level spacing and the bar denoting an average over many resonances] is essentially independent of mass number and excitation energy (see section 5), so that $\overline{(V^{+-})^2}$ scales like d. Comparing $V^{+-}/\Delta E$ for an experiment on a pair of levels in the neutron resonance region with $|\Delta E|\ \cong 1$ eV with the value for an experiment on two low-lying nuclear states with $|\Delta E|\ \cong 100$ keV, we obtain an enhancement factor $(100keV/\mid eV)^{1/2}\ \simeq 3 \cdot 10^2$. (ii) Kinematic enhancement: The neutron decay of the p-wave resonance through its s-wave resonance admixture is enhanced in amplitude over parity-allowed neutron decay by the factor $(\Gamma_s/\Gamma_p)^{1/2}$ where the $\Gamma's$ are the partial neutron widths for the two resonances. Replacing the widths by the penetration factors, we obtain $(\Gamma_s/\Gamma_p)^{1/2}\ \simeq\ (kR)^{-1}$. With a nuclear radius $R\ \simeq\ 5$ fm and a neutron energy $(\hbar k)^2/(2m) = 1eV$, this becomes $\simeq 30$. Combining the two enhancement factors, we obtain 10^4, at least an order of magnitude less than the observed enhancement. [In cases other than the 0.7 eV p-wave resonance in La, our result is not far off the mark set by the experimental data]. We return below [section 5, end] to the discrepancy and note here that the enhancement is at least qualitatively understandable.

All of the above-mentioned tests of parity violation are sensitive to the expectation value of $(\vec{\sigma}\ \vec{k})$, a pseudoscalar, with $\vec{\sigma}$ the neutron spin and $\hbar\vec{k}$ its momentum. Tests of time-reversal symmetry*(without parity violation) either have used detailed balance and compared[14] the cross section for a nuclear reaction like $^{27}Al(p,\alpha)^{24}Mg$ with that of its inverse, $^{24}Mg(\alpha,p)^{27}Al$; or are planned[9] to measure the difference in transmission of polarized neutrons through an aligned target. In the latter case, the observable is proportional to the expectation value of $(\vec{I}\cdot\vec{k})\cdot(\vec{\sigma}\cdot[\vec{I}\times\vec{k}])$ with $\vec{I}$ the spin of the target. Finally, a simultaneous test of both parity and time-reversal symmetry (from the point of view of weak interaction physics a more promising candidate for a positive signal than the previous one would involve the expectation value of $(\vec{\sigma}\ [\vec{I}\times\vec{k}]\ )$[15]. The last experiment is particularly difficult.

3. INFORMATION CONTENT OF DATA IN THE CN REGIME

The analysis of experimental data on level spacings[16] and partial width amplitudes[17] of resonances of the same spin and parity in the CN regime (with no attention paid to symmetry violation) has convincingly shown that these data display stochastic fluctuations which can be described in terms of random matrix models. Studies of completely chaotic (or globally instable) classical systems and of their quantum counterparts strongly suggest that such stochastic behaviour is generic for systems which are chaotic in the classical limit[18]. In consequence, an observable like $\vec{\sigma}\cdot\vec{p}$ in the case of parity violation [or any of the other observables testing symmetry violation and mentioned under 2.] itself acquires stochastic features in the CN regime: The matrix elements of the observable in the space of CN levels are random variables with mean value zero. Relevant information on the strength of the symmetry-breaking interaction is contained in the <u>mean square</u> of the matrix elements; in the absence of further information the measurement of

* Somewhat beyond the title of this paper but relevant to its topic is the analysis of the level-spacing distribution of isolated neutron resonances with regard to a bound on time-reversal symmetry violation[27].

an individual matrix element (for instance, V^{+-} in the p-wave resonance in ^{140}La) carries very little information content - beyond the statement that the symmetry is broken. Likewise, a null measurement in an individual case does not permit a meaningful upper bound on symmetry breaking to be deduced. These statements bring to the fore the question in the title of this paper.

Three ways out of this impasse have been suggested. (i) For time-reversal symmetry breaking, a non-zero result would in itself be most exciting. Therefore, it is of interest to identify those properties of pairs of resonances which would give a maximum sensitivity for the effect independent of statistical considerations. An experiment on such a pair is a shot in the dark but benefits from enhancement factors that are bigger by at least an order of magnitude than in cases (ii) and (iii) below. (This approach is not useful for parity violation where positive evidence exists already, and where quantitative information is looked for.) The conditions optimizing the search for symmetry violation have been worked out[6,7]; it is not very useful to spell them out in detail here. One difficulty with this approach is that it requires specification of the resonance parameters (energies, widths). These are often not known experimentally. (ii) The mean square of the matrix elements of the relevant observables should be determined experimentally by measuring as many cases as possible (at different bombarding energies and/or on different target nuclei); the value obtained should be compared with the theoretical average. The latter can be worked out reliably in terms of a statistical model (section 4). It does not depend on detailed knowledge about energies or partial widths of individual resonances and depends only on the spreading width $\Gamma^{\downarrow}$ for the relevant symmetry-breaking interaction, and on average properties of the CN like transmission coefficients and level density which can be determined reliably. For this reason, a detailed knowledge of resonance parameters is likewise not needed in the experimental data which simplifies the analysis considerably. The advantage of the approach is its reliability and the fact that it yields a value for or an upper bound on $\Gamma^{\downarrow}$. Its main drawback is that many data are needed. (iii) This last difficulty can be overcome[2] by using the statistical model mentioned under (ii) to generate not only the mean square observable but its entire probability distribution. Then a meaningful comparison even with a small data base can be made. The distribution cannot be generated analytically with methods that are available at present; one must resort instead to Monte- Carlo simulations. This approach has not been implemented yet.

Important questions relating to approaches (ii) and (iii) are the following. (i) Does resonant enhancement survive the averaging process? (ii) Which of several observables sensitive to the same symmetry violation shows the biggest enhancement? Taking tests of time-reversal symmetry as an example, I think here not only of a comparison between tests of detailed balance and the scattering of polarized neutrons by aligned targets. I also have in mind[1–4] the optimum choice of normalization factors. Since absolute cross-sections can never be determined with sufficient accuracy in any of the tests mentioned above, the observables must be expressed in terms of relative cross-sections. Is it more advantageous to normalize in terms of the average, or of the root of the variance of the cross-section for the symmetry-allowed process? In section 4, I outline the formulation of the statistical model and in section 6 answer these questions.

4. STATISTICAL MODEL

For pedagogical reasons, I proceed in two steps. First, I present the model of the stochastic S-matrix for pure CN scattering in the absence of symmetry breaking[19]. Next, I indicate the modifications which arise when some symmetry is broken[20,5]. I confine myself to the simplest case: No direct reactions, i.e. pure resonance scattering, and I also leave out the elastic phase shifts. An element of the scattering matrix with fixed spin and parity connecting channels a and b is given by

$$S_{ab} = \delta_{ab} - 2i\pi \sum_{\mu,\nu=1}^{N} W_{a\mu}(D^{-1})_{\mu\nu}W_{\nu b} \tag{1}$$

where the inverse propagator is given by

$$D_{\mu\nu} = E\delta_{\mu\nu} - H_{\mu\nu} + i\pi \sum_c W_{\mu c} W_{c\nu} \quad . \tag{2}$$

Eq. (1) is an N-level Breit-Wigner formula for resonance scattering; the matrix S is unitary. The amplitudes $W_{a\mu} = W^*_{\mu a}$ are partial width amplitudes coupling channels and CN levels. In the N-dimensional space of CN levels, the Hamiltonian matrix is written as $H_{\mu\nu}$, while E is the energy. Time-reversal symmetry allows us to choose $W_{a\mu} = W_{\mu a}$ and $H_{\mu\nu} = H_{\nu\mu}$.

The stochastic features of S are simulated by replacing the actual Hamiltonian of the nucleus by an ensemble of Hamiltonian matrices of dimension N. One uses the Gaussian Orthogonal Ensemble (GOE): All matrices in the ensemble are symmetric and real; the elements in and above the diagonal are uncorrelated real Gaussian distributed random variables with zero mean value and a second moment $\overline{(H^{GOE}_{\mu\nu})^2} = (1+\delta_{\mu\nu})\lambda^2/N$. The bar denotes the average over the ensemble. (The bar was used in section 2 to denote the running average over the levels of an actual nucleus.There is no danger of confusion here since by virtue of an ergodicity argument[21], the ensemble average and the running average over a member of the ensemble are equal; in statistical theory, the nuclear Hamiltonian is assumed to be a member of the ensemble). The parameter λ has the dimension energy and fixes the mean level spacing in the ensemble; the fluctuations of the eigenvalues and eigenvectors of the ensemble are then determined in a parameter-free fashion. The GOE is[21] the generic ensemble for time-reversal symmetric systems.

The GOE is invariant under orthogonal transformations of the basis of states μ. Average quantities like $\overline{S_{ab}}$ or higher moments of S can therefore only depend on orthogonal invariants constructed from the amplitudes $W_{a\mu}$. This fact reduces the number of independent parameters in the model (1) enormously since the only such invariants are $\sum_\mu^N W_{a\mu} W_{\mu b}$. In the absence of direct reactions, $\overline{S_{ab}} = \delta_{ab}\overline{S_{aa}}$, we have $\sum_\mu^N W_{a\mu} W_{\mu b} = \delta_{ab} \sum_\mu^N W^2_{a\mu}$. The value of each dimensionless sum $\lambda^{-1} \sum_{\mu}^{N} W_{a\mu}^2$ is uniquely determined by $\overline{S}_{aa}$ (or, equivalently by the optical-model potential in channel a). We see that the input parameters for the stochastic model in eqs. (1,2) are completely determined by the optical model; the model predictions on higher moments of S (needed to evaluate averages and variances of cross-sections) are then parameter-free.

To account for time-reversal-symmetry breaking without violation of parity, the model is extended by replacing the GOE Hamiltonian by an ensemble H of matrices having the form

$$H = H^{GOE} + \alpha H^{GUE} \quad . \tag{3}$$

Here, H^{GUE} stands for the time-reversal non-invariant Gaussian Unitary Ensemble, the generic ensemble for systems without time-reversal symmetry[21]. It consists of Hamiltonian matrices $H_{\mu\nu}^{GUE} = H_{\nu\mu}^{GUE\,*}$. The elements in and above the diagonal are uncorrelated complex Gaussian distributed random variables with zero mean value and a second moment $\overline{|H_{\mu\nu}^{GUE}|^2} = \frac{\lambda^2}{N}$. The dimensionless parameter α measures the relative strength of the GUE.

For parity violation, eq. (1) is generalized so that S_{ab} (and the channel indices a,b) refers to states of fixed spin but either parity. The matrix H acquires block form, the two blocks referring to states with positive and negative parity, respectively, and in obvious notation has the form

$$H = \begin{pmatrix} H^{++} & \alpha H^{+-} \\ \alpha H^{-+} & H^{--} \end{pmatrix} \quad . \tag{4}$$

To implement stochasticity, both H^{++} and H^{--} are taken to be (uncorrelated) GOE's. Parity violation is due to $H^{+-} = (H^{-+})^*$, with H^{+-} an ensemble of matrices with elements that are uncorrelated Gaussian distributed random variables, with mean value zero and a second moment scaled to that of H^{++} or H^{--}. The variables are real (complex) in the case without (with) simultaneous violation of time- reversal symmetry. The dimensionless parameter α measures the strength of parity violation. (In section 2, a single element of αH^{+-} in a given nucleus was written as V^{+-}).

In the extensions (3,4) of the statistical model, symmetry breaking may also occur in the matrix elements $W_{a\mu}$. In the case of time-reversal violation, the condition $W_{\mu a} = W_{a\mu}(= W^*_{a\mu})$ may be violated. In the case of parity violation, $W_{a\mu}$ may connect a channel a and a level μ of opposite parity. Symmetry violation in the partial width amplitudes $W_{a\mu}$ does not enjoy any enhancement, however, and is therefore discarded[20,5].

It is the aim of the statistical model to convert experimental data for (or bounds on) symmetry violation into values of (or bounds on) $|\alpha|$. To this end, it is necessary to calculate the mean value (or, more likely, the r.m.s. value) of the relevant observable from the model. Since $|\alpha| << 1$, this is done by expanding in powers of α, and keeping the lowest-order nonvanishing term.

In all cases of interest, it has been possible[1–5,20,22] to calculate this lowest-order term exactly in the limit $N \to \infty$. There is no space here to describe the method[23]. But I do wish to stress the importance of this fact. In the entire CN regime, from the domain of poorly overlapping resonances with mean width $\Gamma << d$ all the way to strongly overlapping resonances where $\Gamma >> d$, closed-form expressions for the relevant observables are available the evaluation of which only requires the numerical calculation of three-dimensional integrals. This makes it possible to decide which observable, and which domain of CN scattering, provide the largest enhancement factors.

5. THE SPREADING WIDTH

It was mentioned under 2. that the spreading width $\Gamma_2^{\downarrow} = 2\pi\overline{V^2}/d$ where $\overline{V^2}$ is the mean square matrix elements of the symmetry-breaking interaction is a useful measure of symmetry breaking: $\Gamma^{\downarrow}$ changes little with excitation energy or mass number, while $\overline{V^2}$ alone changes very strongly.

Empirical evidence supporting this statement comes from the analysis of isospin-symmetry breaking in CN reactions. The statistical model[24] for this process is almost identical to that of parity violation in eq. (4), states of opposite parity being replaced by states of isospin quantum numbers T_1 and $T_2 = T_1 + 1$. With d_1 the average level spacing of the T_1-states and $\alpha^2\overline{(H_{\mu\nu}^{(12)})^2} = \overline{V_c^2}$ the mean square Coulomb matrix element, $\Gamma_2^{\downarrow} = 2\pi\overline{V_c^2}/d_1$ has been deduced for many nuclei (with mass number A) from a large body of experimental data. The results are shown in figure 1. I mention that the experiments used in figure 1 have been done in a wide range of nuclear excitation energies. While the r.m.s Coulomb matrix element $\overline{V_c^2}^{1/2}$ ranges over many orders of magnitude, we see that $\Gamma_2^{\downarrow}$ is nearly constant around 10 keV, with fluctuations in either direction by a factor 5 or so.

A theoretical understanding of this nearly constant behaviour of the spreading width, which is expected to prevail also for any other symmetry-breaking interaction, was obtained by French et al.[25-27]. The parameter of fundamental interest is the relative strength of the symmetry-breaking term in the effective nucleon-nucleon interaction of the shell model. The investigations addressed the problem: How do matrix elements of the effective two-body interaction in the shell model "propagate" into large spaces spanned by many-particle configurations, such as the ones appropriate for random-matrix models? A simple order-of-magnitude estimate yields[27]

$$\Gamma^{\downarrow} \cong 2\pi \cdot 10^5 \cdot \alpha_0^2 \ eV \tag{5}$$

where α_0 is the relative strength of the symmetry-breaking effective nucleon- nucleon interaction (in comparison to the symmetry-conserving one). Eq. (5) does evidently not depend on mass number and excitation energy. Eq. (5) has a simple intuitive interpretation. We estimate, roughly, the spreading width $\Gamma^{\downarrow}_{strong}$ for the strong interaction to have the order of magnitude 1 MeV, as evidenced, f.i. by the spreading widths of $E1$ giant resonances. Putting more precisely $\Gamma^{\downarrow}_{strong} = 2\pi\ 10^5$ eV, we find from eq. (5) that $\Gamma^{\downarrow}/\Gamma^{\downarrow}_{strong} \cong \alpha_0^2$. This explains the order of magnitude of the estimate (5).

The deviations from the mean of the spreading widths in figure 1 are significant, both for devising experiments with optimum enhancement, and for a precise reduction of values of or bounds on spreading widths to data on the effective nucleon-nucleon interaction. Unfortunately it is an open question to what extent the departures of $\Gamma^{\downarrow}$ from its mean value can be understood theoretically. This is all the more regrettable since fluctuations of similar size seem to appear in the data on parity violation in a series of nuclei[4].

6. RESULTS

Let us begin with time-reversal symmetry tests not involving parity violation. The present upper bound on α_0 (99% confidence level) is about 10^{-3}. An improvement by up to two orders of magnitude of the sensitivity of the test (which uses detailed balance on a pair of optimally chosen close-lying resonances in the regime $\Gamma << d$) is possible[7]. - For an averaged observable, the optimal choice in the case of detailed balance is[2,4] the ratio $[\overline{(\sigma_{ab} - \sigma_{ab})^2}]\ /\ \overline{(\sigma_{ab}} \cdot \overline{(\sigma_{ba}})$. It has been known for some time[20,27] that in the regime $\Gamma >> d$, this ratio is given by $\Gamma^{\downarrow}/\Gamma$ where $\Gamma = 2\pi \sum_c T_c/d$ is the average CN width given in terms of the sum of the transmission coefficients. The question whether this form would prevail also in the regime $\Gamma << d$ (where Γ is much smaller and the enhancement therefore much bigger) has, to within a numerical factor which is of order unity, been answered in the affirmative[2-4]. Limitations of the enhancement arise from the finite energy resolution δE of the experiment: For $\delta E > \Gamma$, the ratio $\Gamma^{\downarrow}/\Gamma$ must be replaced by

$\Gamma^{\downarrow}/\delta E$. With an optimal energy resolution of proton beams of 300 eV and alpha beams of $\simeq 1$ keV, a detailed-balance experiment on the (p,α) and (α,p) reaction in the regime of non-overlapping resonances could lower the existing upper bound on α_o by an order of magnitude. Candidates for this experiment are the reactions $^{37}Cl(\rho,\alpha)^{34}S$ and $^{39,41}K(p,\alpha)^{36,38}Ar$. - For the test involving the transmission of polarized neutrons through an aligned target, i.e. the observable $(\vec{I}\cdot\vec{k})$ $(\vec{\sigma}[\vec{I}$ x $\vec{k}])$ or, experimentally, a quantity proportional to $(N_+ - N_-)$, the optimal choice of normalization[2] uses average cross-sections. The mean- square observable is thus $\overline{\sigma_+ - \sigma_-)^2}/(\overline{\sigma_+}\cdot\overline{\sigma_-})$ with $\sigma_\pm$ the total cross-section for neutrons with helicity $\pm$, and fixed target alignment. From the Ericson regime ($\Gamma >> d$) all the way down to the domain of isolated resonances ($\Gamma << d$), this quantity is[1-4] again proportional to $\Gamma^{\downarrow}/\Gamma$, suggesting large enhancement in the latter domain. However, the two momentum factors $\hbar\vec{k}$ in the observable $(\vec{I}\cdot\vec{k})$ $(\vec{\sigma}[\vec{I}$ x $\vec{k}])$ show that a non-zero signal requires either p-waves (or higher waves) in both entrance and exit channels, or an s-wave d-wave (or higher-wave) pair. In either case, the penetration factors for angular momentum $l > 0$ strongly suppress the signal (and, thereby, the enhancement factor) near threshold. This suppression actually outweighs[2,4] the enhancement as energy is lowered, suggesting the experiment to be performed sufficiently far above threshold for penetration effects to be negligible. For the actual experiment considered at Los Alamos, the energy resolution unfortunately becomes rapidly poorer with increasing neutron energy. The optimum choice lies somewhere in the interval 70-400 eV, with enhancement factors not much in excess of ten[4]. However, the accuracy attainable in this kind of test is expected to be much higher than in the case of detailed balance. As a result it appears that both tests hold about the same promise, i.e., to lower the bound on α_0 by an order of magnitude.

I turn to measurements of parity violation without time-reversal symmetry violation. Work on this problem is still in progress[5]; here I give some preliminary results. The observable is always proportional to S_{ab}^{+-}, the S-matrix element connecting channels of opposite parity. For $\Gamma >> d$, the average $\overline{\mid S_{ab}^{+-} \mid^2}$ can be easily worked out and is given by $[T_a^+ T_b^- / \sum_c T_c](\Gamma^\downarrow/\Gamma)$ where I have assumed $d_+ \cong d_- = d, \Gamma^+ \cong \Gamma^- = \Gamma = \frac{d}{2\pi}\sum_c T_c$. Therefore, $\overline{| S_{ab}^{+-} |^2} / (\overline{| S_{bb}^{++} |^2 | S_{bb}^{--} |^2})^{1/2}$ has the value $\Gamma^\downarrow/\Gamma$ where $\Gamma^\downarrow$ is the spreading width for the parity-violating interaction. (I have neglected shape-elastic scattering and elastic enhancement factors in the denominator). This form suggests strong enhancement, particularly for isolated resonances. There, a confirmation of the form $\Gamma^\downarrow/\Gamma$ is still lacking; however, the experience gained in the discussion of time-reversal symmetry [1-4] suggests that the extrapolation is correct. The enhancement would then again probably be limited by experimental resolution, and by the p-wave penetration factor.

A theoretical analysis of a test for simultaneous violation of parity and time-reversal symmetry has not yet been undertaken but should proceed very much along the same lines as sketched in the previous paragraph.

7. CONCLUSIONS

The question raised in the title, in the Introduction, and at the end of section 3, have all been answered at least in part. A search for a symmetry violation may profit from the particularly large enhancement factors which come with the optimum choice of a pair of resonances. To obtain reliable mean values or upper bounds on the strength of the symnmetry-violating interaction, it is advantageous to consider average (or r.m.s.) values of observables, generated from a sufficiently large data base. Such averages enjoy considerable enhancement factors, too, depend only on well-known average properties of the CN, and it is possible to choose the observable so as to maximize the enhancement. The data can thus be related to $\Gamma^\downarrow$, the spreading width for the symmetry-violating interaction, which is a universal measure of symmetry violation in the CN regime, and which in turn relates to the strength of the symmetry-breaking part of the effective nucleon-nucleon interaction in the shell model. All this shows that the CN is an excellent laboratory for symmetry tests and studies of broken symmetries: Compared with studies of isolated low-lying nuclear levels, the large enhancement factors lead to higher sensitivity, and the smooth dependence of averaged observables on (known) average CN properties, to a more reliable evaluation of $\Gamma^\downarrow$.

There are some shadows on this bright picture. Assembling a sufficently large experimental data base is difficult, and the influence of missing resonances or of resonances with incorrect spin/parity assignments seems not to have been assessed. On the theoretical side, the precise connection between the spreading width $\Gamma^{\downarrow}$ and the corresponding part of the effective nucleon-nucleon interaction [which would go beyond the simple estimate (5)] is missing, not to speak of the connection between the latter and the bare nucleon-nucleon interaction. It is not clear whether and how details of the symmetry-breaking force relating to its range, its dependence on spin etc. can be tested in the CN regime. Finally, experimental tests of parity violation will always use known p-wave resonances, and not look for an effect at energies different from the p-wave resonance energy. This procedure introduces a bias which further enhances the effect of symmetry violation but has so far not been considered theoretically.

Nonetheless, taking up the point of view mentioned in the Introduction, I believe that further symmetry tests in the CN regime will and should be carried out in the future. They should yield valuable information on parity violation, appreciably lower the bound on time-reversal symmetry violation and, if all goes well, shed light on the interaction which breaks both parity and time-reversal symmetry.

Acknoledgments

In writing this paper, I have benefitted greatly from the work done and the insights gained by D. Davis. I am grateful to him, to V.E. Bunakov, to H.L. Harney and to A. Müller for many informative and stimulating discussions on the subject. The manuscript was prepared while I visited the Lawrence Livermore and Berkeley National Laboratories. I am grateful to M. Blann for his thoughtful and generous hospitality.

References

1. E.D. Davis and U. Hartmann, submitted to Phys. Rev. C
2. E.D. Davis, Phys. Lett. **226** (1989) 197
3. E. D. Davis, in Proc. Workshop on "The breaking of fundamental symmetries in nuclei", Santa Fe, 1988
4. E.D. Davis, in Proc. Workshop on "Spin and Symmetries", TRIUMF, 1989 (to be published)
5. H.L. Harney and A. Müller, private communication
6. V.E. Bunakov, Phys. Rev. Lett **60** (1988) 2250
7. V.E. Bunakov and H.A. Weidenmüller, Phys Rev. **C39** (1989) 70
8. V.P. Alfimenkov, Sov. Phys. Usp. **27** (1984) 797, and references therein
9. J.D. Bowman, contribution to this Conference and references therein to earlier work at Los Alamos
10. B. Heckel et al., Phys. Rev. C **29** (1984) 2389
11. V.P. Alfimenkov et al., JETP Lett. **39** (1984) 416
12. Y. Masuda et al., to be published
13. O.P. Suskov and V.V. Flambaum, JETP Letters **32** (1980) 352; V.E. Bunakov and V.P. Gudkov, Nucl. Phys. **401** (1983) 93
14. E.P. Blanke et al., Phys. Rev. Lett. **51** (1983) 355
15. L. Stodolsky, Phys. Lett B **172** (1986) 5
16. R.V. Haq, A. Pandey and O. Bohigas, Phys. Rev. Lett. **48** (1982) 1086
17. G.E. Mitchell et al., Phys. Rep. **117** (1985) 1
18. O. Bohigas and M.-J. Giannoni, in: Lecture Notes in Physics, Springer-Verlag (1984) 209 p.1
19. H.A. Weidenmüller, Ann. Phys. (N.Y.) **158** (1984) 120
20. D. Boosé, H.L. Harney and H.A. Weidenmüller, Phys. Rev. Lett. **56** (1986) 2012
21. T.A. Brody et al., Rev. Mod. Phys. **53** (1981) 385
22. D. Boosé, H.L. Harney and H.A. Weidenmüller, Z. Phys. A **325** (1986) 363
23. J.J. M. Verbarschot, H.A. Weidenmüller and M.R. Zirnbauer, Phys. Rep. **129** (1985) 367
24. H.L. Harney, A. Richter and H.A. Weidenmüller, Rev. Mod. Phys. **58** (1986) 607
25. J.B. French et al., Ann. Phys. (N.Y.) **181** (1988) 198
26. J.B. French et al., Ann. Phys. (N.Y.) **181** (1988) 235
27. J.B. French et al., in "Tests of time reversal invariance in neutron physics", edited by N.R. Robertson, World Scientific, Singapore 1987

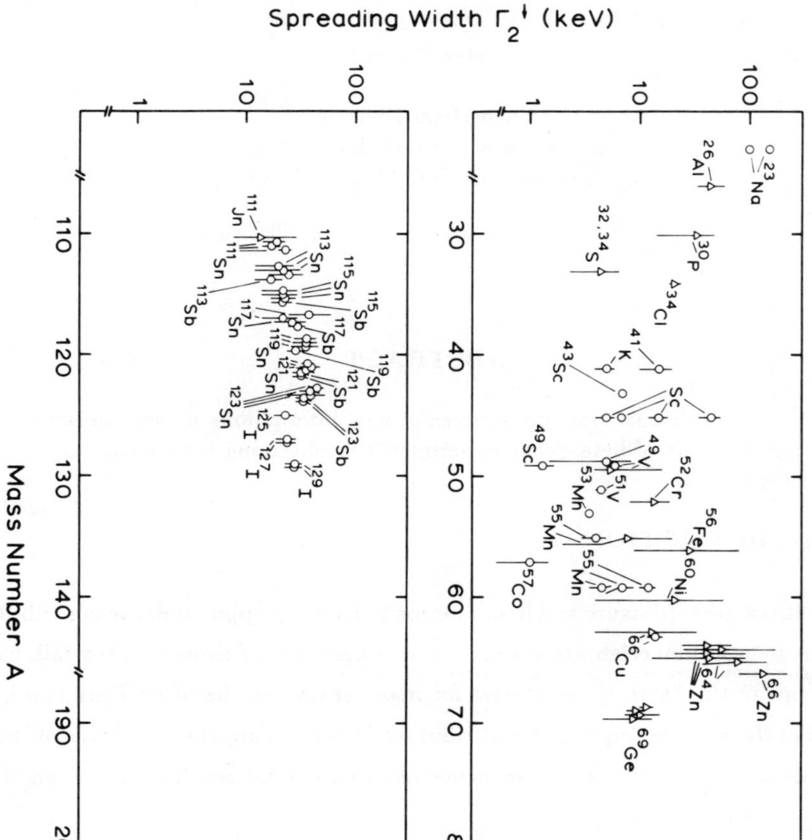

*Fig.*1. The spreading width $\Gamma_2^{\downarrow}$ versus mass number A for both isolated and mutually overlapping isobaric analogue resonances. Taken mostly from ref. (23) with new data supplied by H.L. Harney, private communication (1989)

BETA-DECAY BEYOND THE STANDARD MODEL

Peter Herczeg

Theoretical Division
Los Alamos National Laboratory
Los Alamos, New Mexico 87545

ABSTRACT

We discuss possible sources of new interactions in beta-decay and the role of beta-decay experiments in obtaining information on them.

1. INTRODUCTION

It is a great pleasure and honor for me to have the opportunity to contribute to this Symposium celebrating Felix. The subject I have chosen for my talk has been one of the fields of his interest for many years. In the fifties Felix and his collaborators made important contributions[1)] concerning the properties of the beta-decay interaction to the sequence of experimental results that paved the way to the $V-A$ theory.

In 1968, when I came to Caltech as a postdoctoral fellow, associated half-time with the group led by Felix, the $V-A$ theory was, as ten years earlier, consistent with all pertinent data. Nevertheless, there was strong motivation to search for deviations from the $V-A$ structure in beta-decay (and other processes). One of the reasons was CP-violation, discovered in 1964. The $V-A$ theory could not account for this effect. While it was recognized that the observed CP-violation could be due to a new force, it was equally possible that with appropriate modifications the weak interaction could account for it. Another motivation to search for new interactions was the non-renormalizability of the $V-A$ theory. Felix was keenly interested in the implications of the new theoretical models, and devoted his time to experiments that probed further the nature of the weak interactions and the origin of CP-violation.

Today the motivation to search for new interactions is not weaker. Although the minimal standard model is consistent with all observations, for many theoretical reasons, and especially because of the large number of undetermined parameters in the model, the presence of new physics is expected at some level.

In this talk we shall discuss possible sources of new interactions in beta-decay, and the role of beta-decay experiments in obtaining information on them.

In the next section we discuss beta-decay in left-right symmetric models, and then in models with new fermions. Sections 3 and 4 deal with scalar and tensor interactions, respectively. In Section 5 we analyze the effects of leptoquark exchange. The last section contains our conclusions.

In considering new contributions to beta-decay we shall assume that the quarks and leptons are elementary, and that their interactions are described by a renormalizable gauge theory. Then the $d \to ue^-\overline{\nu}_e$ (and $u \to de^+\nu_e$) transition underlying beta-decay will be described by an interaction, which is included in the most general local (assuming that the masses of the exchanged particles are much larger than the momentum transfer) nonderivative four-fermion interaction. Neglecting induced form factors and the q^2 – dependence of the remaining form factors, the general beta-decay interaction has then the familiar form[2)]

$$\begin{aligned} H_\beta = & \,\overline{e}\gamma_\lambda(C_V + C_V'\gamma_5)\nu_e\overline{p}\gamma^\lambda n \\ & + \overline{e}\gamma_\lambda\gamma_5(C_A + C_A'\gamma_5)\nu_e\overline{p}\gamma^\lambda\gamma_5 n \\ & + \overline{e}(C_S + C_S'\gamma_5)\nu_e\,\overline{p}n + \overline{e}\gamma_5(C_P + C_P'\gamma_5)\nu_e\,\overline{p}\gamma_5 n \\ & + \overline{e}\frac{\sigma_{\lambda\mu}}{\sqrt{2}}(C_T + C_T'\gamma_5)\nu_e\,\overline{p}\frac{\sigma^{\lambda\mu}}{\sqrt{2}}n + H.c. \end{aligned} \tag{1}$$

The constants C_K, C_K' are real if time-reversal (T)-invariance holds. The terms proportional to C_P and C_P' do not contribute in the nonrelativistic approximation for the nucleons.

In writing the Hamiltonian (1) we have assumed that only one neutrino state is involved in beta-decay, but we shall consider where important the more general case. We shall always assume that the kinematic effects of the masses of the neutrinos that can be produced in beta-decay can be neglected.[3]

In the minimal standard model the effective interaction describing the transition $d \to u e^- \overline{\nu}_e$ is the V-A interaction

$$H = -\frac{G}{\sqrt{2}} U_{ud}\, \overline{e}\gamma_\lambda (1-\gamma_5)\nu_e\, \overline{u}\gamma^\lambda(1 - \gamma_5)d + H.c. \quad , \tag{2}$$

where $G/\sqrt{2} = g^2/8M_W^2$, and $U_{ud} = \cos\theta_1$ is the ud-element of the Kobayashi-Maskawa matrix. The neutrino is a massless two-component particle. The beta-decay Hamiltonian corresponding to (2) is of the form (1) with

$$C_V/g_V = -C_V'/g_V = C_A/g_A = -C_A'/g_A = -(G/\sqrt{2})U_{ud} \tag{3}$$

and all the other constants equal to zero. In Eq. (3) $g_V = g_V(0), g_A = g_A(0)$, where $g_V(q^2)$ and $g_A(q^2)$ are the usual vector and axial-vector form factors for the $n \to p$ transition ($g_V = 1$ in the CVC approximation, and $g_A \simeq +1.25$).

Although the minimal standard model interaction dominates beta-decay, the data still allow sizable deviations from the V-A structure.[4] Recent experimental searches for deviations from the predictions of the minimal standard model focused among T-even observables on the longitudinal polarization of beta-rays from unpolarized nuclei. For the Hamiltonian (1), assuming that all the neutrino mass-eigenstates can be emitted in beta decay, the $e^\mp$ longitudinal polarization P_L in the direction parallel to $\overrightarrow{p_e}$ in allowed transitions is [5]

$$P_L = G\frac{p_e}{E_e}/(1 + b\frac{m_e}{E_e}) \quad , \tag{4}$$

where

$$\begin{aligned} G\xi &= 2|M_F|^2[\mp Re(C_S C_S'^* - C_V C_V'^*) - \frac{\alpha Z m_e}{p_e} Im(C_S C_V'^* + C_S' C_V^*)] \\ &+ 2|M_{GT}|^2[\mp Re(C_T C_T'^* - C_A C_A'^*) \\ &+ \frac{\alpha Z m_e}{p_e} Im(C_T C_A'^* + C_T' C_A^*)] \quad , \end{aligned} \tag{5}$$

$$\begin{aligned} \xi &= |M_F|^2 (|C_S|^2 + |C_S'|^2 + |C_V|^2 + |C_V'|^2) \\ &+ |M_{GT}|^2(|C_A|^2 + |C_A'|^2 + |C_T|^2 + |C_T'|^2) \end{aligned} \tag{6}$$

and b is the Fierz interference term, given by

$$b\xi = \pm 2(1 - \alpha^2 Z^2)^{1/2}\, Re\big[|M_F|^2(C_S C_V^* + C_S^{'} C_V^{'*}) - |M_{GT}|^2(C_T C_A^* + C_T^{'} C_A^{'*})\big] \quad . \tag{7}$$

From T-odd observables experimental information is available on the coefficients D and R of the correlations $< \vec{J} > \cdot \vec{p}_e \times \vec{p}_\nu / J E_e E_\nu$ and $\vec{\sigma} \cdot < \vec{J} > \times \vec{p}_e / J E_e$, respectively ($\vec{\sigma}$ = electron spin, $\vec{J}$ = nuclear spin).[6]. D and R can be written as $D = D_t + D_f$, $R = R_t + R_f$, where D_t, R_t represent the T-violating contributions and D_f, R_f are the T-invariant contributions due to electromagnetic final state interactions. D_t and R_t are given by[5]

$$D_t\xi = 2\delta_{J'J} M_F M_{GT}\Big(\frac{J}{J+1}\Big)^{1/2} Im(C_S C_T^* + C_S^{'} C_T^{'*} + C_V C_A^* + C_V^{'} C_A^{'*}) \quad , \tag{8}$$

$$\begin{aligned} R_t\xi = &\pm 2\lambda_{J'J}|M_{GT}|^2\, Im(C_T C_A^{'*} + C_T^{'} C_A^*) \\ &+ 2\delta_{JJ'} M_F M_{GT}\Big(\frac{J}{J+1}\Big)^{1/2} Im(C_S C_A^{'*} \\ &+ C_S^{'} C_A^* + C_V C_T^{'*} + C_V^{'} C_T^*) \quad , \end{aligned} \tag{9}$$

where $\lambda_{J'J}$ is an angular momentum factor, defined in Ref. 5.

In observables we shall keep only the lowest order terms in new beta-decay couplings. Note that then

$$\begin{aligned} C_S C_A^{'*} + C_S^{'} C_A^* &\simeq -(C_S C_A^* + C_S^{'} C_A^{'*}) \simeq g_A(C_S C_V^{'*} + C_S^{'} C_V^*) \\ &\simeq -g_A(C_S C_V^* + C_S^{'} C_V^{'*}) \\ &- \big[(g^2/8M_W^2)U_{ud}\big]^2 g_A\,(\overline{C}_S - \overline{C}_S^{'}) \quad , \end{aligned} \tag{10}$$

$$\begin{aligned} C_T C_A^{'*} + C_T^{'} C_A^* &\simeq -(C_T C_A^* + C_T^{'} C_A^{'*}) \simeq g_A(C_T C_V^{'*} + C_T^{'} C_V^*) \\ &\simeq -g_A(C_T C_V^* + C_T^{'} C_V^{'*}) \\ &- \big[(g^2/8M_W^2)U_{ud}\big]^2 g_A\,(\overline{C}_T - \overline{C}_T^{'}) \quad , \end{aligned} \tag{11}$$

where we have introduced the notation $\overline{C}_K \equiv C_K/(-g^2/8M_W^2)U_{ud}$, $\overline{C}_K^{'} = C_K^{'}/(-g^2/8M_W^2)U_{ud}$.

2. NEW V,A INTERACTIONS

2.1 Model Independent Considerations

The most general interaction for $d \to ue^-\overline{\nu}$ constructed from vector and axial-vector currents can be written in the form

$$\begin{aligned} H^{V,A} &= \overline{e}\gamma^{\lambda}(1-\gamma_5)(\Sigma_i U_{ei}\nu_i)\,\big[a_{LL}\overline{u}\gamma_{\lambda}(1-\gamma_5)d \\ &\quad + a_{LR}\overline{u}\gamma_{\lambda}(1+\gamma_5)d\big] \\ &\quad + \overline{e}\gamma^{\lambda}(1+\gamma_5)\,(\Sigma_i V_{ei}\nu_i)\big[a_{RL}\overline{u}\gamma_{\lambda}(1-\gamma_5)d \\ &\quad + a_{RR}\overline{u}\gamma_{\lambda}(1+\gamma_5)d\big] \;+\; H.c. \end{aligned} \tag{12}$$

where u, d are mass-eigenstates, U_{ei} and V_{ei} are, respectively, elements of the left-handed and the right-handed leptonic mixing matrices, and ν_i are the light neutrino mass-eigenstates (i.e. the neutrino mass-eigenstates that can be produced in beta-decay). Apart from U_{ei} and V_{ei}, the Hamiltonian (12) contains seven independent parameters (four complex numbers minus an overall phase). The observables are quadratic functions of the a_{ik}'s; a set of independent parameters $|a_{ik}|^2 (i,k = L,R)$ and the phases of $a_{LL}a^*_{LR}$, $a_{RR}a^*_{RL}$, and $a_{LL}a^*_{RR}$. The last of these appears only in observables proportional to the neutrino masses, which we shall not consider. For beta-decay into ν_i the coupling constants C_V, C_V', C_A and C_A' can be read off from Eq. 12. Thus $C_A = g_A[(a_{LL} - a_{LR})U_{ei} - (a_{RL} - a_{RR})V_{ei}]$, etc. Normalized observables (such as polarizations and asymmetries) depend only on the ratios $\eta_{ik} = a_{ik}/a_{LL} (ik = LR, RL, RR)$.

As the neutrinos are not observed, the beta-decay probability is the sum of the probabilities of decays into the energetically allowed neutrino mass-eigenstates. The effect of this is taken into account (assuming, as we do, that the kinematic effects of the neutrino masses can be neglected) by replacing in observables a_{LL}, η_{RL} and η_{RR} by $a^{(e)}_{LL} \equiv a_{LL}\sqrt{u_e}$, $\eta^{(e)}_{RL} \equiv \eta_{RL}\sqrt{\widetilde{v}_e}$, and $\eta^{(e)}_{RR} \equiv \eta_{RR}\sqrt{\widetilde{v}_e}$, respectively, where $u_e = \Sigma'_i\, |U_{ei}|^2, v_e = \Sigma'_i |V_{ei}|^2, \widetilde{v}_e = v_e/u_e$ (the summation is over the matrix elements associated with the light neutrinos).

In pure transitions normalized observables are either independent of the η_{ik}'s or proportional to (see Ref. 7)

$$1 - 2|\eta_{RR}^{(e)} + \eta_{RL}^{(e)}|^2 \quad \text{(Fermi transitions)} \quad ,$$

and

$$1 - 2|\eta_{RR}^{(e)} - \eta_{RL}^{(e)}|^2 \quad \text{(Gamow – Teller transitions)} \quad .$$

The rates involve $|a_{LL}|^2$, and in Gamow-Teller decays and in mixed transitions also g_A; normalized T-even observables in mixed transitions depend on $(g_A M_{GT}/g_V M_F)\mathrm{Re}[(1 - \eta_{LR})/(1 + \eta_{LR})]$, and some also on $\mathrm{Re}[(\eta_{RR}^{(e)} + \eta_{RL}^{(e)})(\eta_{RR}^{(e)} - \eta_{RL}^{(e)})*]$, in addition to $|\eta_{RR}^{(e)} + \eta_{RL}^{(e)}|^2$ and $|\eta_{RR}^{(e)} - \eta_{RL}^{(e)}|^2$. The parameter η_{LR} enters in lowest order always in the above combination, so that even in cases where the nuclear matrix elements are known, uncertainties in the theoretical value of g_A prevent the possibility of setting stringent limits on $\mathrm{Re}\eta_{LR}$ and $|\eta_{LR}|$.

The T-violating component of the D-coefficient is given by (see Eq. 8)

$$D_t = a_D \ \mathrm{Im}\ (\eta_{LR} + \eta_{RL}^{(e)}\eta_{RR}^{(e)}*) \quad . \tag{13}$$

The constant a_D contains the nuclear matrix elements for ^{19}Ne-and for n-decay $a_D \simeq -1.03$ and $a_D \simeq 0.87$, respectively.

What information is available on the parameters η_{ik}?

From experimental results on pure Gamow-Teller transitions[8] and on the e^+–longitudinal polarization in a Fermi transition[9] one can deduce[10]

$$|\eta_{RR}^{(e)}| < 0.13 \tag{14}$$

and

$$|\eta_{RL}^{(e)}| < 0.13 \ . \tag{15}$$

Comparisons of e^+–polarization in Fermi and Gamow-Teller decays[11,12] yield $P_L^F/P_L^{GT} = 1.00 \pm 0.003$ (see Ref. 12), implying $(P_L^F/P_L^{GT} - 1 \simeq -8\mathrm{Re}\eta_{RR}^{(e)}\eta_{RL}^{(e)*})$

$$-5.2 \times 10^{-4} < (-Re\eta_{RR}^{(e)}\ \eta_{RL}^{(e)}*) < 7.4 \times 10^{-4} \qquad \text{(90\% c.l.)}. \tag{16}$$

The best limit on D_t/a_D comes from ^{19}-Ne-decay. Barring a cancellation between the two terms in D (see Eq. 13), the experimental value $D = (0.1 \pm 0.6) \times 10^{-3}$ [Ref. 13] yields

$$-0.86 \times 10^{-3} < \mathrm{Im}\ \eta_{LR} < 1.05 \times 10^{-3} \qquad (90\%\ \text{c.l.}), \qquad (17)$$

and

$$-0.86 \times 10^{-3} < \mathrm{Im}\ \eta_{RL}^{(e)} \eta_{RR}^{(e)*} < 1.05 \times 10^{-3} \qquad (90\%\ \text{c.l.}). \qquad (18)$$

From Eqs. (16) and (18) one obtains

$$|\eta_{RR}^{(e)}|\ |\eta_{RL}^{(e)}| < 1.3 \times 10^{-3} \quad . \qquad (19)$$

There are further constraints on $\eta_{RL}^{(e)}$ and $\eta_{RR}^{(e)}$ from beta-ray asymmetries in polarized ^{19}Ne-and n-decays. In the case when $\eta_{RL}^{(e)}, \eta_{RR}^{(e)}$ are real and $\eta_{RR}^{(e)}$ positive, the combined beta-decay constraints yield $(-0.07) \lesssim \eta_{RL}^{(e)} \lesssim 0.05$ and $\eta_{RR}^{(e)} \lesssim 0.13$ (Ref. 14). For the general case ($\eta_{RL}^{(e)}, \eta_{RR}^{(e)}$ complex) the constraints have not been investigated yet.

A stringent limit on $|\eta_{RL}^{(e)}|$ and $|\eta_{RR}^{(e)}|$ has been derived from the energetics of Supernova 1989A. For right-handed neutrinos lighter than about 10 MeV the limit obtained (from the $e_R^- p \to \nu_e^{(R)} n$ reaction) is (Ref. 15; see also Ref. 16)

$$|\eta_{RL}^{(e)}| < 3 \times 10^{-5} \qquad (20)$$

and

$$|\eta_{RR}^{(e)}| < 3 \times 10^{-5} \quad , \qquad (21)$$

unless $|\eta_{RL}^{(e)}| > 2 \times 10^{-2}$ or $|\eta_{RR}^{(e)}| > 0.054$. The large values of $|\eta_{RL}^{(e)}|$ or $|\eta_{RR}^{(e)}|$, required to avoid the constraints (20) and (21) are not ruled out yet. A new class of beta-decay experiments involving measurements of the longitudinal positron polarization from polarized nuclei[17]), with a new technique for e^+-polarization measurements[18]) is expected to improve the limits on $|\eta_{RL}^{(e)}|$ and $|\eta_{RR}^{(e)}|$ (Ref. 19).

For η_{LR} the best limit is

$$|\eta_{LR}| < 0.1 \qquad (22)$$

provided by data on inclusive neutrino and antineutrino scattering.[20)]

Right-handed interactions in beta-decay can arise if there are new charged gauge bosons which have right-handed couplings to the usual leptons and/or quarks, or if there are new quarks and leptons which mix with the usual ones and which have right-handed couplings to the W. Below we consider such possibilities. Still another possible source of right-handed interactions is leptoquark-exchange, which we shall consider in Section 5.

2.2 Left-Right Symmetric Models

Left-right symmetric models are attractive extensions of the standard electroweak model, which provide an understanding of the origin of parity-violation in the weak interactions.[21)] The simplest models are based on the gauge group $SU(2)_L\times SU(2)_R\times U(1)$ (Ref. 22). These contain two distinct charged gauge boson fields W_L and W_R, which couple to left-handed and right-handed fermion currents, respectively. The gauge boson mass-eigenstates W_1 and W_2 are related to W_L and W_R as

$$\begin{aligned} W_L &= \cos\zeta\, W_1 + \sin\zeta W_2 \\ W_R &= e^{i\omega}\,(-\sin\zeta W_1 + \cos\zeta W_2) \end{aligned} \tag{23}$$

where ζ is a mixing angle and ω is a CP-violating phase. The exchange of W_1 and of W_2 generate an effective beta-decay interaction of the form 12 (Ref. 23).

Let us consider first the simplest version of $SU(2)_L\times SU(2)_R\times U(1)$ models, where the $SU(2)_L$ and $SU(2)_R$ gauge coupling constants g_L and g_R are equal ($g_L = g_R \equiv g$), the left-handed and the right-handed quark mixing matrices U_L and U_R are the same, $\widetilde{v}_e, \widetilde{v}_\mu, ... = 1$, and CP-invariance holds (manifestly left-right symmetric models). In these the parameters $a_{LL}^{(e)}$, η_{LR}, $\eta_{RL}^{(e)}$ and $\eta_{RR}^{(e)}$ are given by

$$\begin{aligned} a_{LL}^{(e)} &= a_{LL} \simeq (g^2\cos^2\zeta/8m_1^2)\cos\theta_1^L \\ \eta_{LR} &= \eta_{RL}^{(e)} \simeq -\zeta \\ \eta_{RR}^{(e)} &= \eta_{RR} \simeq m_1^2/m_2^2 \equiv \delta \quad , \end{aligned} \tag{24}$$

where m_1 and m_2 are the masses of W_1 and W_2, and $\cos\theta_1^L$ is the $ud-$element of U_L.

The parameters ζ and δ are subject to the constraints described in Section 2.1. In addition, one has $|\zeta| < 0.033, \delta < 0.04$ from muon decay data, and the more stringent (but less reliable) bounds $|\zeta| \lesssim 4 \times 10^{-3}, \delta \lesssim 3 \times 10^{-3}$ from nonleptonic transitions.[24] There is also a bound $|\zeta| \lesssim 10^{-2} - 10^{-3}$ implied by charged current universality.[25,26]

The above bounds are sufficient for the supernova limits (20) and (21) to become applicable, so that $|\zeta|$, $\delta \lesssim 3 \times 10^{-5}$. As a consequence, in manifestly left-right symmetric models the right-handed current effects in beta-decay are expected to be too small to be observable.

In the most general version of $SU(2)_L \times SU(2)_R \times U(1)$ models (models with non-manifest left-right symmetry and $\widetilde{v}_e$, $\widetilde{v}_\mu, ... \neq 1$) muon decay gives a limit $|\eta_{RL}^{(e)}| < 0.047$, but no constraint on $\eta_{RR}^{(e)}$ (Ref. 7). The $K^o \to \overline{K}^o$ transition sets a weak upper bound on $|\eta_{RR}^{(e)}|$, not better than 7.5×10^{-2} (Ref. 26). The best limit on $|\eta_{RR}^{(e)}|\ |\eta_{RL}^{(e)}|$ is from beta-decay data (Eq. 19). The upper limits on $|\eta_{RL}^{(e)}|$ and $|\eta_{RR}^{(e)}|$ are already near the values, where the supernova limits (20) and (21) become applicable.

For η_{LR} one has the constraint $|\eta_{LR}| \lesssim 4 \times 10^{-3}$ from nonleptonic $K-$decays and the $D-$coefficient.[7] There is no constraint on η_{LR} for light right-handed neutrinos from charged-current universality (see Ref. 16).

The T-violating component of the $D-$coefficient receives a contribution in first-order in $W_{1,2}-$exchange[27], given by

$$D_t \simeq -a_D \zeta \frac{g_R}{g_L} (\cos \theta_1^R / \cos \theta_1^L) \sin (\alpha + \omega) \quad , \tag{25}$$

where $\cos \theta_1^R e^{i\alpha} = (U_R)_{ud}$. In neglecting the second term in Eq. (13) we have assumed that $g_R^2 m_1^2 / g_L^2 m_2^2$ can be neglected relative to one, as indicated by the analysis in Ref. 16. The phase $(\alpha + \omega)$ contributes also to the electric dipole moment of the neutron D_n and to ϵ' / ϵ (Ref. 27). The limits on $|D_t a_D|$ from these observables are $\simeq 3 \times 10^{-5}$ (Ref. 28), but they are not as reliable as the limit (17).

2.3 Exotic Fermions

The interaction responsible for $d \to ue^-\bar{\nu}$ can contain terms involving right-handed currents even in models where the gauge group remains the standard SU(2) × U(1) group. This happens if new quarks and leptons exist which mix with the usual ones, and whose right-handed components are in non-singlet representations of SU(2) (Ref. 29). If the electric charge and color assignments of the new fermions are the standard ones, the only possibility for the non-singlet right-handed fermions is to be in doublet representations (mirror-fermions or vector doublets).[30] The new fermions have to be heavy (heavier than about 20 GeV) in view of limits from direct production at accelerators (see Ref. 30).

Requiring that the light-heavy fermion mixing does not lead to flavor-changing neutral currents involving the usual quarks and charged leptons (which are severely constrained by experiment), there is a considerable simplification in the form of the mixing matrices.[30] In this framework the parameters $a_{LL}^{(e)}$, η_{LR}, $\eta_{RL}^{(e)}$ and $\eta_{RR}^{(e)}$ are in lowest order in light-heavy mixings given by

$$a_{LL}^{(e)} \simeq (-g^2/8M_W^2)\sqrt{u_e} \quad , \tag{26}$$

$$\eta_{LR} \simeq s_R^u \, s_R^d \, (\widehat{V}_R)_{ud} \quad , \tag{27}$$

$$\eta_{RL}^{(e)} \simeq s_R^e \sqrt{\widetilde{v}_e} \quad , \tag{28}$$

$$\eta_{RR}^{(e)} \simeq \eta_{LR} \, \eta_{RL}^{(e)} \quad . \tag{29}$$

Here $s_R^u \equiv \sin\theta_R^u$, etc.; θ_R^e, θ_R^u are the light-heavy mixing angles associated with the electron, u–quark, and the d–quark; the quantities u_e and $\widetilde{v}_e$ were already defined in Sectin 2.1 (Ref. 31). The matrix $\widehat{V}_R$ is unitary for $n_R^u = n_R^d = m_R^u$, where n_R^u, n_R^d and m_R^u is the number of charge – $\frac{2}{3}$ singlet states, the number of charge – $(-\frac{1}{3})$ singlet states and the number of charge – $\frac{2}{3}$ doublet states, respectively.[30]

A comprehensive analysis of constraints on mixings between the usual and possible new fermions was made in Ref. 30. The results for the parameters involved in (26) - (29) are $|s_R^e \sqrt{\widetilde{v}_e}| < 0.042$ and $|s_R^u\, s_R^d\, \mathrm{Re}\,(\widehat{V}_R)_{ud}| < 6\times10^{-3}$, so that

$$| \eta_{RL}^{(e)} | < 0.042 \tag{30}$$

$$|\mathrm{Re}\; \eta_{LR} | < 6 \times 10^{-3} \quad . \tag{31}$$

The limits (30), (31), and Eq. 29 imply

$$| \mathrm{Re}\; \eta_{RR}^{(e)} | < 2.5 \times 10^{-4} \quad . \tag{32}$$

The elements of $\widehat{V}_R$ are in general complex. The phase in $(\widehat{V}_R)_{ud}$ contributes to D_t (Ref. 32). Writing $(\widehat{V}_R)_{ud} = e^{i\phi}\,(\widehat{V}_R^{'})_{ud}$, where $(\widehat{V}_R^{'})_{ud}$ is real, we have (see Eqs. 27 - 29)

$$\begin{aligned} D_t &\simeq a_D \left[1 - (s_R^e \sqrt{\widetilde{v}_e})^2\right] \mathrm{Im}\eta_{LR} \simeq a_D\; \mathrm{Im}\eta_{LR} \\ &\simeq a_D\; s_R^u s_R^d\; (\widehat{V}_R^{'})_{ud} \sin\phi \quad . \end{aligned} \tag{33}$$

The bounds (17) imply

$$|s_R^u\; s_R^d\; (\widehat{V}_R^{'})_{ud} \sin\phi| < 1\times10^{-3} \quad . \tag{34}$$

An upper limit of $\sim 3\times10^{-5}$ on $|s_R^u\; s_R^d(\widehat{V}_R^{'})_{ud} \sin\phi|$ follows again from D_n and $\epsilon^{'}/\epsilon$.

Eqs. (31) and (34) yield the limit

$$|\eta_{LR}| < 6.1\times10^{-3} \quad . \tag{35}$$

From Eqs. (29), (30) and (34) we have

$$|\mathrm{Im}\; \eta_{RR}^{(e)}| < 4.4\times10^{-5} \tag{36}$$

which, when combined with the bound (32), yields

$$|\eta_{RR}^{(e)}| < 2.6\times10^{-4} \quad . \tag{37}$$

From (35) and (37) on obtains $|\,\mathrm{Re}\eta_{RR}^{*}\;\eta_{RL}| \le |\eta_{RR}^{(e)}|\,|\eta_{RL}^{(e)}| < 1.1\times10^{-5}$.

3. SCALAR INTERACTIONS

The most stringent bound on the T-invariant component of scalar-type beta-decay interactions comes from the value[33)]

$$b_F = (1.2 \pm 5) \times 10^{-3} \tag{38}$$

of the Fierz interference term (Eq. 7), derived from ft-values of superallowed $0^+ \to 0^+$ decays. This implies (see Eqs. 7 and 10)

$$|\mathrm{Re}(\overline{C}_S - \overline{C}'_S)| = |b_F| < 9.4 \times 10^{-3} \qquad (90\%)\ \text{c.l.}). \tag{39}$$

The best limit on T-violating scalar couplings is set by a measurement of the R-coefficient in ^{19}Ne–decay.[34)] The experimental value $R = -0.079 \pm 0.053$ yields (see Eqs. 9 and 10)

$$|\mathrm{Im}(\overline{C}_S - \overline{C}'_S)| < 0.64 \qquad (90\%)\ \text{c.l.}). \tag{40}$$

The limits on $|C_S|$ and $|C'_S|$ are an order of magnitude weaker than (39) (Ref. 4).

Scalar-type couplings can arise from charged Higgs-boson exchange. Charged Higgs bosons are present already in the standard model if the Higgs sector contains more than one Higgs doublet. The couplings of the charged Higgs boson to fermions in such models are not determined. The possibility that the corresponding scalar beta-decay couplings have values near the limits (39) and (40) cannot be ruled out, although there is no known natural model with such features (see Ref. 35).

Another possible source of scalar-type beta-decay interactions is lepto-quark exchange, which we shall discuss in Section 5.

Denoting the coupling of a charged Higgs boson (H) to $\overline{u}d$ and for example to $\overline{\nu}_e e$ by f'' and f', C_S is given by $(f'f''/m_H^2)g_S$, where $g_S \equiv g_S(0)$ is defined by $< p|\overline{u}d|n > = g_S(q^2)\overline{u}_p u_n$. g_S was estimated in Ref. 36 and in Ref. 37, obtaining $g_S \simeq 1$ and $g_S \simeq 0.6$, respectively.

4. TENSOR INTERACTIONS

The experimental bounds on tensor couplings in beta-decay are

$$|\mathrm{Re}(\overline{C}_T - \overline{C}'_T)| < (1.3 \times 10^{-2}) \qquad (95\% \text{ c.l.}), \qquad (41)$$

mainly from the Fierz interference term in ^{22}Na (Ref. 4), and

$$|\mathrm{Im}(\overline{C}_T - \overline{C}'_T)| < (9.1 \times 10^{-2}) \qquad (95\% \text{ c.l.}) \qquad (42)$$

from the result $\mathrm{Im}(\overline{C}_T\overline{C}'^*_A + \overline{C}'_T\overline{C}^*_A) = -0.020 \pm 0.047$ (Ref. 38), derived from data on $e^{\mp}$ longitudinal polarizations. The limits on $|\overline{C}_T|$ and $|\overline{C}'_T|$ are an order of magnitude weaker than (41) (Ref. 4).

A mechanism that could generate tensor couplings would be the exchange of spin-two bosons, coupled to lepton and tensor currents. However, there is no known renormalizable gauge theory involving tensor bosons. The only other possible mechanism for elementary fermions which gives rise to tensor couplings in beta-decay at the tree-level is the exchange of spin-zero leptoquarks (see Section 5).

The constants C_T, C'_T are given in terms of the coupling constants a_{TT}, a'_{TT} of the four-fermion couplings $\overline{e}(\sigma^{\mu\nu}/\sqrt{2})\nu_e\overline{u}(\sigma_{\mu\nu}/\sqrt{2})d$ and $\overline{e}(\sigma^{\mu\nu}/\sqrt{2})\gamma_5\nu_e\overline{u}(\sigma_{\mu\nu}/\sqrt{2})d$ as $C_T = a_{TT}g_T, C'_T = a'_{TT}g_T$, where $g_T \equiv g_T(0)$ is defined by $< p|\overline{u}(\sigma_{\mu\nu}/\sqrt{2})d|n > \simeq g_T(q^2)\overline{u}_p(\sigma_{\mu\nu}/\sqrt{2})u_n$. A calculation[37] yields $g_T \simeq 1.2g_A$.

5. LEPTOQUARK EXCHANGE

Leptoquarks are bosons which induce quark-lepton transitions. They appear in many extensions of the minimal standard model. In models where they do not induce proton decay, leptoquarks could be light enough to cause observable effects in some low-energy processes.[39] Both spin-one and spin-zero leptoquarks occur. The $d \to ue^-\nu$ transition can be mediated by either Q ($\equiv$ electric charge) $= \frac{2}{3}$ or Q$= -\frac{1}{3}$ leptoquarks. We shall denote spin-one and spin-zero leptoquarks of electric charge Q by $X^{(Q)}$ and $Y^{(Q)}$, respectively. The four-fermion interactions $H_{X(Q)}$ for $d \to ue^- \nu_e$ generated by $X^{(Q)}$–leptoquark exchange can be written in the form

$$H_{X(2/3)} = \sum_{i,j=V,A} f_{ij}\, \overline{e}\, \Gamma_i d\, \overline{u}\, \Gamma_j \nu_e + H.c., \tag{43}$$

$$H_{X(-1/3)} = \sum_{i,j=V,A} h_{ij}\, \overline{e}\, \Gamma_i u^c\, \overline{d}^c\, \Gamma_j \nu_e + H.c., \tag{44}$$

where $\Gamma_V = \gamma_\mu$, $\Gamma_A = \gamma_\mu \gamma_5$, and u^c, d^c are the charge-conjugates of u, d. The f_{ij}'s and h_{ij}'s are proportional to the product of the fermion-leptoquark couplings involved, and to m_{LQ}^{-2}, where m_{LQ} is the mass of the leptoquark. We shall assume that the left-handed neutrino state involved in the leptoquark interaction is the same as the one participating in the minimal standard model interaction. In Eqs. (43) and (44) we have taken for simplicity the left-handed and the right-handed neutrino states $\nu_e^{(L)}$ and $\nu_e^{(R)}$ to be the left-handed and right-handed components of the same neutrino field ν_e, and shall assume that ν_e is a mass-eigenstate. For general neutrino states the beta-decay coupling constants associated with interaction terms involving $\nu_e^{(L)}$ ($\nu_e^{(R)}$) have to be multiplied by $\sqrt{u_e}$ ($\sqrt{v_e}$). The four-fermion interactions induced by $Y^{(2/3)}$–and $Y^{(-1/3)}$–leptoquark exchange ($H_{Y(2/3)}$ and $H_{Y(-1/3)}$) have the same form as (43) and (44), respectively, except for $V, A \to S, P$ ($\Gamma_S = 1, \Gamma_P = \gamma_5$).

We shall consider now the effects of the various leptoquark-exchange Hamiltonians on beta-decay.[40] Leptoquarks do not contribute in lowest order to leptonic and nonleptonic processes. Leptoquark contributions to semileptonic processes are therefore not constrained significantly by them.

<u>$X^{(2/3)}$–leptoquarks.</u> After a Fierz transformation the Hamiltonian (43) yields a beta-decay interaction of the form (1) with

$$C_V/g_V = C_A/g_A = \frac{1}{2}\,(f_{VV} + f_{AA}) \quad , \tag{45}$$

$$C'_V/g_V = C'_A/g_A = \frac{1}{2}\,(f_{VA} + f_{AV}) \quad , \tag{46}$$

$$C_S/g_S = -\,C_P/g_P = -\,f_{VV} + f_{AA} \quad , \tag{47}$$

$$C'_S/g_S = -\,C'_P/g_P = -\,f_{VA} + f_{AV} \quad , \tag{48}$$

$$C_T = C_T' = 0 \quad , \tag{49}$$

where g_V, g_A and g_S have been defined earlier, and $g_P \equiv g_P(0)$ is defined by $< p|\, \overline{u}\gamma_5 d\, |n> = g_P(q^2)\overline{u}_p\gamma_5 u_n$.

A stringent constraint on the pseudoscalar-type couplings comes from the experimental result[41)]

$$R_\pi \equiv \frac{\Gamma(\pi^+ \to e^+\nu_e)}{\Gamma(\pi^+ \to \mu^+\nu_\mu)} = (1.218 \pm 0.014) \times 10^{-4} \tag{50}$$

on $\pi \to \ell\nu$ decays. R_π is given by

$$R_\pi = (1.233 \times 10^{-4})\,(Q^{(e)}/Q^{(\mu)}) \quad , \tag{51}$$

where $Q^{(e)}$ and $Q^{(\mu)}$ are sensitive to new semileptonic interactions involving axial-vector and pseudoscalar quark currents[42)] ($Q^{(e)}/Q^{(\mu)} = 1$ in the minimal standard model). The experimental result (50) implies

$$0.9663 < \frac{Q^{(e)}}{Q^{(\mu)}} < 1.0095 \qquad (90\%\ \text{c.l.}). \tag{52}$$

In the limits (52) we have included the theoretical uncertainty of 0.3% in the calculation of the coefficient of $Q^{(e)}/Q^{(\mu)}$ in Eq. 51. Neglecting the leptoquark contribution to a_{LL}, and assuming that there is no cancellation between the contribution of the pseudoscalar couplings involving the electron and the muon[43)], Eq. 52 yields[44)]

$$|\, Re\, (\overline{C}_P - \overline{C}_P')/g_P| < 1.2 \times 10^{-3} \tag{53}$$

$$|\, Im\, (\overline{C}_P - \overline{C}_P')/g_P| < 6.2 \times 10^{-4} \quad . \tag{54}$$

The bounds (53) and (54), and the relations (47) and (48) imply (with $g_S \simeq 1$)

$$|\, Re\, (\overline{C}_S - \overline{C}_S')\,| \lesssim 1.2 \times 10^{-3} \quad , \tag{55}$$

$$|\, Im\, (\overline{C}_S - \overline{C}_S')\,| \lesssim 6.2 \times 10^{-4} \quad . \tag{56}$$

For the Fierz-interference term b_F (Eq. 7) and the T-violating component of R in ^{19}Ne−decay (55) and (56) set the limits

$$|\, b_F| \lesssim 1.2 \times 10^{-3} \tag{57}$$

and

$$|\,(R_t)_{Ne}| \lesssim 1.6 \times 10^{-4} \quad , \tag{58}$$

respectively.

The V, A part of the $d \rightarrow ue^-\nu_e$ interaction (minimal standard model contribution plus the V, A part of the leptoquark contribution) is of the form (12) with

$$a_{LL} = a_{LL}^{(o)} + \frac{1}{4}\,(f_{VV} + f_{AA} - f_{VA} - f_{AV}) \quad , \tag{59}$$

$$a_{RR} = \frac{1}{4}\,(f_{VV} + f_{AA} + f_{VA} + f_{AV}) \quad , \tag{60}$$

$$a_{LR} = 0 \quad , \tag{61}$$

$$a_{RL} = 0 \quad , \tag{62}$$

where $a_{LL}^{(o)}$ is the minimal standard model contribution $[a_{LL}^{(o)} = -(g^2/8M_W^2)U_{ud}]$. As a consequence, the T-violating component of the D-coefficient vanishes. Neglecting the scalar contribution to P_L^F, we have $P_L^F = P_L^{GT} \simeq 1 - 2|\,\eta_{RR}|^2$ (and therefore $P_L^F/P_L^{GT} = 1$).

The only constraints on the V, A−part of the interaction (as well as on the V, A−part of the other leptoquark-exchange interactions to be discussed below) are those described in Section 2.1.

<u>$X^{(-1/3)}$−leptoquarks.</u> The beta-decay interaction resulting from the Hamiltonian (44) is of the form (1) with

$$C_V/g_V = -\, C_A/g_A = -\,\frac{1}{2}\,(h_{VV} + h_{AA}) \quad , \tag{63}$$

$$C_V'/g_V = -\, C_A'/g_A = -\,\frac{1}{2}\,(h_{VA} + h_{AV}) \quad , \tag{64}$$

$$C_S/g_S = -\, C_P/g_P = -\, h_{VV} + h_{AA} \quad , \tag{65}$$

$$C'_S/g_S \;=\; -\,C'_P/g_P \;=\; -\,h_{VA} \,+\, h_{AV} \quad , \tag{66}$$

$$C_T \;=\; C'_T \;=\; 0 \quad . \tag{67}$$

In view of the relations (65) and (66), the bounds (53) - (58) apply also in this case.

The constants a_{ik} of the $V, A-$part of the interaction, with the minimal standard model contribution included, are

$$a_{LL} \;=\; a^{(o)}_{LL} \quad , \tag{68}$$

$$a_{LR} \;=\; -\,\frac{1}{4}(h_{VV} \,+\, h_{AA} \,-\, h_{VA} \,-\, h_{AV}) \quad , \tag{69}$$

$$a_{RR} \;=\; 0 \quad , \tag{70}$$

$$a_{RL} \;=\; -\,\frac{1}{4}(h_{VV} \,+\, h_{AA} \,+\, h_{VA} \,+\, h_{AV}) \quad . \tag{71}$$

The T-violating component of the D-coefficient is now generally non-zero. We have from Eq. (13)

$$D_t \;=\; (a_D/(g^2/8M_W^2)\, U_{ud})\, \frac{1}{4}\, Im\, (h_{VV} \,+\, h_{AA} \,-\, h_{VA} \,-\, h_{AV}) \, . \tag{72}$$

Leptoquark contributions to D_t are constrained only by the experimental limit on D.

For P_L^F and P_L^{GT} we have in this case $P_L^F \;=\; P_L^{GT} \;\simeq\; 1 \,-\, 2|\eta_{RL}|^2$ (so that again $P_L^F/P_L^{GT} \;=\; 1$).

$Y^{(2/3)}$–leptoquarks. A novel feature of the beta-decay interaction generated by spin-zero leptoquark-exchange is the presence of tensor terms, in addition to $V, A, S, P-$couplings.

The $V, A-$part of the beta-decay interaction corresponding to $H_{Y(2/3)}$ is the same as for $H_{X(-1/3)}$, except for $h_{VV} \,+\, h_{AA} \;\rightarrow\; -\frac{1}{2}\,(-f_{SS} \,+\, f_{PP})$ and $h_{VA} \,+\, h_{AV} \;\rightarrow\; -\frac{1}{2}\,(-f_{SP} \,+\, f_{PS})$. Thus D_t is generally non-zero (given by Eq. 72 with the above substitutions), and also for P_L^F, P_L^{GT} the same conclusions hold as for $H_{X(-1/3)}$.

The $S, P, T-$ couplings are related as

$$C_S/g_S \;=\; C_P/g_P \;=\; C_T/g_T \quad , \tag{73}$$

$$C'_S/g_S = C'_P/g_P = C'_T/g_T \quad , \tag{74}$$

where $C_S = -\frac{1}{4}(f_{SS} + f_{PP})$, $C'_S = -\frac{1}{4}(f_{SP} + f_{PS})$. Consequently, the limits (53) and (54) imply

$$|\, b_{GT}| = |\, Re(\overline{C}_T - \overline{C}'_T)/g_A| \lesssim 10^{-3} g_T \tag{75}$$

$$|Im\,(\overline{C}_T - \overline{C}'_T)| < (6.2 \times 10^{-4}) g_T \quad , \tag{76}$$

and also the bounds (55) - (57). For R_t in ^{19}Ne–decay we obtain

$$|(R_t)_{Ne}| \lesssim |1.9 - 1.6(1 - \frac{g_S}{g_T})| \times 10^{-4} |g_T| \quad . \tag{77}$$

<u>$Y^{(-1/3)}$–leptoquarks.</u> The V, A–part of the interaction resulting from $H_{Y(-1/3)}$ is the same as for $H_{X(2/3)}$, except for $f_{VV} + f_{AA} \to -\frac{1}{2}(-h_{SS} + h_{PP}), f_{VA} + f_{AV} \to -\frac{1}{2}(-h_{SP} + h_{PS})$.

The $S, P, T-$ couplings are related as

$$C_S/g_S = C_P/g_P = -C_T/g_T \tag{78}$$

$$C'_S/g_S = C'_P/g_P = -C'_T/g_T \tag{79}$$

and therefore the bounds (55) - (57), (75) and (76) hold. For $(R_t)_{Ne}$ we find

$$|(R_t)_{Ne}| \lesssim |1.9 - 1.6(1 + \frac{g_S}{g_T})| \times 10^{-4} |g_T| \quad . \tag{80}$$

6. CONCLUSIONS

In this talk we discussed possible sources of new interactions in beta-decay[45], their effects on beta-decay observables, and the constraints on the beta-decay coupling constants. We can summarize the main points as follows:

- Right-handed interactions can arise in beta-decay at the tree level in left-right symmetric models (or other models involving new gauge bosons with right-handed couplings), in models with exotic fermions, and in models involving leptoquarks.

An improvement of the existing limit on $|\eta_{RR}^{(e)}|$ through beta-decay measurements would further constrain models with $X^{(2/3)}-$ and $Y^{(-1/3)}$-type leptoquarks, as well as general left-right symmetric models. In the above models $|\eta_{RR}^{(e)}|$ is not constrained by muon-decay data. The $K^o \to \overline{K}^o$ transition does not impose a significant limit on $|\eta_{RR}^{(e)}|$ generated by leptoquark exchange, and constrains $|\eta_{RR}^{(e)}|$ in general left-right symmetric models only weakly. In models with exotic fermions various data combined set already a stringent upper limit ($\simeq 3 \times 10^{-4}$) on $|\eta_{RR}^{(e)}|$.

Improvements of the limit on $|\eta_{RL}^{(e)}|$ through beta-decay measurements would set more stringent constraints on models with $X^{(-1/3)}-$and $Y^{(2/3)}-$type leptoquarks. For general left-right symmetric models the best present limit on $|\eta_{RL}^{(e)}|$ comes from muon-decay, and in models with exotic fermions from a combination of data (including muon-decay).

The beta-decay limits for $\mathrm{Re}(\eta_{RR}^{(e)}\ \eta_{RL}^{(e)*})$ and $\mathrm{Im}(\eta_{RR}^{(e)}\ \eta_{RL}^{(e)})$ combined provide a complementary constraint for $|\eta_{RR}^{(e)}|$ and $|\eta_{RL}^{(e)}|$.

It should be noted that the present limits on $|\eta_{RR}^{(e)}|$ and $|\eta_{RL}^{(e)}|$ are close to the values where the limits derived from the energetics of Supernova 1989A become applicable.

- More sensitive searches for a nonzero $D-$coefficient would be of importance.

  The T-violating component of the $D-$coefficient receives a tree-level contribution both in left-right symmetric models and in models with exotic fermions, and also in models with $X^{(-1/3)}-$ and $Y^{(2/3)}-$type leptoquarks. For D_t in left-right symmetric models and in models with exotic fermions more stringent limits than the present experimental limit on D follow from the experimental results on D_n and ϵ'/ϵ. However these bounds are not rigorous, in view of the uncertainties in the calculations. The contribution to D_t from leptoquark exchange can be as large as the present experimental upper limit for D.

- For scalar type beta-decay interactions generated by charged Higgs-boson exchange values of the T-invariant and T-violating components of $C_S - C_S'$ as large as the present limits cannot be ruled out. Although there is no compelling model where the size of the couplings can be so large, improvements of the present limits would be worthwhile.

Scalar interactions are generated also by leptoquark exchange. However, limits on pseudoscalar couplings derived from the $\pi \to e\nu/\pi \to \mu\nu$ ratio (whose validity require the absence of some cancellations) set stringent limits on such interactions (cf. Eqs. 55 and 56).

- The only possible source of tensor interactions in beta-decay at the tree level appears to be the exchange of spin-zero leptoquarks. For tensor interactions generated this way the stringent limits from $\pi \to e\nu/\pi \to \mu\nu$ apply (see Eqs. 75 and 76). In spite of these points, improved limits on tensor couplings, even above these limits, would be desirable.
- In models where the quarks and leptons are composite, which we did not discuss here, new beta-decay interactions could arise as a result of constituent exchange[46]. The strength of such interactions is of the order of $4\pi/\Lambda^2$, where Λ is the compositeness scale. With $\Lambda \gtrsim (1-3)$ TeV (the current limits from other processes), the corresponding beta-decay constants could have values near the present upper bounds.

ACKNOWLEDGEMENTS

I would like to thank J. Deutsch, S. Hatamian and M. Skalsey for valuable conversations, and to Petr Vogel, who organized this Symposium, for inviting me to give a talk on this occasion. This is a good opportunity also to express my gratitude to Felix Boehm for the countless stimulating coversations over the years on many issues in nuclear and particle physics, and in particular on symmetry tests.

This work was supported by the United States Department of Energy.

Many Happy Returns, Felix!

REFERENCES

1. Boehm, F. and Wapstra, A. H., Phys. Rev. 107, 1202 (1957); Phys. Rev. 109, 456 (1958); Boehm, F., Novey, F. T. B., Barnes, C. A. and Stech, B., Phys. Rev. 108, 1497 (1957).

2. Our metric and $\gamma-$matrices are the same as in Bjorken, J. D. and Drell, S. D., Relativistic Quantum Mechanics (McGraw-Hill, New York, 1964). Our constants C_K, C_K' are the same as those in Jackson, J. D., Treiman, S. B. and Wyld, Jr., H. W., Phys. Rev. 106, 517 (1957), except for the opposite sign of C_V', C_A, C_S', C_P' and C_T'.
3. Some effects of neutrino mass on beta-decay are discussed by Robertson, R. G. H., in these Proceedings.
4. A comprehensive study of constraints on the Hamiltonian (1) from beta-decay experiments, assuming T-invariance, was made in Boothroyd, A. I., Markey, J. and Vogel, P. The bounds obtained for example for tensor interactions are $|C_T| < (9 \times 10^{-2})g_A, |C_T'| < (8.5 \times 10^{-2})g_A$ and $|\mathrm{Re}(C_T - C_T')| < (1 \times 10^{-2})g_A$. These limits are still the most stringent ones on such types of couplings.
5. Jackson, J. D., Treiman, S. B. and Wyld, Jr., H. W., Ref. 2 and Nucl. Phys. 4, 206 (1957).
6. For a recent review of experiments searching for T-violation in beta-decay see Boehm, F., Hyp. Int. 43, 95 (1988).
7. Herczeg, P., in Weak and Electromagnetic Interactions in Nuclei, proceedings of the International Symposium, Heidelberg, West Germany, 1987, edited by Klapdor, H. V. (Springer-Verlag, Berlin, 1986), p. 528.
8. Klinken van, J., Koks, F. W. J. and Behrens, H., Phys. Lett. 79B, 199 (1978).
9. Klinken van, J., Stam, K., Venema, W. Z., Wichers, V. A. and Atkinson, D., Phys. Rev. Lett. 50, 94 (1983).
10. See Ref. 7. The limits on $|\eta_{RR}^{(e)}|$ and $|\eta_{RL}^{(e)}|$ from these measurements may actually be weaker, since the neglect of forbidden matrix elements in the most precise experiments used in Ref. 8 may not be justified [Deutsch, J., in Fundamental Symmetries and Nuclear Structure, proceedings of the Workshop on The Breaking of Fundamental Symmetries in Nuclei, edited by Ginocchio, J. N. and Rosen, S. P., (World Scientific, Singapore, 1989, p. 36].

11. Wichers, V. A., Hageman, T. R., Klinken van, J. and Wilschut, H. W., Phys. Rev. Lett. 58, 1821 (1987).

12. Carnoy, A. S., Deutsch, J., Girard, T. A. and Prieels, R., Univ. Catholique de Louvain preprint, to appear in the Proceedings of the International Symposium on Weak and Electromagnetic Interactions in Nuclei (WEIN), May 1989, Montreal, Canada.

13. Hallin, A. L., Calaprice, F. P., MacArthur, D. W., Piilonen, L. E., Schneider, M. B., and Schreiber, D. F., Phys. Rev. Lett., 52, 337 (1984).

14. See Carnoy, A. S., Deutsch, J. and Holstein, B. R., Phys. Rev. 38, 1636 (1988), and Deutsch, J., Ref. 10. The limits on $\eta_{RL}^{(e)}$ and $\eta_{RR}^{(e)}$ (for $\eta_{RL}^{(e)}, \eta_{RR}^{(e)}$ real, $\eta_{RR}^{(e)} > 0$) are the same as the limits on ζ and δ in these papers.

15. Mohapatra, R. N. and Barbieri, R., Phys. Rev., D39, 1229 (1989); Raffelt, G. and Seckel, D., Phys. Rev. Lett. 60, 1793 (1988).

16. Langacker, P. and Sankar, Uma, S., Phys. Rev. D40, 1569 (1989).

17. Quin, P. A. and Girard, T. A., to appear in Phys. Lett.

18. Skalsey, M., Girard, T. A. and Rich, A., Phys. Rev. C32, 1014 (1985); Skalsey, M., et al., Phys. Rev. C39, 986 (1989); Girard, T. A., Carnoy, A. S., Deutsch, J., Prieels, R. and Schmit, M. L., Z. Phys. A330, 51 (1988); Carnoy, A. S., et al., Ref. 12.

19. Deutsch, J., private communication.

20. Abramowicz, H., et al., Z. Phys. C12, 225 (1982).

21. Pati, J. C. and Salam, A., Phys. Rev. Lett. 31, 661 (1973); Phys. Rev. D10, 275 (1974); Mohapatra, R. N. and Pati, J. C., Phys. Rev. D11, 566 (1975); Phys. Rev. D11, 2558 (1975); Senjanović, G. and Mohapatra, R. N., Phys. Rev. D12, 1502 (1975).

22. For brief reviews of the aspects of $SU(2)_L \times SU(2)_R \times U(1)$ models relevant to our discussion see Langacker, P. and Sankar, Uma, S., Ref. 16, and Herczeg, P., Phys. Rev. D34, 3449 (1986).

23. For discussions of beta-decay in $SU(2)_L \times SU(2)_R \times U(1)$ models and for references to earlier work see Deutsch, J., Ref. 10, and Herczeg, P., Ref. 7.

24. For a review see Herczeg, P., Ref. 22. For the limit on δ from muon decay we used the experimental result given in A. E. Jodidio et al., Phys. Rev. D37, 237(E) (1988).

25. Wolfenstein, L., Phys. Rev. D29, 2130 (1984); Deutsch, J., Univ. Catholique de Louvain preprint, May, 1989, to appear in Acta Physica Hungarica.

26. See Ref. 16, where a comprehensive analysis of constraints on general $SU(2)_L \times SU(2)_R \times U(1)$ models was made.

27. Herczeg, P., Phys. Rev. D28, 200 (1983), and in Neutrino Mass and Low-Energy Weak Interactions, Telemark, 1984, edited by V. Barger and D. Cline (World Scientific, Singapore, 1985), p. 288.

28. We used the formulas for D_n and ϵ'/ϵ in Herczeg, P., in Neutrino Mass...., Ref. 27, based on the calculations of Beall, G. and Soni, A. [Phys. Rev. Lett. 47, 552 (1981)], and Chang, D. [Nucl. Phys. B214, 435 (1983)], respectively. We used the experimental limit $|D_n| < 1.2 \times 10^{-25}$ (95% c.l.) [Smith, K. F., et al., Univ. of Sussex preprint, Oct. 1989], and the result $\epsilon'/\epsilon = (3.3 \pm 1.1) \times 10^{-3}$ [Burkhardt, H., et al., Phys. Lett. B206, 169 (1988)].

29. The term "exotic fermion" refers to fermions with noncanonical $SU(2)_L \times U(1)$ assignments; see Langacker, P. and London, D., Phys. Rev. D38, 886 (1988). This paper contains also references to earlier work on mixings between the usual fermions and various types of exotic ones.

30. Langacker, P. and London, D., Phys. Rev. D38, 886 (1988).

31. In Ref. 30 $\sqrt{u_e}$ and $\sqrt{v_e}$ are denoted by c_L^ν and s_R^ν, respectively.

32. Herczeg, P., to be published.

33. Ormand, W. E., Brown, B. A., and Holstein, B. R., preprint 1989.

34. Schneider, M. B., Calaprice, F. P., Hallin, A. L., MacArthur, D. W. and Schreiber, D. F., Phys. Rev. Lett. 51, 1239 (1983).

35. Herczeg, P., in Tests of Time Reversal Invariance in Neutron Physics, proceedings of the Workshop on Tests of Time Reversal Invariance in Neutron Physics, Chapel Hill, 1987, edited by Roberson, N. R., Gould, C. R. and Bowman, J. D. (World Scientific, Singapore, 1987), p. 24.

36. McWilliams, B. and Li, L.-F., Nucl. Phys. B179, 62 (1981).

37. Adler, S. L., et al., Phys. Rev. D11, 3309 (1975).

38. Skalsey, M. and Hatamian, S., Bull. Am. Phys. Soc. 30, 1277 (1985). See also Skalsey, M. and Hatamian, S., Phys. Rev. C31, 2218 (1985).

39. Some recent papers concerned with leptoquarks and their phenomenology are Buchmüller, W. and Wyler, D., Phys. Lett. B177, 377 (1986); Kizikuri, Y., Phys. Lett. B185, 183 (1987); Shanker, O., Nucl. Phys. B204, 375 (1982). See also references quoted in these papers.

40. Herczeg, P., to be published.

41. Bryman, D. A. et al., Phys. Rev. D33, 112 (1986).

42. For a review see Herczeg, P., in Rare Decay Symposium, proceedings of the Rare Decay Symposium, Nov. 1988, Vancouver, Canada, edited by Bryman, D., et al. (World Scientific, Singapore, 1989), p. 24.

43. If $(\overline{C}_P - C_P')/(C_P^\mu - C_P^{\mu'}) = m_e/m_\mu$, where $C_P^\mu, C_P^{\mu'}$ are coupling constants analogous to C_P, C_P', but associated with couplings involving $\overline{\mu}\nu_\mu$ and $\overline{\mu}\gamma_5\nu_\mu$, one would have $Q^{(e)}/Q^{(\mu)} = 1$ like in the minimal standard model. This happens for Higgs couplings in some models, but here such a cancellation would have to be regarded as accidental.

44. In our paper (Ref. 42) and in O. Shanker (Ref. 39) the assumption was made that the pseudoscalar contributions to $\pi \rightarrow \ell\nu$ are small relative to the minimal standard model contribution. The upper limit in Eq. 53 then becomes 10^{-5}. However, there is no justification for such an assumption. The upper limit in Eq. 49 of Ref. 42 should be 2.9×10^{-5}, instead of 5.7 $\times\ 10^{-5}$.

45. The same mechanisms can, of course, generate new charged current quark-lepton four-fermion interactions involving the other leptons. The corresponding coupling constants do not respect in general $e - \mu - \tau$ universality. $e - \mu - \tau$ universality holds only for left-right symmetric models with $u_\ell = v_\ell = 1 (\ell = e, \mu, \tau)$ and for the contributions of η_{LR} to normalized observables in general left-right symmetric models.

46. For a review see Eichten, E., et al., Rev. Mod. Phys. 56, 579 (1984); see also Jodidio, A., et al., Phys. Rev. D34, 1967 (1986) and references quoted therein.

NUCLEAR ANAPOLE MOMENTS

I.B. Khriplovich
Institute of Nuclear Physics
630090 Novosibirsk, USSR

Abstract

A new electromagnetic moment, so called anapole, arises in a system if the parity is violated. We discuss the anapole moments of heavy nuclei and the perspectives of their measurement in atomic and molecular experiments.

It is great pleasure for me to deliver a talk at the Symposium dedicated to Felix Boehm's birthday. My topic is related to the check of fundamental symmetries in nuclei and atoms. Both these fields have been influenced very much by Felix Boehm's investigations. So I hope that he will find the talk of interest.

To explain what is an anapole moment, let us start in a rather formal way from the multipole expansion for a vector potential $\mathbf{A}(\mathbf{r})$ induced by a current density $\mathbf{j}(\mathbf{r})$ of a stationary system. Expanding the well-known expression

$$\mathbf{A}(\mathbf{R}) = \int \frac{d\mathbf{r}\mathbf{j}(\mathbf{r})}{|\mathbf{R}-\mathbf{r}|} \tag{1}$$

in powers of r/R, we get

$$\begin{aligned} A_i(\mathbf{R}) \;&=\; \frac{1}{R}\int d\mathbf{r} j_i(\mathbf{r}) \;+\; \nabla_k \frac{1}{R}\int d\mathbf{r} r_k j_i(\mathbf{r}) \;+ \\ &+\; \frac{1}{2}\nabla_k\nabla_\ell \frac{1}{R}\int d\mathbf{r}\; r_k r_\ell j_i(\mathbf{r}) \;+\; \ldots \end{aligned} \tag{2}$$

(here and below we put the velocity of light equal to unity, $c = 1$). Everybody knows that the first term of this expansion vanishes, but nevertheless we shall prove it in a somewhat unusual way to demonstrate a trick that will be exploited below. (We follow here Ref.[1]; see also book [2].) In a stationary case, the continuity equation degenerates into

$$\operatorname{div} \mathbf{j}(\mathbf{r}) \;=\; 0. \tag{3}$$

For a system of finite size, the integral $\int d\mathbf{r}\nabla_m(r_i j_m(\mathbf{r}))$ vanishes. So, taking into account eq. (3), we get immediately

$$\int d\mathbf{r} j_i(\mathbf{r}) \;=\; 0. \tag{4}$$

In the same way we see that in the second term of expansion (2) the symmetric part of the integral vanishes:

$$\int dr \nabla_m (r_i r_k j_m) = \int dr (r_k j_i + r_i j_k) = 0. \tag{5}$$

So, we are left here with

$$\frac{1}{2} \int dr (r_k j_i - r_i j_k) \tag{6}$$

which corresponds evidently to the magnetic moment contribution to a vector potential. In fact, it is the third term of the expansion (2) we are interested in. And again, the totally symmetric part of the integral vanishes:

$$\int dr \nabla_m (r_i r_k r_\ell j_m) =$$

$$= \int dr (r_k r_\ell j_i + r_k j_\ell r_i + j_k r_\ell r_i) = 0. \tag{7}$$

So, we are left with an irreducible symmetric second-rank tensor and a vector. The tensor

$$M_{\ell m} = \int dr (r_\ell \epsilon_{mik} + r_m \epsilon_{\ell ik}) r_i j_k \tag{8}$$

is nothing else but the magnetic quadrupole moment (MQM). As to the vector, there are in general two suitable structures, $\int d\mathbf{r}\mathbf{r}(\mathbf{r}\mathbf{j})$ and $\int d\mathbf{r}r^2\mathbf{j}$. But they are related by the identity

$$\int dr (r^2 \mathbf{j} + 2\mathbf{r}(\mathbf{r}\mathbf{j})) = 0 \tag{9}$$

following from the contraction of eq.(7) in a pair of indices. We choose $\int d\mathbf{r}r^2\mathbf{j}$ as the independent structure. Since the indices k and ℓ in $r_k r_\ell j_i$ get contracted (see expansion(2)), no wonder that ∇_k and ∇_ℓ in $\nabla_k \nabla_\ell \frac{1}{R}$ are contracted too which leads to $\delta(\mathbf{R})$. After a little algebra, we come to the following expression for the corresponding contribution into the electromagnetic vector-potential:

$$\mathbf{A}^{(a)}(\mathbf{R}) = \mathbf{a}\delta(\mathbf{R}), \tag{10}$$

$$\mathbf{a} = -\pi \int d\mathbf{r} r^2 \mathbf{j}(\mathbf{r}). \tag{11}$$

The vector $\mathbf{a}$ is just the anapole moment (AM). We have omitted in expression (10) the term $\nabla(\mathbf{a}\nabla)\frac{1}{R}$ which can be eliminated by a gauge transformation.

This way of reasoning may be considered by somebody not too convincing since starting from the multipole expansion for $R >> r$ we have come to $\delta(\mathbf{R})$. A rigorous derivation of equations (10), (11) can be found, e.g., in book [2].

Why are not MQM and AM usually discussed among static electromagnetic multipoles? Because they violate some fundamental symmetries. Indeed, the MQM $M_{\ell m}$ of a quantum-mechanical system, say, of a nucleus, should be by usual arguments proportional to the symmetric traceless tensor $Q_{\ell m} = I_\ell I_m + I_m I_\ell - \frac{2}{3}\delta_{\ell m} I(I+1)$, the spin of a nucleus $\mathbf{I}$ being its only vector characteristic (in rest frame). But while $M_{\ell m}$ is odd both under time reversal T and space inversion P, $Q_{\ell m}$ is evidently even under both operations. So, the existence of MQM violates both T- and P-invariance. We wish to restrict discussion to T-even phenomena, so MQM will not be dealt with here anymore.

As to AM, it should be, by the same line of reasoning, proportional to the spin $\mathbf{I}$. According to (11), an anapole moment $\mathbf{a}$ is a polar T-odd vector, while $\mathbf{I}$ is an axial T-odd one. Therefore, a system where P-invariance only is violated, can acquire an anapole moment.

But how does it happen? The point is that such a system possesses helical spin structure [2,3,4]. If, without taking into account the parity nonconservation, the system is polarized along the z axis, then the P-odd effects induce the spin projection on the plane xy tangent to the circle r=const. (see Fig. 1). The same xy components are acquired naturally both by the magnetic moment of such a system and the magnetic field created by it. As to the current

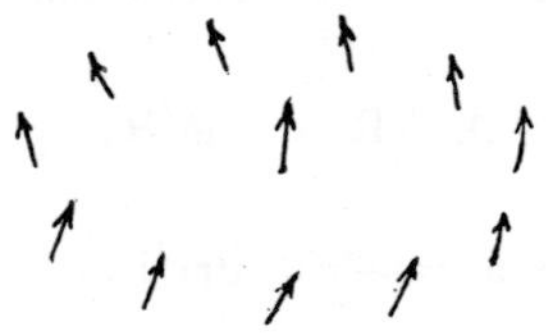

Fig.1 Spin Helix

distribution, alongside with the ordinary circular current in the plane xy there is a component corresponding to the current in a winding coiled on a torus[3]. The fact is of importance that the corresponding magnetic field is confined inside the torus. No wonder that the interaction with the vector potential (10) is of a contact type, take note of a δ – function in it. The interaction occurs provided the wave function of a probe particle penetrates the torus.

V.G. Vaks and Ya.B. Zel'dovich were the first to predict the existence of such a source of electromagnetic field in parity nonconserving systems [3], the term "anapole" for this source was suggested by A.S. Kompaneyets.

In fact, there exists a whole family of higher P-odd electromagnetic moments corresponding to the modulation of the confined magnetic field in polar and azimuthal angle. By virtue of the same arguments, the interaction with each of them is of a contact type. The corresponding contributions to the vector potential contain gradients of $\delta(\mathbf{R})$. The classification of these moments and the references to numerous papers where this classification has been derived can be found, e.g., in book [2].

One more remark perhaps would be appropriate here. It should not be thought that an AM can arise only in those somewhat exotic cases when parity nonconserving weak interactions are of importance. More commonplace example of a system with AM is a chiral radical, a molecule that does not coincide with its mirror image and possesses a non-paired electron. There is no parity nonconservation in this case since the AM is proportional here to the internal pseudoscalar corresponding to the co-ordinate helix of the molecule. It has been shown by M.E. Pospelov and myself that, curiously enough, a conducting substance with such molecules can be magnetized by an applied voltage. The point is that the voltage induces here a current, and the latter, interacting with the molecular AM orients it and in this way the angular momentum and magnetic moment as well. As to the T-violation inherent in the proportionality between the angular momentum and electric field, $\mathbf{I} \sim \mathbf{E}$, one should not be surprised by it since dissipative processes are involved here (just compare with the usual Ohm law $\mathbf{j} = \sigma \mathbf{E}$).

2. Now we come to atoms and muclei, and to parity nonconservation (PNC) in them. The existence of PNC in atoms is firmly established at present. The effect has been observed already by various groups in bismuth, thallium, cesium and lead. It is no accident that all these atoms are heavy. Just in heavy atoms, the P-odd weak electron-nucleus interaction is strongly enhanced, roughly as Z^3.

To be precise, the nuclear-spin independent PNC effects have been observed in atoms, namely these effects increase as Z^3. And one power of Z here is in fact the so-called weak nuclear charge Q which is in the standard electroweak model numerically close to $-\mathbf{N}$, where $N = A - Z \sim Z$ is the neutron number. The factor Q arises since in the nuclear-spin independent phenomena all nucleons act coherently.

As to the atomic PNC effects dependent on nuclear spin, they have not been observed until recently due to their smallness. Indeed, here only the unpaired nucleon is operative, so we lose the factor $|Q| \simeq N >> 1$ in heavy atoms. Moreover, within the standard model, the corresponding dimensionless constant is not of the order of unity, but very small numerically, $\simeq 0.05$.

However, some years ago it was recognized[5,6] that the main source of the nuclear-spin dependent PNC effects in atoms is perhaps not the weak electron-nucleon interaction, but P-odd nuclear forces. They create a nuclear anapole moment, and the latter, interacting electromagnetically with electrons, induces P-odd effects in atoms. At first sight, these effects are extremely small since in this electromagnetic interaction there is the factor $e^2 = \alpha = 1/137 << 1$ (we put c=1, $\hbar = 1$). But let us recall the toroidal winding picture of an anapole. It is only natural that the magnitude of the anapole moment $\mathbf{a}$ is proportional to the magnetic flux through the torus cross-section. In this way, the nuclear radius squared enters the result leading to enhancement factor $A^{2/3} >> 1$.

Curiously enough, the anapole moment can be calculated analytically for a heavy nucleus within shell model, under some simplifying assumptions. We start from the following expression for the effective P-odd weak interaction potential for an external nucleon:

$$W = \frac{G}{\sqrt{2}}\frac{g}{2m_p}\{\sigma\mathbf{p}, \rho(r)\}. \tag{12}$$

Here $G = 10^{-5}/m_p^2$ is the Fermi constant of weak interaction, m_p is the proton mass, σ and $\mathbf{p}$ are the spin and momentum operators of the valence nucleon, $\rho(r)$ is the spherically symmetric density of the nucleons of the core normalized with the condition $\int d\mathbf{r}\rho(r) = A$ (the atomic number is assumed to be large, $A >> 1$.) The pseudoscalar $\sigma\mathbf{p}$ guarantees the P-odd character of W, being evidently the only pseudoscalar we can construct within nuclear shell model. The anticommutator combination $\{\sigma\mathbf{p}, \rho(r)\}$ makes the operator W hermitian. The dimensionless constant g characterizes the P-odd interaction of the valence nucleon with the nuclear core. It is estimated to be $g_p \simeq 4$ for an external proton, and $g_n \leq 1$ for a neutron.

To calculate the AM of a heavy nucleus analytically, we assume the density $\rho(r)$ to be constant throughout space and to coincide with the mean nuclear density ρ_o. This approximation is quite reasonable if the wave function of an external nucleon is localized mainly in the region of the core. The Schroedinger equation

$$[-\frac{1}{2m_p}\Delta + U(r) + W(\mathbf{r})]\psi(\mathbf{r}) = E\psi(\mathbf{r}) \tag{13}$$

at $\rho(r) = \rho_o = \text{const}$ has to first order in the perturbation W the solution

$$\psi(\mathbf{r}) = (1 - i\frac{G}{\sqrt{2}}g\rho_o\sigma\mathbf{r})\psi_o(\mathbf{r}). \tag{14}$$

Here ψ_o is the unperturbed wave function of an external nucleon. It might seem that now interaction (12), being equivalent to the electromagnetic interaction with a constant vector potential $\mathbf{A} = -(G/\sqrt{2})(g/e)\rho_o\sigma$, should not result in any physical effects whatsoever. Indeed, the contribution into the nuclear AM from the orbital part of the current density vanishes in this approximation. However, the spin part

$$\mathbf{j}^s = \frac{e\mu}{2m_p} \nabla \times (\psi^\dagger \sigma \psi) \tag{15}$$

of the current density is operative even in this approximation, due to the non-commutativity of the σ – matrices. (In formula (15), μ is the magnetic moment of the valence nucleon.) Simple calculations by means of formulas (11), (14), (15) yield

$$a = \frac{e\mu}{2m_p} \frac{G}{\sqrt{2}} \frac{gA^{2/3}}{r_o} \frac{9}{5} \frac{(I+1/2)(-1)^{I+1/2-\ell}}{I(I+1)} \mathbf{I}. \tag{16}$$

We have taken here for the mean squared radius of the external nucleon, entering the calculations, the value $< r^2 >= \frac{3}{5} r_o^2 A^{2/3}$, $r_o = 1.2 \cdot 10^{-13}$ cm, and set $\rho_o = (\frac{4\pi}{3} r_o^3)^{-1}$. ℓ in formula (16) is the orbital angular momentum of the valence nucleon.

Estimates show that the contribution of the core excitations to the nuclear AM is small[7]. There are thus reasons to believe that formula (16) is a good approximate expression for the AM of a heavy nucleus. The remarkable enhancement factor $A^{2/3}$ should be noted in particular.

Since interaction (12) ensures the basic contribution to the AM of a heavy nucleus, it is worth to be calculated beyond the constant density approximation. In Ref.[6], the numerical calculations of the AM of heavy nuclei were performed with the use of a realistic description of the core density $\rho(r)$. The Woods-Saxon potential with the account for the spin-orbit interaction was used. The agreement of the approximate analytical calculation with the numerical one proved to be fairly satisfactory (see below). Anyway, the discrepancy in the results is comparable with the expected accuracy of the nuclear shell model used in both variants of the calculation.

The main uncertainty in the calculation of the AM of heavy nuclei is related to the dimensionless constant g which characterizes P-odd nuclear forces. The value $g_p = 4$ may well differ from the true one, say, by a factor of 2, the constant

g_n is known even less reliably. The point is that theoretical interpretation of other P-odd nuclear effects, which gives the information on the constants $g_{p,n}$, is very difficult. It is clear thus that the stability of the calculation of AM in terms of g makes anapole moments a source of a valuable information about P-odd nuclear forces.

The electromagnetic interaction of the electron current with the vector potential created by nuclear AM is of a contact type and in this respect does not differ from the neutral current weak interaction. It is natural therefore to measure this electromagnetic interaction in units of the Fermi weak interaction constant G. The corresponding dimensionless parameter κ_a, as obtained by means of (16), is

$$\kappa_a = \frac{9}{10} g \frac{\alpha\mu}{m_p r_o} A^{2/3} . \tag{17}$$

The values of κ_a calculated for nuclei of an experimental interest are[6] (at $g_p = 4,\ g_n = 1$)

	Cs	Tl	Pb	Bi
analytical result	0.33	0.44	−0.08	0.45
numerical result	0.25	0.36	−0.09	0.31

The AM of the ^{133}Cs nucleus has been calculated also in a recent work [22]. In that paper only the π − meson exchange contribution to the P-odd nuclear forces has been taken into account. This contribution constitutes roughly a half of the value $g_p = 4$ accepted by us. No wonder therefore that their result 0.14 for $\kappa_a(^{133}\mathrm{Cs})$ constitutes roughly a half of the corresponding our value 0.25-0.33. The result of the Ref.[22] not only confirms in fact ours, but, being obtained in a quite different approach, demonstrates once more the stability of the calculation of nuclear AM in terms of the constant $g_{p,n}$.

One more remark would be appropriate here. The nuclear AM can be considered as a peculiar radiative correction to the, say, electron-nucleus scattering amplitude (see diagram 2). But being of a contact nature, this contribution to the scattering amplitude is in general indistinguishable from other radiative corrections described, e.g., by diagrams 3 and 4. And in a gauge theory of electroweak interactions, only the total scattering amplitude, i.e., the sum of all

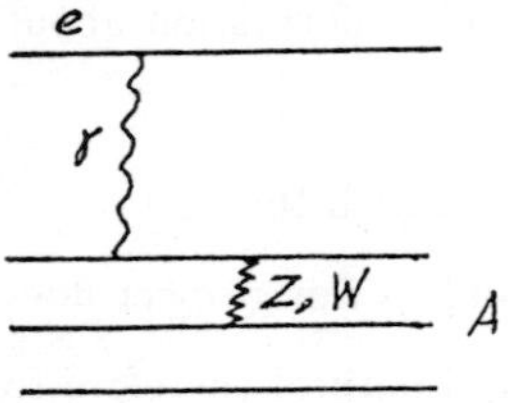

Fig. 2

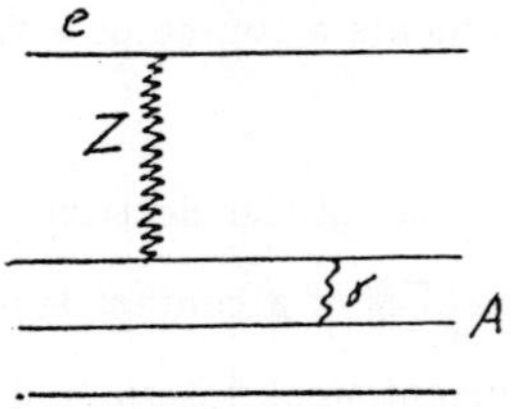

Fig. 3

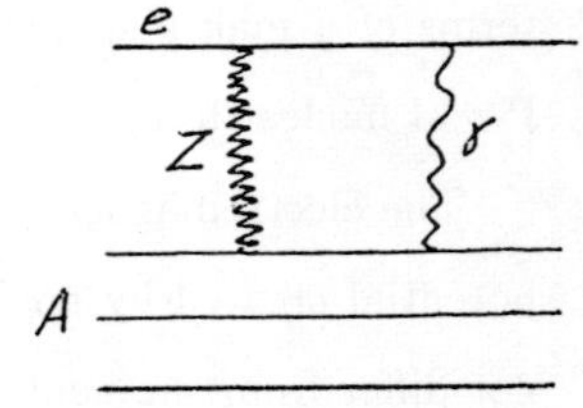

Fig. 4

diagrams, is gauge-invariant, independent of the gauge choice for the Green's functions of heavy vector bosons. No wonder that, generally speaking, the AM of an elementary particle or a nucleus is not a gauge-invariant, i.e., physically well-defined, quantity.

However, our result (16) for the AM of a heavy nucleus is gauge invariant since only the diagram 2 is enhanced as $A^{2/3}$.

One more case when the notion of AM makes sense, is the deuteron in the zero-range approximation, $\kappa r_o << 1$ ($\kappa = \sqrt{m_p \epsilon}$, ϵ is the binding energy). Here[5)]

$$\mathbf{a} = \frac{\pi e}{m_p}\{c(\mu_p - \mu_n - 1/3) + \lambda(\mu_p + \mu_n)\}\mathbf{I}\,. \tag{18}$$

The parameter c refers to the P-odd np scattering amplitude that conserves the total spin S, but changes the isotopic spin T. λ refers to the amplitude that violates S, but conserves T. At $c \sim \lambda \sim 10^{-6} m_p^{-1}$, the corresponding dimensionless constant κ_d is small, about 10^{-2}. However, within the standard model the hadronic axial current is an isotopic vector, so for the isoscalar deuteron it vanishes. But in practice, the zero-range parameter is not so small here, $\kappa r_o \sim 1/3$, so it is only natural that other radiative corrections are in this case numerically quite comparable to the AM contribution.

Even the nucleon AM can make sense, but in a (theoretical) chiral limit $m_\pi \to 0$ only, m_π being the π – meson mass. It was shown by A.I. Vainshtein and myself to be given in this limit by the diagrams 5 and 6. The circle on these

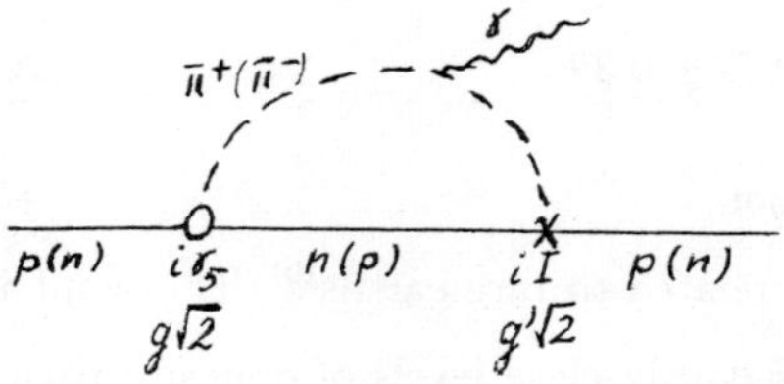

Fig. 5

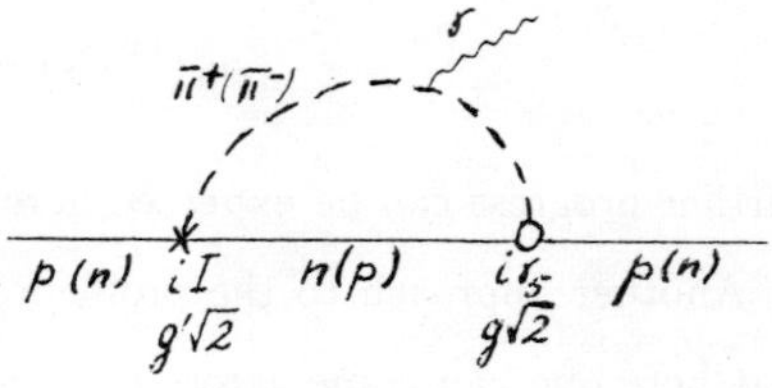

Fig. 6

diagrams refers to the usual strong interaction πNN vertex, the cross describes the P-odd weak πNN interaction. The result is

$$a_{p,n} = -\frac{egg'}{12 m_p m_\pi}\sigma. \tag{19}$$

The diagrams discussed lead to the same result for a proton and neutron since under the permutation p $\longleftrightarrow$ n the strong coupling constant g does not change, and the weak one g′ changes sign together with the charge e of the π – meson. Here we disagree with recent paper[22] according to which the same diagrams lead to $a_p \neq a_n$. One more disagreement refers to the dependence on the π – meson mass m_π. Our result is m_π^{-1}, and theirs $\ln m_\pi$.

However, in practice numerically there is nothing special about this contribution to the weak eN scattering amplitude, other radiative corrections are of the same magnitude.

3. We return now to the AM's of heavy nuclei. How can they be measured? The first way is to look for a difference in the magnitude of atomic P-odd effects at different hyperfine components of optical transitions [8]. The mutual orientation of the nuclear and electron spins varies from one hyperfine level to another, no wonder therefore that the magnitude of nuclear spin-dependent P-odd effects varies as well. The effects can be looked for both in the optical activity experiment of the bismuth - lead type and in the studies of P-odd correlations in an external electric field (cesium - thallium type experiment).

Just in the experiment of the latter kind, with cesium beam, the first evidence has been found recently of the AM of the nucleus ^{133}Cs[9]:

$$\kappa_a(^{133}\mathrm{Cs}) = 0.72 \pm 0.39. \tag{20}$$

Further progress can be expected here soon.

Another approach to the problem is related to rare earths[10]. The point is that here one can come across the anomalously close levels of opposite parity with the total angular momentum of electrons differing by unity. These levels can mix only by a nuclear-spin dependent P-odd interaction which is a vector in electron variables. As a result, the effects of this interaction could appear to be considerably enhanced, both absolutely and relatively, as compared with the effects independent of the nuclear spin. Both the optical rotation experiments and the cesium-like ones could be performed here. It should be kept in mind, however, that nuclear-spin dependent effects vanish when averaged over hyperfine structure (HFS). So HFS should be resolved in such experiments. And it is sometimes very small in rare earths.

One more suggestion on P-odd nuclear-spin dependent effects is to study optical activity in microwaves in the transitions between the components of HFS of heavy atoms[11]. Perhaps the major obstacle in this case is the almost equal population of the HF levels. A natural way out would be to lower the temperature of the gas to some Kelvins. The idea is not as crazy as it looks. Active atoms could be implanted into a matrix of frozen noble gases at Kelvin temperatures[12]. The densities obtained in this way are quite reasonable, but the matrices themselves are very thin films, say, about 10 microns thick. Anyway, the size of the waveguides and resonators is much larger.

Another suggestion here refers to the searches for P-odd effects in the microwave transitions in an atomic beam[13].

And now we turn to molecules. A heavy diatomic molecule has a quantum number Ω, the projection of the total electron angular momentum $\mathbf{j}$ on the molecular axis $\mathbf{n}$. Under the reflection in a plane that contains $\mathbf{n}$, Ω changes sign. So the states $|\Omega>$ and $|-\Omega>$ are degenerate. The degeneracy is lifted by the molecular rotation. The level splitting arising is evidently very small in atomic scale

$$\Delta E \sim \frac{m}{M} Ry, \tag{21}$$

M being the reduced mass of a molecule. And the stationary states $\frac{1}{\sqrt{2}}(|\Omega> + |-\Omega>)$ and $\frac{1}{\sqrt{2}}(|\Omega> - |-\Omega>)$ have opposite parities. It can be easily seen that the nuclear-spin independent weak interaction, being a pseudoscalar in electron variables, cannot mix these states. But the nuclear-spin dependent one, being an electronic vector, has the nonvanishing matrix element $<\Omega = \frac{1}{2}|H|\Omega = -\frac{1}{2}>$. So the opposite-parity levels are mixed by it at $|\Omega| = 1/2$.

This discussion refers to the heteronuclear molecules only. In homonuclear ones, a new quantum number arises, (molecular) parity. The corresponding states, "gerade" and "ungerade", are mixed by a P-odd interaction, but the energy interval between u- and g-states is generally speaking of the usual atomic scale. W. Demtroeder has noted however that even in homonuclear molecules one can come across an accidental closeness of g- and u-states with $\Delta\Omega = 0, \pm 1$ and the corresponding enhancement of parity-nonconserving effects.

Therefore, in a heavy diatomic heteronuclear molecule nuclear-spin dependent P-odd effects are drastically enhanced, due to the small energy intervals[14-18]. The corresponding experiments are being discussed for a long time already, but unfortunately there is still no experimental progress in this direction.

And finally, one more topic concerning molecules. Many years ago, it was recognized that parity nonconservation will cause the splitting in energy between optical isomers, i.e., molecules that are mirror reflections of each other. (The full list of references is too long to be presented here, see e.g., book[2].) The origin of the effect becomes quite evident if one recalls the spin helix discussed above. It is only natural that the energy of a molecule is different for the two cases: when the spin helix is of the same sign as the co-ordinate helix, and when their signs are opposite. This energy difference is of course quite tiny. And although the calculations of this splitting is being carried out by various groups, the experiments here are perhaps of unsurmountable difficulty now. Hopefully, there is however an exception to the last assertion. I mean the search for the splitting of HF constants in the ground state of optical isomers[19–21]. But of course it should be a radical. And for the effect to be appreciable, there should be a heavy atom in it. And this atom should be sufficiently close to the optically active group, and the unpaired electron should be close to both. The effect arises of course due to the nuclear-spin dependent P-odd interaction. Numerically, the energy splitting constitutes

$$\delta \mathrm{E} \sim (10 - 100)\kappa_{\mathrm{a}} \text{ Hz}. \tag{22}$$

The observation of such a splitting does not seem by itself to be a fantastic task.

4. Sixty-five years ago, the studies of atomic hyperfine structure gave the first clue to the existence of nuclear electromagnetic moments. Since then, atomic and molecular spectroscopy serves as a source of invaluable information on nuclear properties, such as multipole moments and radii of nuclei. Now we are at the threshold of the new turn of events in this story: optical spectroscopy starts to bring data of a quite new kind on nuclear structure, data on P-odd nuclear forces. Moreover, the discovery and study of a new electromagnetic moment, the anapole one, is of major interest by itself.

References

1. O.P. Sushkov, V.V. Flambaum and I.B. Khriplovich, Zh. Eksp. Teor. Fiz. 87, 1521 (1984) (Sov. Phys. JETP 60, 873 (1984)).
2. I.B. Khriplovich, Parity Nonconservation in Atomic Phenomena (Nauka, Moscow, 1988). (Its English translation is being published by Gordon and Breach.)
3. Ya.B. Zel'dovich, Zh. Eksp. Teor. Fiz. 33, 1531 (1957) (Sov. Phys. JETP 6, 1184 (1957)). (This paper contains also the mention of the analogous results obtained by V.G. Vaks.)

4. L.M. Barkov, I.B. Khriplovich and M.S. Zolotorev, Comments Atom. Mol. Phys. 8, 79 (1979).

5. V.V. Flambaum and I.B. Khriplovich, Zh. Eskp. Teor. Fiz. 79, 1656 (1980) (Sov. Phys. JETP 52, 835 (1980)).

6. V.V. Flambaum, I.B. Khriplovich and O.P. Sushkov, Phys. Lett. B146, 367 (1984).

7. V.V. Flambaum, I.B. Khriplovich and O.P. Sushkov, Phys. Lett. B162, 213 (1985); Nucl. Phys. A449, 750 (1986).

8. V.N. Novikov, O.P. Sushkov, V.V. Flambaum and I.B. Khriplovich, Zh. Eskp. Teor. Fiz. 73, 802 (1977) (Sov. Phys. JETP 46, 420 (1977)).

9. M.S. Noecker, B.P. Masterson and C.E. Wieman, Phys. Rev. Lett. 61, 310 (1988).

10. V.A. Dzuba, V.V. Flambaum and I.B. Khriplovich, Z. Phys. D 1, 243 (1986).

11. V.N. Novikov and I.B. Khriplovich, Pis'ma Zh. Eksp. Teor. Fiz. 22, 162 (1975) (Sov. Phys. JETP Lett. 22, 74 (1975)).

12. E.L. Al'tshuler et al., Zh. Tekhn. Fiz. 54, 1956 (1984) (Sov. Phys. JTP).

13. V.G. Gorshkov, V.F. Yezhov, M.G. Kozlov and A.I. Mikhailov, Yad. Fiz. 48, 1363 (1988) (Sov. J. Nucl. Phys.).

14. L.N. Labzovsky, Zh. Eksp. Teor. Fiz. 75, 1208 (1978) (Sov. Phys. JETP 48).

15. O.P. Sushkov and V.V. Flambaum, Zh. Eksp. Teor. Fiz. 75, 1208 (1978) (Sov. Phys. JETP 48, 608 (1978)).

16. V.G. Gorshkov, L.N. Labzovsky and A.N. Moskalev, Zh. Eksp. Teor. Fiz. 76, 414 (1979) (Sov. Phys. JETP 49).

17. V.V. Flambaum and I.B. Khriplovich, Phys. Lett. A110, 121 (1985).

18. M.G. Kozlov, Zh. Eksp. Teor. Fiz. 89, 1933 (1985) (Sov. Phys. JETP 62.

19. I.B. Khriplovich, Zh. Eksp. Teor. Fiz. 79, 354 (1980) (Sov. Phys. JETP 52 177 (1980)).

20. V.G. Gorshkov, M.G. Kozlov and L.N. Labzovsky, Zh. Eksp. Teor. Fiz. 82, 1807 (1982) (Sov. Phys. JETP 55).

21. I.B. Khriplovich, Z. Phys. A322, 507 (1985).

22. W.C. Haxton, E.M. Henley and M.J. Musolf, Phys. Rev. Lett. 63, 949 (1989).

NEUTRINO MASS -- RECENT RESULTS

R. G. H. Robertson

Physics Division, Los Alamos National Laboratory,

Los Alamos, NM 87545, U. S. A.

ABSTRACT

Some recent developments in the experimental search for neutrino mass are discussed. Simpson and Hime report finding new evidence for a 17-keV neutrino in the β decay of ^{3}H and ^{35}S. New data from Los Alamos on the electron neutrino mass as measured in tritium beta decay give an upper limit of 13.5 eV at the 95% confidence level. This result is not consistent with the long-standing ITEP result of 26(5) eV within a "model-independent" range of 17 to 40 eV. It now appears that the electron neutrino is not sufficiently massive to close the universe by itself.

1. INTRODUCTION

It is a great pleasure for me to be able to join in honoring Felix Boehm at this symposium. From my earliest days as a graduate student, I recognized Felix as the doyen of the kind of physics I would like to do, and so the honor is really mine. I was never a Boehm student, which is something I regret, but then I was never a Boehm competitor, either, something for which I am extremely grateful! Today I would like to discuss some topics in neutrino mass, a field to which Felix Boehm has made many important contributions.

The continuing intensive experimental search for neutrino mass is motivated by the profound implications for cosmology and for particle physics. As is well known, the universe would be gravitationally closed by a neutrino having a mass of a few tens of eV, and the 1980 report[1] by the group at the Institute for Theoretical and Experimental Physics (ITEP) in Moscow of a 35 eV electron neutrino mass therefore aroused great interest. In the intervening 9 years, the ITEP group have improved their apparatus, taken more data, refined their analysis, and still find qualitatively the same result. New experiments at the Los Alamos National Laboratory on the beta decay of free molecular tritium[2] now give an upper limit of 13.5 eV, in disagreement with the ITEP result.

While there has been no recent work on the mass of the μ neutrino, a vigorous program of research on the τ continues. The Argus collaboration at DESY has observed[3] several examples of the decay of the τ to 5 charged pions, and can set an upper limit of 35 MeV on the mass of ν_τ. This represents a substantial advance over the previous limit (also Argus[4]) of 70 MeV, but at present one can see no laboratory technique that would probe μ and τ neutrino masses to the levels allowed by cosmology.

The new work of Simpson and Hime[5,6] appears to show evidence for a small admixture of a 17-keV neutrino in the electron neutrino, in contradiction with a number of other experiments.

2. THE 17-keV NEUTRINO

In 1985 Simpson[7] reported that there was at the low-energy end of the tritium beta spectrum a distortion indicative of a 3% admixture of a 17.1-keV antineutrino with the dominant electron antineutrino. It was subsequently shown by Lindhard and Hansen[8] and by Eman and Tadic[9] that about 67% of the distortion could be explained by Simpson's use of an incorrect screening potential. A similar effect, exchange between the orbital electrons and the outgoing beta particle, was noted by Haxton[10] to be responsible for another 15% of the distortion. The remaining evidence for a 17-keV neutrino from beta decay of tritium in Si seemed too model-dependent to be conclusive.

The 18.6-keV Q-value for tritium beta decay makes tritium a poor candidate for revealing a 17-keV neutrino, and several groups took up the search in ^{35}S (Q = 167 keV) and ^{63}Ni (Q = 67 keV). In ^{35}S five groups claimed[11-15] to find no evidence for a 17-keV neutrino at levels below that found by Simpson, but Simpson[16] rebutted three of those claims by pointing out that the analyses had been done incorrectly (the data sets had only been fitted under the assumption of no second neutrino, and not with full variation of all allowed parameters). Indeed, in one case, there was a better fit when a 1 to 2% admixture of a 17-keV neutrino was included, and in other cases the upper limits were no longer sufficiently stringent to be interesting. Most difficult to dismiss, however, is the very detailed study of ^{63}Ni carried out by Hetherington et al.[17], in which an upper limit of 0.3% (90% CL) was set on the heavy neutrino admixture.

Hime and Simpson[6] now report not only that the beta spectrum of tritium implanted in Ge shows the effect of a heavy neutrino admixture, but also that there is strong evidence in ^{35}S for the same admixture.[5] From the tritium-in-germanium data, Hime and Simpson conclude that there is a 1.1(5)% admixture of a 16.9(1)-keV neutrino, in good agreement with the revised values from the earlier experiment[7] on tritium implanted in Si. The ^{35}S result is lower, but still consistent, at 0.73(9,6)% admixture of a 16.9(4)-keV neutrino. The uncertainties in parentheses are statistical and systematic, respectively. While an effect is unmistakably present (i.e., not merely a statistical fluctuation), it is puzzling that it does not show up in the ^{35}S and ^{63}Ni experiments that have sufficient sensitivity to see it. If the effect is due to a heavy neutrino admixed with the electron neutrino, one would, of course, expect to see it in all beta decays, provided that the energy is high enough.

In Table I we summarize the relevant measurements, to the best of our knowledge.

Table I. Measurements and calculations relating to the 17-keV neutrino proposed by Simpson.

Authors (Reference)	Source	m_ν (eV)	$\sin^2\theta$
Simpson (7)	T in Si	17.1(2)	0.03
Haxton (10)	Exchange corrections		
Lindhard & Hansen (8)	Screening corrections		
Eman & Tadic (9)	Screening corrections		
Simpson (6,16)	(revised)	17.1(2)	0.011(3)
Altzitzoglu et al. (11)	^{35}S		<0.004 (99% CL)
Ohi et al. (12)	^{35}S		<0.0015 (90% CL)
Apalikov et al. (14)	^{35}S		<0.0017 (90% CL)
Datar et al. (13)	^{35}S		<0.006 (90% CL)
Markey & Boehm (15)	^{35}S		<0.003 (90% CL)
Hetherington et al. (17)	^{63}Ni		<0.003 (90% CL)
Hime & Simpson (6)	T in Ge	16.9(1)	0.011(5)
Simpson & Hime (5)	^{35}S	16.9(4)	0.0073(9,6)

3. TRITIUM BETA DECAY EXPERIMENTS

It has been known for more than 50 years that the energy spectrum of electrons emitted in beta decay yields information about the mass of the electron neutrino.[18] Tritium is especially suitable for this work in view of its large matrix element, low energy release, simple atomic structure and convenient half-life. The beta spectrum may be written:

$$N(E) = C\,F(Z_f,R,E)\,p_e\,E\,\Sigma_i w_i(E_0-E_i-E)[(E_0-E_i-E)^2-m_\nu{}^2c^4]^{1/2}$$

$$\times\,[1+\alpha_1(E_0-E)+\alpha_2(E_0-E)^2] \quad ; \quad E \le E_0-E_i-m_\nu c^2$$

where $F(Z_f,R,E)$, a smoothly varying function of energy, is the Fermi function that corrects for Coulomb distortion of the outgoing wave. The total energy is E_0. Small theoretical and experimental corrections are absorbed in the polynomial with coefficients α_1 and α_2. Weak magnetism and nuclear recoil give[19] α_1 a value of 2.312×10^{-9} eV^{-1}. The summation is over all final states of the daughter system. Each final state has a different energy, and calculating the energies E_i and branching ratios w_i to the final states is a matter of fundamental importance in all tritium experiments. Equally important, but more amenable to experimental checks, are energy loss as the electron traverses the source material, instrumental resolution, and backscattering.

In 1981, a group at the Institute for Theoretical and Experimental Physics (ITEP) in Moscow reported[1] from their study of the tritium spectrum that ν_e had a mass of 35 eV, with revolutionary implications for particle physics and cosmology. More recent ITEP work[20] has reduced this value slightly to 26(5) eV, with a "model-independent" range of 17 to 40 eV. Fritschi et al.[21] found in a similar type of experiment at the University of Zürich an upper limit of 18 eV, and other measurements have neither confirmed nor contradicted these works.[22,23] Both the Zürich and ITEP experiments have very high statistical accuracy, and the difference between the two results must be a consequence of systematic effects. A probable origin for such effects is the solid source materials used, for which molecular structure calculations are difficult to carry out to the necessary precision.

Unlike other experiments presently in operation, the experiment[22] at the Los Alamos National Laboratory makes use of a gaseous source of T_2 to capitalize on the simplicity of the two-electron system. The final-state calculations can be carried out with some confidence for atomic and molecular tritium, but with less certainty for solid sources like those used by ITEP and Zürich. Use of a gaseous source also confers the advantages of minimal and well understood energy-loss corrections, and no backscatter corrections. Thus the gaseous source minimizes systematic uncertainties, but it is technically more difficult, and statistical accuracy can be hard to obtain.

3.1 The Los Alamos Experiment

The Los Alamos group reported[22] in 1987 an initial result, $\nu_e <$ 27 eV at 95% confidence level (CL), which showed that it was possible to perform experiments with gaseous sources at a useful level of sensitivity. The precision of the result was limited almost entirely by statistical accuracy, systematic effects being very much smaller. Sensitivity to neutrino mass increases extremely slowly with data acquisition time, roughly as the fifth root, so it was clear that significant improvement in the limit could only transpire through an increase in the data rates. To this end, a number of improvements were made, the principal one being the replacement of the simple single-element proportional counter in the spectrometer with a 96-pad Si microstrip detector array.

The new detector is an octagonal array of 300-μm-thick planar passivated Si wafers (n-type) each with a sensitive area of 7 x 10 mm^2. The sensitive area is subdivided into 12 strips on 0.83-mm centers by readout pads. There are thus 96 microstrips tiling the surface of a 2-cm diameter cylinder. For 23-keV electrons, resolutions of 2.5 to 4.0 keV FWHM are observed.

Numerous other improvements have been made. An axial gradient has been superimposed on the magnetic field of the source to eliminate the trapping of electrons, which previously necessitated a large and complex correction. Baffles were installed in the spectrometer, and the acceptance reduced by half, to improve the lineshape. A getter pump was added to remove atmospheric gases that caused extra energy loss when Kr was being recirculated. The stability and background of the Si detector that monitors the source strength was improved, and the source density was stabilized with the addition of a servo regulator. Notwithstanding the loss in spectrometer acceptance, the gross data rate is 8 times higher than previously, and the signal-to-background about the same.

The beta spectrum is formed by setting the spectrometer to analyze a fixed momentum (equivalent to an energy of 23 keV) and scanning the accelerating voltage on the source. Before and after a tritium data set, the 17820-eV K-conversion line of $^{83}Kr^m$ is recorded two or three times to determine the instrumental resolution and energy scale.

Analysis of the data begins with manual creation of a set of 87 "windows" on the energy spectra from the individual pads. Each pad receives counts corresponding to a slightly different momentum, the total range being about 100 eV in energy from one end of the detector to the other. The data are thus organized by summing counts from the same pad numbers on each wafer to form 12 spectra, each independently calibrated by a $^{83}Kr^m$ spectrum similarly formed. The "raw" tritium spectra can be compared to the theoretical spectrum modified by corrections for energy loss, instrumental resolution, apparatus efficiency, and the final-state spectrum. The neutrino mass and its variance are then estimated[22] from a plot of the sum of Ξ^2 for all pads against m_ν^2.

Electrons lose energy by inelastic scattering as they spiral through the source gas. The cross section differential in energy has been constructed from various data[22] The total inelastic cross section is very tightly constrained by the Liu sum rule[24] to be $\sigma_0 = 3.474(11) \times 10^{-18}$ cm^2 at 18.5 keV. The gas-density profile in the source is determined by kinetic theory from the measured throughput of gas scavenged by pumps into a calibrated volume, given the dimensions and temperature (130K) of the source tube. The probability of an electron interacting is 8.5 %, and the number of interactions per decay is 9.1 %, an indication of the generally small scattering probability and the very minor role played by plural interactions. The stopping power computed with our differential cross section (which satisfies the Liu sum rule) is 0.44×10^{-16} eV-cm per atom, 18% below the Bethe stopping power.[25] This difference represents the major uncertainty in the energy-loss estimates.

Measurement of the instrumental resolution is accomplished by circulating $^{83}Kr^m$ (from the decay of ^{83}Rb) through the source and recording the nominally monoenergetic K-conversion line at 17820(3) eV. This single calibration is sufficient because, in the Los Alamos apparatus, the spectrometer is always set to analyze the same momentum, and spectra are obtained simply by scanning the acceleration voltage applied to the source. Conversion lines are accompanied by shakeup and shakeoff satellites, and, rather than rely on calculations for their positions and intensities, a direct measurement of the equivalent Kr K-shell photoionization spectrum was made at the Stanford Synchrotron Radiation Laboratory.[26] Excellent agreement between the shapes of the spectra is obtained when the slightly better-resolution photoionization spectrum is convoluted with a Gaussian to match the internal-conversion data. Most important, a long tail (2×10^{-4} eV^{-1}) observed in the data but not predicted by theory is shown to be a part of the Kr spectrum (and not instrumental). A more detailed description of this work is given elsewhere.[26,27] A spectrum of thermal electrons from the source region accelerated to 19 keV showed evidence for a weak tail of 7×10^{-6} eV^{-1}, and the Kr data also shows evidence of marginal statistical significance for a residual tail at about this level. This residual tail being presumably of instrumental origin, the Los Alamos group took the instrumental resolution to be given by a Gaussian with an added flat tail of 7×10^{-6} eV^{-1} extending to 350 eV, and extracted the Gaussian parameters by fitting to the diagram line and first satellites of the Kr spectrum, for which the theoretical description appears to be good. The Gaussian standard variances for the 12 spectra averaged about 120 eV^2. The instrumental linewidth is sufficiently narrow that details of its shape have only a small influence on the results. The effect of the added tail on m_ν^2 is 15 eV^2, with a 15 eV^2 uncertainty.

There are contributions to the tritium linewidth not contained in the Kr calibration. The partition of recoil energy between internal and translational degrees of freedom of the THe^+ ion contributes[28] a variance of 9×10^{-2} eV^2. Zero-point vibrational motion in the T_2 molecule[29] and thermal motion create Doppler broadenings of variance 4×10^{-4} and 4×10^{-2} eV^2, respectively. These contributions are negligible. Also, no significant trapped-ion density was found.[30]

The small variation of apparatus efficiency with acceleration voltage introduces a spectral distortion that can influence the neutrino mass derived. It is customary to parametrize this with empirically determined linear and quadratic correction terms α_1 and α_2 in the spectrum. In the Los Alamos apparatus both the spectrometric data and the monitor data are subject to efficiency corrections. The monitor efficiency function may easily be measured by plotting its rate, corrected for source pressure, against acceleration voltage, but there is no comparable method for the spectrometric data. However, it is possible to measure the spectrometric efficiency by recording other lines in the $^{83}Kr^m$ spectrum and by recording the tritium spectrum over a wider range of energies. This work is still in progress at Los Alamos. Until that work is complete, the Los Alamos group considers both linear and quadratic terms in the fit separately, adopts a mean value and treats the total spread in neutrino mass squared (105 eV^2) as a systematic uncertainty.

The final-state spectrum (of the THe^+ ion) has the most important influence on the tritium spectrum. Calculations have been reported for the decay of T_2 in the sudden approximation. The Martin-Cohen (MC) calculation[31] is truncated at 94 eV excitation, and the Quantum Theory Project (QTP) calculation[32] at 164 eV. The MC and QTP calculations are in very good accord, the latter (the one we adopt) giving an m_ν^2 8 eV^2 larger owing to its greater range. The MC calculation omits 1.3% of the strength, while the QTP one omits 0.5%, and the distribution of this strength is responsible for the difference between the variances, 545 eV^2 and 617 eV^2, respectively, and the sum-rule result of Kaplan and Smelov,[33] 1110 eV^2. Despite this alarming discrepancy, the effect on neutrino mass is actually rather small, as was demonstrated by simulating the missing 0.5 % of strength with discrete and continuous distributions that satisfy the sum rule. An upward correction to m_ν^2 of 20(10) eV^2 for the strength missing in the QTP calculation results.

The validity of the sudden approximation, on which all these calculations rest, has not seriously been questioned, largely because of the work of Williams and Koonin[34] (WK), who claimed that the rescattering contributions (i.e., the interaction of the beta directly with orbital electrons) were less than 10^{-3} in the case of the atom. WK, however, treated only s-wave final states, arguing that other partial waves would each contribute of order $(1/pa_0)^2 = (1/36)^2$, where p is the beta momentum, and a_0 the Bohr radius. They then invoked Intemann's argument[35] that the highest partial wave of interest would have an ℓ of order pa_0, but erroneously found this quantity to be 1/36, whereas it is actually 36. A complete calculation appears to be very difficult, but Friar[19] has obtained a closed-form expression for the p-wave bound and continuum strengths. The Los Alamos group doubled this contribution to make an estimate for the T_2 molecule, and found m_ν^2 shifts upward by 6 eV^2. Friar showed that higher ℓ contributions fall off very rapidly for bound states, and a semiclassical model indicates that this is true for the continuum as well. The total effect was estimated to be not more than about 3 times the p-wave, but the need for a complete continuum calculation for the T_2 molecule is manifest.

The tritium data set obtained (13400 measurements) spanned an interval of about 9 days and led to spectra totalling 8000 counts in the last 100 eV, of which 1400 were background. The spectra covered the range 16500 - 19200 eV. The maximum-likelihood procedure[22] was used to obtain values for m_ν^2, E_0, amplitude, background, α_1 and α_2. These results, and their 1-σ statistical uncertainties, are listed in Table II. In Table III we list the estimated uncertainties (1-σ) in m_ν^2 from all sources. The uncertainty in the endpoint energy, E_0, has not yet been evaluated.

Los Alamos took the conservative viewpoint that they did not know which is the correct description of the curvature, and that the two choices are but a selection from a large variety of possible efficiency functions. The best estimate is then the average of the α_1 and α_2 fits, and the uncertainty associated with efficiency correction is the difference between them.

Table II. Results from analysis of August, 1988 data; uncertainties are one standard deviation statistical.

	α_1 only	α_2 only	
m_ν^2	-250(90)	-145(90)	eV^2
E_0	18567.2	18569.2	eV
α_1	$-2.6(2) \times 10^{-5}$	--	eV^{-1}
α_2	--	$-8.8(5) \times 10^{-9}$	eV^{-2}
Ξ^2 (ave.)	945.8	945.1	(892 D.O.F)

Table III. Contributions (eV^2) to the uncertainty in m_ν^2 at one standard deviation.

Analysis:		
Statistics		90
Beta monitor statistics, dead time		5
Energy Loss:		
18% in theoretical spectrum shape:		15
5% Uncertainty in source density		4
Resolution		
Width		12
Skewness		6
Tail		15
Final States		
Differences between theories		8
Region above truncation point		10
Rescattering		20
Apparatus Efficiency		
Linear vs Quadratic		105
	Total	143

In Figure 1 are plotted the residuals for the fit near the endpoint for m_ν = 0 and 30 eV, from which it may be seen qualitatively that a 30-eV mass is rejected. That conclusion is borne out quantitatively when all uncertainty components are considered. As discussed elsewhere,[18] setting confidence levels on quantities

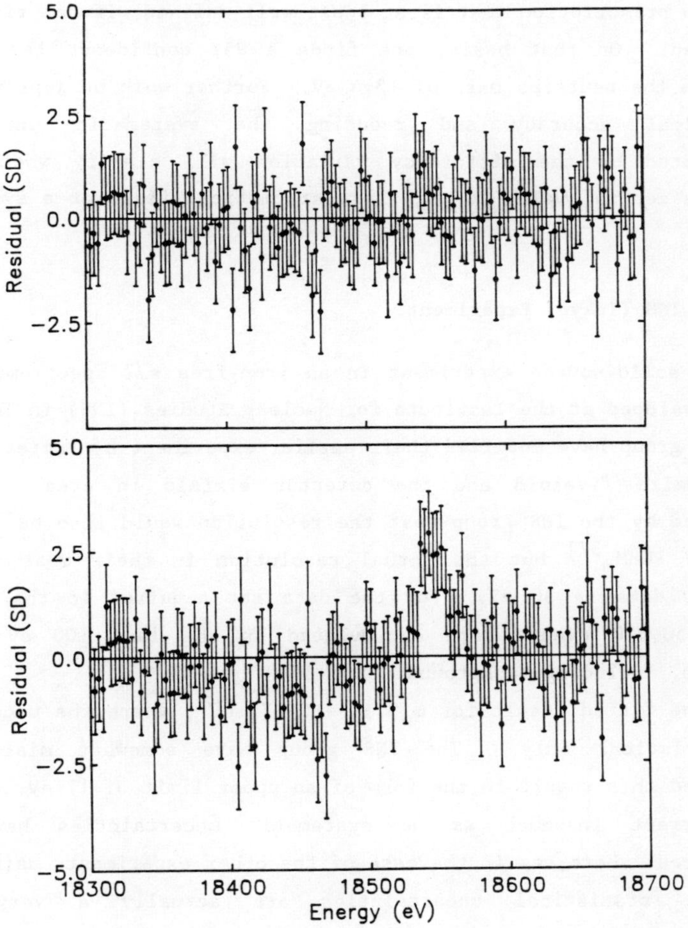

Fig.1. Residuals in fits to neutrino masses of 0 (top) and 30 eV (bottom). All other parameters including α_1 and α_2 have been allowed to vary.

physically forbidden from having negative values is a complex issue, especially when the measured value (through normal statistical fluctuations) falls in the non-physical regime. The final result (from the average of the fits with α_1 and α_2 separately), m_ν^2 = -198(143) eV^2, would have arisen in less than 8% of trials from a neutrino mass greater than 0 eV. To derive a confidence level on the mass is less straightforward. The Particle Data Group[36] sets forth a Bayesian prescription that is at least well-defined, if not rigorously justified. On that basis, one finds a 95% confidence level upper limit on the neutrino mass of 13.4 eV. Further work on improving the statistical accuracy and reducing the systematic uncertainty contributed by the efficiency variation will clarify whether the negative central value for m_ν^2 is a statistical fluke or a systematic effect.

3.2 The INS (Tokyo) Experiment

A solid-source experiment in an iron-free $\pi\sqrt{2}$ spectrometer has been developed at the Institute for Nuclear Studies (INS) in Tokyo.[23] The INS group have modified their earlier experiment by increasing the source size fivefold and the detector sixfold in area. It was predicted by the INS group that the resolution would also be improved to 2 eV FWHM,[37] but the actual resolution in their most recently reported data is 30 eV. For the data set acquired in the fall of 1988, 150,000 events were accumulated in the last 100 eV of the spectrum, including background.

The fitted result for m_ν^2 is -82(87) eV^2, where the uncertainty is statistical only. The INS group have somewhat misleadingly presented this result in the form of an upper limit of 11 eV, which is not correct inasmuch as no systematic uncertainties have been considered. Here, as in the case of the other experiments using solid sources, statistical uncertainties are actually a very minor component, and it is systematic errors that overwhelmingly dominate the result. In particular, the INS group have considered neither the final state spectrum nor the ^{109}Cd shakeup and shakeoff spectrum in estimating their errors. Until that has been done, no conclusion about the neutrino mass can be drawn from the INS work.

4. CONCLUSION

The new data from Los Alamos cannot be reconciled with the ITEP claims of a non-zero electron antineutrino mass, and it appears likely that the systematic errors associated with the final-state spectrum or energy loss for the solid source used by ITEP have been underestimated. The upper limit of 13.4 eV at 95% confidence level means that the electron neutrino is insufficiently massive to close the universe by itself. A mass of 22 eV is required for that, unless the Hubble is less than 50 km/sec/Mpc. We note, finally, that this low limit on the mass of ν_e should permit a new analysis of the neutrino data from the supernova SN1987a largely free of the time-dispersive effects introduced by neutrino mass.

REFERENCES

1. Lubimov, V. A., Novikov, E. G., Nozik, V. Z., Tret'yakov, E. F., Kozik, V. S., Phys. Lett. 94B, 266 (1980).
2. Bowles, T. J., Robertson, R. G. H., Wark, D. L., Wilkerson, J. F., Stephenson, G. J., Friar, J. L., and Knapp, D. A., XXIII Yamada Conference, Osaka, Japan, June 12-16, 1989, unpublished (LA-UR-89-2010).
3. Albrecht, H., Binder, U., Böckmann, P., Gläser, R., Harder, G. et al. Phys. Lett 202B, 149 (1988).
4. Albrecht, H., Binder, U., Harder, G., Philipp, A., Schmidt-Parzefall, W. et al., Phys. Lett. 163B, 404 (1985).
5. Simpson, J. J. and Hime, A., Phys. Rev. D39, 1825 (1989).
6. Hime, A., and Simpson, J. J., Phys. Rev. D39, 1837 (1989).
7. Simpson, J. J., Phys. Rev. Lett. 54, 1891 (1985).
8. Lindhard, J. and Hansen, P. G., Phys. Rev. Lett. 57, 965 (1986).
9. Eman, B. and Tadic, D., Phys. Rev. C33, 2128 (1986).
10. Haxton, W. C., Phys. Rev. Lett. 55, 807 (1985).
11. Altzitzoglou, T., Calaprice, F., Dewey, M., Lowry, M., Piilonen, L., et al., Phys. Rev. Lett. 55, 799 (1985).
12. Ohi, T., Nakajima, M., Tamura, H., Matsuzaki, T., Yamazaki, T., Phys. Lett. 160B, 322 (1985).
13. Datar, V. M. et al., Nature 318, 547 (1985).
14. Apalikov, A., Boris, S., Golutvin, A., Laptin, L., Lubimov, V. et al., JETP Lett. 42, 289 (1985).
15. Markey, J. and Boehm, F., Phys. Rev. C32, 2215 (1985).
16. Simpson, J. J., Phys. Lett. B174 (1986).
17. Hetherington, D. W., Graham, R. L., Lone, M. A., Geiger, J. S., Lee-Whiting, G. E., Phys. Rev. C36, 1504 (1987).
18. Robertson, R. G. H. and Knapp, D. A., Ann. Rev. Nucl. Part. Sci. 38, 185 (1988).
19. Friar, J. L. (unpublished).

20. Boris, S., et al., Phys. Rev. Lett. 58, 2019 (1987); Boris, S. D., et al., Pis'ma Zh. Eksp. Teor. Fiz. 45, 267 (1987) [Sov. Phys. JETP Lett. 45, 333 (1987)].
21. Fritschi, M., et al., Phys. Lett 173B, 485 (1986).
22. Wilkerson, J. F., et al., Phys. Rev. Lett. 58, 2023 (1987).
23. Kawakami, H., Nisimura, K., Ohshima, T., Shibata, S., Shoji, Y., et al. Phys. Lett. 187B, 198 (1987).
24. Liu, J. W., Phys. Rev. A 7, 103 (1973).
25. Inokuti, M., Rev. Mod. Phys. 43, 297 (1971).
26. Bartlett, R. J., Brown, G. S., Bowles, T. J., Crasemann, B., Henderson, J., Knapp, D. A., Robertson, R. G. H., Schaphorst, S. Sorenson, S. L., Trela, W. J, Wark, D. L., and Wilkerson, J. F. (unpublished).
27. Robertson, R. G. H., Bowles, T. J., Wark, D. L., Wilkerson, J. F., and Knapp, D. A.,in "Tests of Fundamental Laws in Physics," Proceedings of the XXIVth Moriond Workshop, Les Arcs, France, January 1989, edited by O. Fackler and J. Tran Thanh Van (Editions Frontières, Gif-sur-Yvette, France) p. 255.
28. Kolos, W., et al., Phys. Rev. A 31, 551 (1985).
29. Cohen, J. S., private communication.
30. Staggs, S. T., et al., Phys. Rev. C 39, 1503 (1989).
31. Martin, R. L. and Cohen, J. S., Phys. Lett. 110A, 95 (1985).
32. Fackler, O. et al., Phys. Rev. Lett. 55, 1388 (1985).
33. Kaplan, I. G. and Smelov, G. V., in Proceedings of the International Symposium on Nuclear Beta Decays and Neutrino, edited by T.Kotani, H. Ejiri and E. Takasugi (World Scientific, Singapore 1986), p. 347.
34. Williams, R. D. and Koonin, S. E., Phys. Rev. C 27, 1815 (1983)
35. Intemann, R. L., Phys. Rev. A 26, 3012 (1982).
36. Particle Data Group, Phys. Lett. 170B, 1 (1986).
37. Kawakami, H., et al., in "Neutrino Mass and Related Topics," Proc. XVI INS International Symposium, Tokyo, March 16-18, 1988, edited by S. Kato and T. Ohshima (World Scientific, Singapore, 1988) p. 82.
38. Kawakami, H., et al., INS preprint INS-Rep.-758, KEK preprint 89-60.

NEUTRINO OSCILLATIONS: AN ESSAY IN HONOR OF FELIX BOEHM

S. P. ROSEN
T-DIVISION, LOS ALAMOS NATIONAL LABORATORY
LOS ALAMOS, NEW MEXICO 87545

1 ABSTRACT

We briefly review the theory of neutrino oscillations and the MSW effect and report on new calculations by Rosen and Gelb for solar neutrino-electron scattering. The aim of these calculations is to try to use the scattering process as a means of choosing between the three types of MSW solutions for the ^{37}Cl experiment. Both the efficiency and the resolution of the Kamiokande II detector are taken into account and the ratio R of the MSW prediction to the standard solar model prediction is calculated for different cuts on the minimum electron energy. We find that the adiabatic solution requires R to be less than 1/3, the large angle one requires it to be less than 2/3, and the nonadiabatic one restricts it to a value close to 1/2. The central value of the published KII data is close to 1/2, but the errors are too large to exclude the other solutions.

2 INTRODUCTION

I would like to wish Felix 'Many Happy Returns' on the occasion of his 65th birthday and to express the hope that he will have many more birthdays to come! There is still much work to be done in the field of neutrino oscillations and we shall need all of his insight and skill to complete it.

It is entirely fitting to begin a discussion of neutrino oscillations, one which may ultimately turn out to be nothing more than wishful thinking on the part of this particular theorist, with a quotation attributed to Felix at a recently held international conference, and with the principal result of his work.

The quotation was made by Martin Perl in his review of neutrino properties at the 1989 International Symposium on Lepton and Photon Interactions at High Energies, held at SLAC in August. 'Neutrino Oscillations?' asked Perl rhetorically, 'There is *NO* confirmed evidence for neutrino oscillations!' and he cited a talk by

Felix at the Yamada Conference held at Osaka in June of this year[1)].

In a similar vein, we can refer to the beautiful series of reactor experiments led by Felix. These experiments[2)], carried out at the Goesgen reactor over a period of several years, provide the most sensitive laboratory limits on the disappearance of $\bar{\nu}_e$ in the parameter domain of small Δm^2 and large $\sin^2\theta$. Other experiments at accelerators[3)] provide rather tight limits on the disappearance and appearance of different types of neutrino in the domain of small mixing angle and large mass differences.

Given the weight of negative evidence, why should we pursue the idea of neutrino oscillations? Let me give you my own reasons for doing so:

1. Because the idea[4)] is a beautiful one, especially when coupled with the elegant matter enhancement mechanism of Mikheyev and Smirnov[5)], and Wolfenstein[6)]. Invoking a Diracian principle, I would like to suggest that whatever is beautiful should be true.

2. Because the observation of oscillations is a definitive signal for the existence of neutrino mass[4)], and it is the most sensitive method of searching for masses well below $1eV$.

3. Because there are significant regions of parameter space remaining to be explored, especially with low energy neutrinos over very long distances[2,3)].

4. Because there exists the *possibility* that we are seeing oscillations with solar neutrinos. Indeed the Davis[7)] and Kamiokande II[8)] experiments provide the best (remember that 'best' does not necessarily imply 'good') evidence for neutrino oscillations but they are not definitive.

We appear to have a conflict between theoretical desires on the one hand and hard experimental facts on the other. Now it happens that on my journey to Pasadena I was reading a book by an expert negotiator who believes in looking for 'win-win' resolutions to conflicts between people. So here I am going to look for a 'win-win' resolution of the physics.

3 QUICK REVIEW OF THE MSW EFFECT

As you might gather from the first point, I have a strong predilection for the Mikheyev-Smirnov-Wolfenstein effect (MSW)[5,6,9)]; indeed my hope and my prejudice are that it will provide us with the solution to the solar neutrino problem. The basis for this effect, and also for the Kamiokande II experiment, is to be found in the two lowest order diagrams for neutrino-electron scattering. One diagram involves the exchange of a neutral Z^0 boson between the electron and the neutrino, and in the standard GWS model, it has the same strength for all types of neutrino; the other is a 'charge-exchange' process involving the exchange of a charged W^+ boson, and it comes into play only for the electron-type neutrino ν_e. The existence of this charged-current diagram means that ν_e has a different refractive index, or effective mass, in matter as compared with the other types of neutrino, ν_μ and ν_τ, and that it has a much larger cross-section for scattering from electrons than do the other neutrinos. Both of these effects will play important parts in this discussion.

The phenomenon of neutrino oscillations is a purely quantum mechanical effect involving two almost degenerate neutrino mass eigenstates[4,10], ν_1 and ν_2, with masses m_1 and m_2 respectively, and with a common momentum p which is much greater than both masses. The energies of the two states are then given by the approximate formula

$$E_i = p + m_i^2/2p \;\; (i = 1, 2) \; . \tag{1}$$

As these mass eigenstates evolve in time, they acquire the appropriate phase factors

$$exp(-iE_j t)(j = 1, 2) \tag{2}$$

and hence the phase difference between them oscillates with time. Whenever we have new states defined as coherent combinations of these two states with definite phase relations between them, the character of the new states will oscillate in time along with the phase difference.

Let us define the electron neutrino ν_e and the muon neutrino ν_m as the orthogonal combinations:

$$\nu_e = \cos\theta \nu_1 + \sin\theta \nu_2$$

$$\nu_m = \cos\theta \nu_2 - \sin\theta \nu_1. \tag{3}$$

By electron neutrino and muon neutrino we mean those flavor eigenstates that take part in weak interactions with the electron and muon respectively. As these states evolve in time, the relative phase between ν_1 and ν_2 will change, and what was initially a pure electron neutrino or a pure muon neutrino will become an admixture of the two flavor states.

To explore the MSW effect, we write down the Schroedinger-like time development equation for the oscillation amplitudes[6]:

$$i\frac{dA}{dt} = HA \tag{4}$$

where A represents a column vector of the probability amplitudes for ν_e to remain ν_e and for ν_e to turn into another neutrino type ν_m, a_e and a_m respectively,

$$A = \begin{pmatrix} a_e \\ a_m \end{pmatrix} . \tag{5}$$

The Hamiltonian H is given by:

$$H = \begin{pmatrix} XY \\ YZ \end{pmatrix} \tag{6}$$

with

$$\begin{aligned} X &= \frac{m_1^2c^2 + m_2^2s^2}{2p} + 2^{1/2}G_F N_e \\ Z &= \frac{m_1^2s^2 + m_2^2c^2}{2p} \\ Y &= \frac{m_2^2 - m_1^2}{2p}cs = \frac{\Delta m^2}{2p}cs \\ c &= \cos\theta \\ s &= \sin\theta\ . \end{aligned} \tag{7}$$

For neutrinos propagating through the sun, matter oscillations are incorporated through the last term in the expression for X, which depends upon the Fermi constant G_F and the density of electrons, N_e. All the other terms in the expressions for X, Y, Z in eq (7) correspond to neutrinos propagating in vacuo. Adding the matter term can have powerful consequences: for, under the right conditions, it gives us a Hamiltonian matrix which is not only symmetric, but also one which has equal elements down the diagonal. The eigenvectors of such a matrix are equal admixtures of electron- and muon- neutrino, and for a given off-diagonal element, the separation between the eigenvalues is minimal. In other words the extra term in X gives us a chance to progress from non-maximal mixing in vacuo to maximal mixing in matter.

The condition for equal diagonal elements, $X = Y$ can be written as:

$$2^{1/2} G_F N_e = \frac{\Delta m^2 \cos 2\theta}{2p} \quad . \tag{8}$$

Now the electron density N_e is inherently positive, and the Fermi constant G_F, since it arises from the exchange of a gauge boson, is also positive: therefore the product of Δm^2 and $\cos 2\theta$ must also be positive if the condition (8) is to be satisfied. It is not difficult to show that this requires the electron-neutrino to be dominantly composed of the lighter of the two mass eigenstates, and the muon-neutrino of the heavier.

We take the neutrino masses to be so small that the particles travel with the speed of light; in units where c=1 we can then equate the time t to the distance R travelled. For oscillations which occur in vacuo rather than in a material medium, the probabilities for the survival of the original flavor and the appearance of a new flavor can be written in the standard forms:

$$P(\nu_e, \nu_e; R) = 1 - \sin^2 2\theta \sin^2(\pi R/L) \tag{9}$$

$$P(\nu_m, \nu_e; R) = \sin^2 2\theta \sin^2(\pi R\mathrm{L}), \tag{10}$$

where the oscillation length L is:

$$\begin{aligned} L &= 2\pi/(E_2 - E_1) = 4\pi p/(m_2^2 - m_1^2) = 4\pi p/\Delta m^2 \\ &= 2.5(p/MeV)/(\Delta m^2/eV^2) meters. \end{aligned} \tag{11}$$

For oscillations in a medium with a constant density of electrons, we can describe the oscillation problem in the same way as in the in vacuo case except that the mixing angle and the oscillation length are both modified. The survival probability for an electron neutrino is now:

$$P(\nu_e, \nu_e; R) = 1 - \sin^2 2\theta_M \sin^2(\pi R/L_M). \tag{12}$$

The new oscillation parameters are obtained by diagonalising the equations of motion in eq.(4-7) above:

$$\begin{aligned}
\sin^2 2\theta_M &= \sin^2 2\theta/[\sin^2 2\theta + (L/L_0 - \cos 2\theta)^2] \\
L_M &= L/[sin^2 2\theta + (L/L_0 - \cos 2\theta)^2]^{1/2} \\
L &= 4\pi p/\Delta m^2 \\
L_0 &= 2\pi/2^{1/2} G_F N_e.
\end{aligned} \tag{13}$$

The formula for the new mixing angle in matter has some important properties: no matter how small the in vacuo angle θ may be, the matter angle θ_M reaches its maximal value when

$$L/L_0 = \cos 2\theta \tag{14}$$

which is just another way of writing the equal diagonal element condition of eq(8). Therefore, as long as θ is different from zero, there is always a density for which the neutrino will oscillate with maximal mixing.

For a medium of varying density such as the sun, the enhancement condition of eq. (8) can be satisfied at different locations for neutrinos of different oscillation lengths. Thus all neutrinos in an energy band which is determined by the density profile of the medium and the value of Δm^2 will pass through an enhancement point somewhere in the medium.

To gain some insight into this situation, we express the electron density in units of Avogadro's Number and L_0 in meters. The condition (8) becomes

$$\frac{p}{\Delta m^2} = \frac{7 * 10^6 \cos 2\theta}{\rho_e} \quad . \tag{15}$$

The values of ρ_e encountered in the sun vary from about 150 in the core to close to zero at the edge and so the band of oscillation lengths covers several orders of magnitude:

$$10^4 \leq \frac{p}{\Delta m^2} \leq 10^8. \tag{16}$$

For neutrino momenta in the range of 1-10 MeV, typical of the ^{8}B neutrinos to which the Davis experiment is sensitive, and a small mixing angle, condition (15) corresponds to Δm^2 in the range of 10^{-7} to 10^{-3} eV^2.

In the earth, ρ_e varies from about 3 at the surface to 13 at the center, and so the range of Δm^2 which will give rise to enhanced oscillations for ^{8}B neutrinos is 10^{-8} to 10^{-6} eV^2. We may therefore anticipate that for certain sets of parameters, neutrinos can undergo enhanced oscillations in the earth as well as the sun. This in turn can give rise to a "day-night" effect in which electron neutrinos are converted to another type in the sun and rejuvenated in their passage through the earth[11)].

When we apply the MSW effect to the ^{37}Cl experiment of Ray Davis and his collaborators, we find that there are three types of solution in the parameter space of Δm^2 and $\sin^2 2\theta$: the *adiabatic* solution[5,12)], in which 'low' energy solar neutrinos remain as electron neutrinos while 'high' energy ones are almost completely converted to muon or other neutrino type; the *nonadiabatic* solution[13)], in which the 'low' energy neutrinos are completely converted to another neutrino type while the 'high' energy ones have about a 50% of remaining as electron neutrinos; and the *large angle* solution[14)], in which the probability for the solar neutrinos to remain as electron neutrinos is independent of energy. The dividing line between 'low' and 'high' is in the neighborhood of 6-8 MeV.

These solutions arise from the different ways in which the spectrum of solar neutrinos can overlap the curve of probability for ν_e to remain ν_e as a function of the oscillation length parameter $l = p/\Delta m^2(MeV/ev^2)$. The characteristic behavior of the probability is that it is close to unity for $l \lesssim 3 \times 10^4$ and then falls rapidly to the value of $\sin^2\theta$ in the neighborhood of $l = 10^5$; this is the adiabatic part of the curve. It remains at this value for a decade or more, depending on the magnitude of the mixing angle, and then begins an exponential, nonadiabatic rise back to unity as l approaches 10^8.

For small mixing angles, the probability in the central region is close to zero and so for values of Δm^2 in the range 10^{-4} to 10^{-7}, we find that the probability curve divides the spectrum of ^{8}B neutrinos into two parts, one with a large probability and the other with a small one. In the adiabatic solution the large probability occurs for low energies, and in the nonadiabatic one, it occurs for high energies.

For large mixing angles with $\sin^2\theta = 1/4 - 1/2$, the central region is large enough to contain the entire solar neutrino spectrum. Thus we obtain a solution in which the probability for ν_e to remain ν_e is independent of energy. This solution spans the entire range of Δm^2 for which MSW comes into play, and it includes the values for which the day-night effect occurs.

It is clear that in order to distinguish between these solutions, we must find a way of measuring the probability as a function of energy. One way is to look at the low energy pp neutrinos: for the adiabatic solution they remain as electron neutrinos, whereas for the nonadiabatic one they can be almost entirely converted to another type[13)]. Another way of making the distinction is to look at the scattering of solar neutrinos by electrons.

4 SOLAR NEUTRINO-ELECTRON SCATTERING AND THE KII EXPERIMENT

Solar neutrino-electron scattering is sensitive to all types of neutrino except sterile ones[9)]. Electron neutrinos have the largest cross-section, by a factor between 6 and 7, because they interact with electrons through both charged- and neutral-currents; the muon- or tau-type neutrinos into which they might oscillate interact only through the neutral-current. Should solar neutrinos oscillate into sterile types, then the sterile neutrinos will not scatter from electrons at all.

Total cross-sections for solar neutrino-electron scattering in the context of the standard particle physics model and the MSW mechanism have been calculated by Bahcall, Gelb, and Rosen[15)]. Subsequently, Gelb and Rosen[16)] have examined the implications of the different MSW solutions of the ^{37}Cl experiment for electron scattering. They found that there is a definite relationship between each solution and the ratio R of the observed signal to the signal expected on the basis of the standard solar model in the absence of neutrino oscillations. They found that the adiabatic solution predicts values of R less than, or equal to 1/3, while the large angle solution allows values up to approximately 2/3; by contrast the non-adiabatic solutions predicts that R should be in a narrow range around 50%. In addition R can never fall below 1/6 unless solar neutrinos oscillate into sterile types. This summer Gelb and I have re- examined our calculations taking into account certain properties of the KII detector that were ignored in earlier work. I would like to describe the results here.

To use solar neutrino-electron scattering as a tool for distinguishing between different MSW solutions of the Davis experiment we follow Bahcall, Gelb, and Rosen[15)] and write the differential cross-section for producing a recoil electron with kinetic energy T as:

$$\langle\frac{d\sigma}{dT}\rangle = \frac{1}{\phi_{\text{total}}}\int dq\phi(q)\left[\frac{d\sigma}{dT}(\nu_e e)P_{ee}(q) + \frac{d\sigma}{dT}(\nu_\mu e)(1 - P_{ee}(q))\right] \tag{17}$$

where $\phi(q)$ is the spectrum of neutrinos of energy q produced in the sun, $\frac{d\sigma}{dT}$ is the differential cross-section for neutrino electron scattering, and $P_{ee}(q)$ is the probability for an electron-type neutrino of energy q to remain an electron neutrino. The basic differential cross- section $\frac{d\sigma}{dT}(\nu e)$ depends upon q, T, and the flavor of the incident neutrino.

Since we actually measure the product of the basic differential cross-section times the flux, we must assume that the neutrino flux $\phi(q)$ is given by the standard solar model[9)] in order to extract information about MSW from the data. In addition we must take into account the detector efficiency for recording electrons of an apparent energy T_a and its resolution function, which gives the probability that the true electron energy is T_t when the apparent energy is T_a. In our new calculations, Gelb and I use the prescriptions for efficiency and resolution given in various publications of the Kamiokande collaboration[17)] and described by Bahcall and Haxton[18)] in their recent study of MSW and the standard solar model.

The efficiency function for the Kamioka detector is:

$$\begin{aligned} f(T_a) &= 1 - e^{-[(\frac{T_a - T_0}{4})]^{2.03}} \quad T_a \geq T_0 \\ f(T_a) &= 0 \quad T_a \leq T_0 \\ T_0 &= 4.2MeV. \end{aligned} \tag{18}$$

The detector resolution is approximately given by a Gaussian:

$$\begin{aligned} \rho(T_a, T_t) &= N\exp^{-(T_a - T_t)^2/2\sigma^2(T_a)} \\ \sigma(T_a) &= 0.22T_a\sqrt{\frac{10MeV}{T_a}}, \end{aligned} \tag{19}$$

where N is chosen so that the integral of $\rho(T_a, T_t)$ over apparent energies from the electron mass to infinity is equal to unity.

To obtain the total cross-section, we perform a double integral over both the apparent and true electron energies of the differential cross-section in eq. (17)

weighted by the product of efficiency and resolution functions. The integration runs from the experimentally required minimum energy $T_{\min}$ to the kinetically allowed maximum $T_{\max}$:

$$\langle\sigma(\nu e)\rangle = \int_{T_{\min}}^{T_{\max}} \int_{T_{\min}}^{T_{\max}} dT_a dT_t f(T_a)\rho(T_a, T_t)\langle\frac{d\sigma(T_t)}{dT_t}\rangle \ . \tag{20}$$

It follows from eqs. (17) and (20) that $\langle\sigma(\nu e)\rangle$ is bounded from above by the cross-section for pure electron-neutrino scattering, and from below by that for pure muon-neutrino scattering:

$$\sigma(\nu_\mu e) \leq \langle\sigma(\nu e)\rangle \leq \sigma(\nu_e e) \ . \tag{21}$$

A signal below this lower bound would indicate that the oscillation must take place into a sterile neutrino.

The effect of detector efficiency and resolution upon the magnitude of the cross-section in the standard solar model varies with the minimum energy cut $T_{\min}$: they tend to increase the cross-section when the cut is large, and decrease it when the cut is relatively low. For example, when $T_{\min} = 9.3MeV$, the cross-section is 0.014 SNU with perfect efficiency and resolution and 0.02 SNU with the actual properties given in eqs. (18,19) above. By contrast, for $T_{\min} = 7.5MeV$, the cross-section is 0.042 SNU in the perfect case and 0.037 SNU in the actual one. Here SNU stands for one solar neutrino-electron scattering event per 10^{36} electrons per second. Given a fiducial volume of approximately 680 tons for the KII detector, these cross- sections correspond to event-rates of 0.27, 0.38, 0.82, and 0.72 events per day respectively.

We can understand the proportionately large effect for the higher energy cut on the basis of the behavior of the basic cross-section and the energy spectrum of the ^8B solar neutrinos[19)]. The cross-section increases linearly with energy while the spectrum behaves approximately quadratically, rising to a maximum in the vicinity of 6.5 MeV and then declining to zero near the end-point of 14 MeV. Now the resolution function spreads an apparent energy over a range of roughly 3-4 MeV and, at the higher cut it tends to bring in many more neutrinos with slightly lower energies and cross-sections; the net effect is to give a significantly larger signal. As the cut moves back towards the peak of the spectrum, the reverse tends to happen: the resolution spreading tends to bring in fewer neutrinos, but with higher energies and cross-sections, and the overall effect is a slightly smaller signal.

The results of our new calculations are shown in Figure (1) where we plot the MSW predictions for the cross-section as a function of Δm^2 for a series of values of $\sin^2 2\theta$. We express the cross-section as a fraction of the cross-section predicted by the standard solar model for the same minimum energy cut $T_{\rm min}$ and the same efficiency and resolution functions as given above, but with no oscillations. Two energy cuts are considered: $T_{\rm min} = 9.3$ and 7.5 MeV.

Imposing the restrictions corresponding to the MSW solutions of the ^{37}Cl experiment, we find that the results for the *fraction* of MSW to standard model cross-section are very little changed from the case when efficiency and resolution were not taken into account. The adiabatic solution still implies that the fraction R must be less than 1/3 and the large angle one restricts it to 2/3 or less. For the nonadiabatic solution, however, R tends to be slightly larger than before: whereas in the original calculation R hovered between 0.5 and about 10% *below* it, R now hovers between 0.5 and 10% *above* it.

The value R = 0.5 in the case of the nonadiabatic solution and its relative constancy are not difficult to understand. Solar neutrino-electron scattering in the KII experiment is sensitive to neutrinos in the higher energy half of the spectrum, a region for which the nonadiabatic solution gives about a 40% survival probability to electron-type neutrinos[13)]; thus 40% of the neutrinos will scatter with the maximal electron-type neutrino cross-section, while the other 60% will scatter with the neutral-current cross-section which is a factor of 6 smaller. Simple arithmetic then yields a value of 0.5 for R.

At the time of writing, the KII collaboration has published the results of 450 days of observation between January, 1987 and May, 1988, an interval which overlaps with the period during which the ^{37}Cl experiment has been yielding a signal[7)] of about 4 SNU rather than the overall average of 2.1 SNU. The value of R that they obtain, namely[8)]

$$R = 0.46 \pm 0.13(stat) \pm 0.08(syst), \tag{22}$$

tends to favor the nonadiabatic solution, but the errors are sufficiently large that it does not exclude the adiabatic one. In addition, the result is also consistent with the large angle solution. As more data is accumulated, one may hope that the central value will remain unchanged, while the errors shrink down to a level where one can make a more definite statement about the adiabatic solution; however to distinguish between the other solutions one must turn to other experiments, especially the ^{71}Ga

one.

In so far as the large angle solution is concerned, the conversion of electron-type neutrinos into other types is independent of energy and it predicts that the signal for the ^{71}Ga experiment, which is predominantly sensitive to the low energy pp solar neutrinos, will be suppressed by the same factor of 1/4 - 1/2 as in the ^{37}Cl one. The nonadiabatic solution, on the other hand, yields a much stronger conversion for low energy neutrinos than it does for high energy ones and so it can yield a much greater suppression of the ^{71}Ga; in fact, the signal could be as low as 10 or 20 SNU[13)] instead of the 130 SNU predicted by the standard model. This is illustrated in Figures (2a, 2b and 4e) of a recent paper of Bahcall and Haxton[18)]. We look forward to the results of this experiment with great anticipation.

5 CONCLUSION: THE WIN-WIN SOLUTION

We also look forward to a new *reactor* experiment recently proposed by Felix and his colleagues[20)]. It will be performed with a kiloton detector to be located in a railway tunnel 13 kilometers from the Goesgen reactor, and it will probe the region of parameter space corresponding to large mixing angles and Δm^2 of $10^{-4} eV^2$.

If Felix and colleagues find an effect, then we shall be blessed with a major discovery of profound importance for physics beyond the standard model. If they find no effect, then perhaps they will be confirming, indirectly, the scenario preferred by this author.

Happy Birthday Felix, and please give us the answer when we come to celebrate your 70th !

6 REFERENCES

1. Felix Boehm, Proceedings of the XXIII Yamada Conference, Osaka, Japan, June 1989; and CalTech Preprint CALT-63-546.

2. G. Zacek *et al.* Phys. Rev.**D34**,2621 (1986).

3. R. Eichler, Proceedings of the 1987 Symposium on Lepton-Photon Interactions at High Energies.

4. B. Pontecorvo, Sov. Phys. - JETP**6**,429 (1958) and *ibid* **26**,984 (1968); Z.

Maki, M. Nakagawa, and S. Sakata, Prog. Theoret. Phys. **28**, 870 (1962); V. Gribov and B. Pontecorvo, Phys. Lett. **B28**,493 (1969).

5. S. P. Mikheyev and A. Yu. Smirnov, Sov. J. Nucl. Phys.**42**,913 (1986); Nuovo Cimento **9C**, 17 (1986); and Sov. Phys. Usp. **30**, 759 (1987).

6. L. Wolfenstein, Phys. Rev.**D17**,2369 (1978).

7. R. Davis Jr. *et al.*, in *NEUTRINO '88*, The Proceedings of the 13th International Conference on Neutrino Physics and Astrophysics, Boston (Medford), June 5-11, 1988, edited by J. Schnepps, T. Kafka, W.A. Mann, and Pran Nath (World Scientific, Singapore, 1989) p.518.

8. K. Hirata *et al.* Phys. Rev. Lett.**63**,16 (1989).

9. See John N. Bahcall, *NEUTRINO ASTROPHYSICS* (Cambridge University Press, Cambridge and New York, 1989) for a comprehensive discussion of the solar neutrino problem.

10. B. Kayser, Phys. Rev.**D24**,110 (1981).

11. A.J. Baltz and J. Weneser, Phys. Rev.**D35**,528 (1987) and *ibid* **D37**,3364 (1988); J. Bouchez *et al.*, Z. Phys.**C32**,499 (1986); and E.D. Carlson, Phys. Rev.**D34**,1454 (1986).

12. H.A. Bethe, Phys. Rev. Lett.**56**,1305 (1986).

13. S.P. Rosen and J.M. Gelb, Phys. Rev.**D34**,969 (1986); W.C. Haxton, Phys. Rev. Lett.**57**,1271 (1986); E.W. Kolb, M. Turner, and T.P.Walker,Phys. Lett.**B175**,478 (1986); and S.J. Parke, Phys. Rev. Lett.**57**,1275 (1986).

14. S.J. Parke and T.P. Walker, Phys. Rev. Lett.**57**,2322 (1986); **57**,3124(E) (1986).

15. J.N. Bahcall, J.M. Gelb, and S.P. Rosen, Phys. Rev.**D35**,1976 (1987).

16. S.P. Rosen and J.M. Gelb, Phys. Rev.**D39**,3190 (1989).

17. M. Nakahata, University of Tokyo Preprint UT-ICEPP-88-01; E.W. Beier, Proceedings of the Seventh Workshop on Grand Unification *ICOBAN '86*, edited by J. Arafune (World Scientific, Singapore, 1987), p.79; Y. Totsuka, *ibid*, p.118.

18. J. N. Bahcall and W. C. Haxton, Institute for Advanced Study Preprint, IASSNS AST 89/14 and Phys. Rev. **D40**, 931 (1989).

19. See reference (9) Table 6.3, p.156; Fig. 6.6, p.159; Table 8.6, p.221; and Fig. 8.1, p.222.

20. Felix Boehm, in *NEUTRINO '88*, The Proceedings of the 13th International Conference on Neutrino Physics and Astrophysics, Boston (Medford), June 5-11, 1988, edited by J. Schnepps, T. Kafka, W.A. Mann, and Pran Nath (World Scientific, Singapore, 1989) p.490.

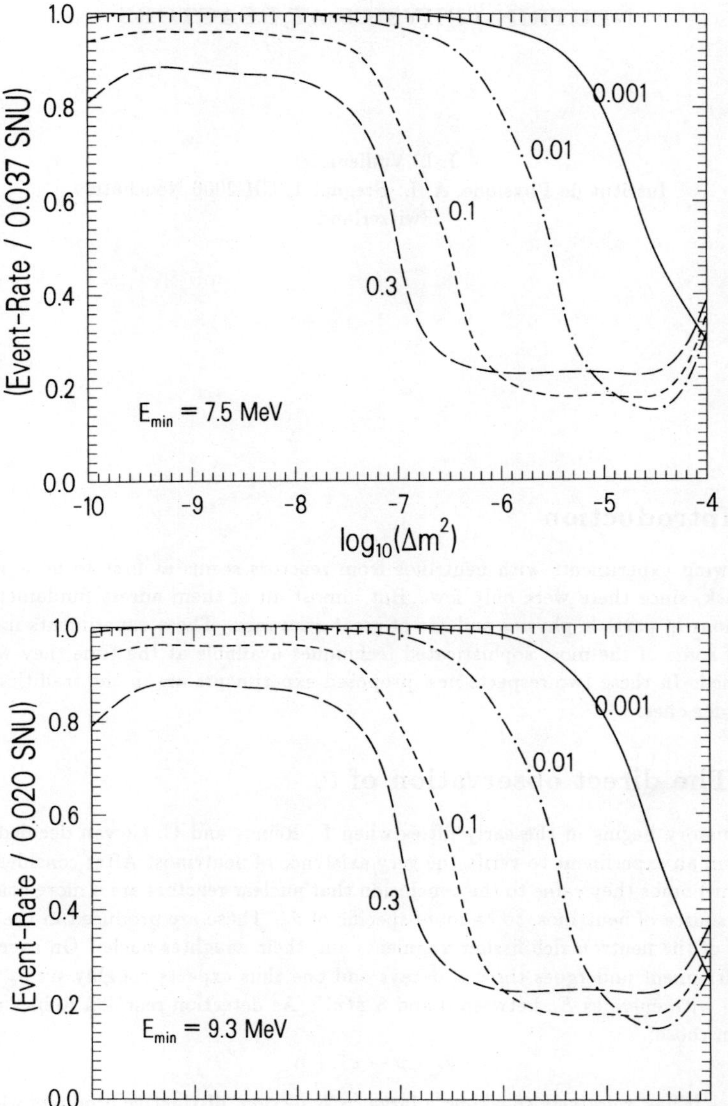

Fig. 1 The ratio of MSW to standard model predictions for solar neutrino electron scattering taking into account detector efficiency and resolution for two different minimum electron energy cuts. Each curve corresponds to the value of $\sin^2 2\theta$ indicated beside it.

NEUTRINO PHYSICS AT REACTORS

J.-L. Vuilleumier
Institut de Physique, A. L. Breguet 1, CH-2000 Neuchâtel
Switzerland

1 Introduction

Reviewing experiments with neutrinos from reactors seems at first to be a simple task, since there were only few. But almost all of them adress fundamental questions in particle physics and are of great relevance. These experiments made use of some of the most sophisticated techniques available at the time they were designed. In these two respects new proposed experiments are in the tradition of the older ones.

2 The direct observation of $\overline{\nu}_e$

The history begins in the early fifties when F. Reines and C. Cowan decided to perform an experiment to verify the very existence of neutrinos. After considering nuclear bombs they came to the conclusion that nuclear reactors are a more convenient source of neutrinos, to be more specific of $\overline{\nu}_e$. These are produced in the β^- decay of the neutron rich fission fragments and their daughter nuclei. On average each fragment undergoes three β decays and one thus expects roughly six $\overline{\nu}_e$ per fission with energies E_ν between 0 and 8 MeV. As detection reaction Reines and Cowan chose:

$$\overline{\nu}_e + p \rightarrow e^+ + n \tag{1}$$

because it has, relatively speaking, a large cross section, distinctive products which allows a good signature and because organic scintillators contain a large amount of protons. The neutron only takes away around 20 keV of energy so that the positron kinetic energy E_{e^+} in good approximation is given by

$$E_{e^+} = E_\nu - E_{threshold}, \quad \text{with} \quad E_{threshold} = 1.804\ MeV. \tag{2}$$

A first attempt was made in 1953 at a reactor in Hanford. The detector was heavily shielded and consisted of an 0.3 m^3 tank filled with Cd loaded liquid scintillator and viewed by 90 photomultipliers. Coincidences were looked for between a prompt e^+ signal and a delayed signal from γ's produced in the capture after thermalization of the neutron on a Cd nucleus. A hint of the signal was seen [1]. Another detector, with better particle identification, was then built. It consisted of two tanks containing 200 l each of H_2O loaded with $CdCl_2$ and sandwiched between three 1400 l liquid scintillators, a rather large scale apparatus at the time. Positrons were identified by the detection in coincidence of two 511 keV annihilation γ rays in two adjacent scintillators. A delayed coincidence was then sought with the Cd capture γ ray signal (fig. 1). The detector was placed in a well shielded location at 11 m from the core of a reactor in Savannah River having, at the time, a power of 700 $MWth$. A reactor-on minus reactor off signal of 3.0 ± 0.2 event$\cdot$hour^{-1} was seen in 1956 [2]. The elaborate signature resulted in a good signal to background ratio of 2.04 but also in a low acceptance of 2.6 ± 1.0 %. The existence of $\overline{\nu}_e$ was thus established and a cross section for reaction 1 averaged over the reactor spectrum of $\overline{\sigma}_{exp} = (12^{+7}_{-4}) \cdot 10^{-44}$ cm^2 was derived. It was in agreement with the value predicted by the $V - A$ theory $\overline{\sigma}_{calc} = (10 \pm 1.7)10^{-44}$ cm^2.

At the same time R. Davis and collaborators looked for the reaction $\overline{\nu}_e + ^{37}Cl \rightarrow e^- + ^{37}Ar$ near another of the Savannah River reactors using a 4000 l tank filled with CCl_4. No signal was seen. The reason is that either neutrinos are Dirac particles with $\nu_e \neq \overline{\nu}_e$, and this particular reaction is entirely forbidden by lepton number conservation, or neutrinos are Majorana particles and then it is is highly suppressed by helicity. Double beta decay experiments are attempting to find which of these explanations is correct. Later R. Davis and collaborators used the same technique to successfully detect solar ν_e: $\nu_e + ^{37}Cl \rightarrow e^- + ^{37}Ar$ [3].

3 The determination of the $\overline{\nu}_e$ spectrum

Subsequent experiments were much more precise. A better knowdledge of the reactor $\overline{\nu}_e$ spectrum soon turned out to be necessary to fully exploit the data. First, one has to know what fissile isotopes contribute in a given reactor. These contributions are time dependent. In a Pu production reactor like the ones in Savannah River for instance ^{235}U dominates largely ($\sim$ 88% on average), followed far behind by ^{239}Pu and ^{238}U. In a pressurized water reactor (PWR) the contribution of ^{235}U, ^{239}Pu, ^{238}U and ^{241}Pu are 55 %, 33 %, 7 % and 5 % respectively on average.

Then one must know the $\overline{\nu}_e$ spectrum for each one of these isotopes. Two approaches have been taken. One is to calculate the spectrum from the yields and decay schemes of the fission fragments [4], [5]. At low energy long lived low endpoint fragments with well known yields and decay schemes dominate so that the calculations are relatively straightforward. At higher energy however short lived nuclei with poorly known or totally unknown yields and decay schemes become more important. Models have to be used which limits the accuracy. The more recent and more elaborate of these calculations achieve a precision of a few percent below 2 MeV and around 20% at 8 MeV [5], [6], [7]. The general feature is a rather smooth distribution peaking around 0.5-1 MeV and extending up to 8-10 MeV.

The other more pragmatic approach consists in measuring the β spectrum of the fission fragments of a given isotope, and to use that information to derive the corresponding $\overline{\nu}_e$ spectrum. The best measurements to date have been performed at the Institut Laue-Langevin in Grenoble. Thin targets of fissile isotopes sandwiched between Ni foils to contain the fission fragments were exposed to a high flux of thermal neutrons near the core of the reactor. The energy of the β's was measured using the BILL magnetic spectrometer. The major isotopes ^{235}U and ^{239}Pu, as well as ^{241}Pu have been investigated this way [8], [9], [10]. The precision on the β spectra is around 2 % for kinetic energies above 2 MeV. Essentially these spectra are superpositions of a large number of β decays with allowed shape P_β. They can however be approximated well by a superposition of a limited number of them, say 30:

$$N_\beta(E_e) = \sum a_i P_\beta(E_e, E_0^i, Z^i). \tag{3}$$

The endpoint energies E_0^i are chosen at regular intervals. The distribution of Z^i, the charge of the fission fragment, as a function of the endpoint energy is taken from experimental data. The yields a_i are varied until good agreement is found with the measured spectrum. The $\overline{\nu}_e$ spectrum is then obtained from the same set of a_i:

$$N_\nu(E_\nu) = \sum a_i P_\beta(E_0^i - E_\nu, E_0^i, Z^i). \tag{4}$$

This procedure was shown to introduce very little additional uncertainty. The spectra determined this way have uncertainties around 3 % between 2 and 8 MeV, much better than those of calculated spectra, with which they agree within errors.

For ^{238}U which has a large fission cross section for fast neutrons no such measurements exist and one has to rely on calculations. Fortunately ^{238}U does not contribute much ($\sim$ 7% in a PWR) to the total spectrum and adds only little to the error. That total reactor spectrum is the weighted average of all the spectra of the individual fissile isotopes. Its uncertainty between 2 and 8 MeV is thus also around 3%. This energy range is adequate for reaction 1 which has a threshold of 1.8 MeV. For reactions with lower threshold, for instance $\overline{\nu}_e e^- \rightarrow \overline{\nu}_e e^-$, one may need to complement these spectra with the calculated ones.

The energies released per fission have been measured with good precision [11], but there is a further source of uncertainty on the normalization of the $\overline{\nu}_e$ reactor spectrum: the power of the reactor in general is known to within 2 % only.

4 The measurement of $\overline{\nu}_e e^- \rightarrow \overline{\nu}_e e^-$ scattering

The reactions

$$\overline{\nu}_e e^- \rightarrow \overline{\nu}_e e^- \text{ and } \nu_e e^- \rightarrow \nu_e e^- \tag{5}$$

are fundamental and their study provides information on basic features of the weak interaction. Both charged (CC) and neutral currents (NC) are involved. They are expected to interfere if the NC and CC final state neutrinos are identical [12]. A measurement of the differential cross section of either reaction allows, in principle, to determine the Weinberg angle $sin^2\theta_W$ and to observe the interference which is expected to be destructive for reasonable values of $sin^2\theta_W$. Practically however $\overline{\nu}_e e^- \rightarrow \overline{\nu}_e e^-$ only has a good sensitivity to both effects, while $\nu_e e^- \rightarrow \nu_e e^-$ essentially only probes the interference. Indeed a measurements of $\nu_e e^- \rightarrow \nu_e e^-$ scattering has been performed at the LAMPF beam dump [13]. In spite of a limited statistics the experiment showed that there was no room for a constructive interference, and tentatively confirmed the destructive interference. As expected the experiment did not give a precise value for the Weinberg angle.

Let us be more specific before looking at experimental results on $\overline{\nu}_e e^- \to \overline{\nu}_e e^-$ scattering. Under the assumption of identical final states for CC and NC interactions and provided vector and axial vector currents only are involved the differential cross section is [12], [7]:

$$\frac{d\sigma}{dT} = \tag{6}$$

$$\frac{G_F^2 m_e}{2\pi}\left[(g_V + x + g_A)^2 + (g_V + x - g_A)^2\left(1 - \frac{T}{E_\nu}\right)^2 + (g_A^2 - (g_V + x)^2)\frac{m_e T}{E_\nu^2}\right]$$

$$+ \frac{\pi\alpha^2\mu_\nu^2}{m_e^2}\frac{1 - T/E_\nu}{T}$$

where T is the electron kinetic recoil energy, E_ν the neutrino energy; in the standard model one has:

$$g_V = 2sin^2\theta_W + \frac{1}{2},\ g_A = \begin{cases} \frac{1}{2} & \text{for } \nu_e \\ -\frac{1}{2} & \text{for } \overline{\nu}_e \end{cases} .$$

Here x is related to the square charge radius $< r^2 >$ of the neutrino

$$x = \frac{2M_W^2}{3} < r^2 > sin^2\theta_W \text{ for } \nu_e,\ x \to -x \text{ for } \overline{\nu}_e \tag{7}$$

and μ_ν is the neutrino magnetic moment. These stem from higher order processes but are non vanishing for Dirac neutrinos only [14]. They are exactly zero for Majorana neutrinos. In the standard model they are due in lowest order to the graphs in fig. 2 and are negligibly small. In particular the magnetic moment is non zero only if the neutrino can flip its spin, in other words if it has a non zero mass m_ν:

$$\mu_\nu = \frac{3G_F e m_\nu}{8\sqrt{2}\pi^2} = 3.2 \cdot 10^{-19} \cdot m_\nu(eV). \tag{8}$$

Here μ_ν is given in units of the electron Bohr magneton μ_B. In models with right handed currents μ_ν can be enhanced by a large factor. In any event the relative contribution of the last term containing μ_ν in eq. 6 is larger the lower the energy. Reactor experiments are thus ideally suited to look for it, in particular if the electron recoil spectrum can be measured down to very low energies (fig. 3).

There is a renewed interest in the magnetic moment of the neutrino which might be responsible for the observed low flux of ^8B ν_e from the sun in the ^{37}Cl (see ref. [3]) and Kamiokande experiments [15]. Voloshin, Vysotskii and Okun [16] have pointed out that the sun, in solar active years, has a nearly toroidal magnetic field above and below the equator. The ^8B ν_e, which are produced very near the center of the sun, may flip their spin on their way out when crossing the magnetic field and become sterile. One then expects the solar ^8B flux to be modulated with the 11 year solar cycle, which is in fact suggested by the data (fig. 4). This requires $\mu_\nu \sim 10^{-10}$, which is achievable in some models, but is in some disagreement with limits from astrophysics and cosmology (see ref. [7]). For instance stellar cooling requires $\mu_\nu < 10^{-11}$ while the observed duration (several seconds) of the SN1987A neutrino burst implies $\mu_\nu < 1 - 10 \cdot 10^{-13}$. Limits of this kind however are model dependant and it seems important to look for the magnetic moment of the neutrino in a lab experiment.

Only one experiment on $\overline{\nu}_e e^- \rightarrow \overline{\nu}_e e^-$ scattering has been performed so far, by Reines, Gurr and Sobel [17] at the Savannah River reactor in the seventies. The aim was, at that time, to verify the predictions of the standard model. The detector consisted of a 15.9 kg plastic scintillator coarsly segmented surrounded by a NaI counter, a Pb shield and a liquid scintillator to veto the cosmics (fig. 5). The signature for a good event was given by a single count in one of the elements of the plastic scintillator with nothing in coincidence. The NaI gave a good anti-Compton efficiency allowing an efficient suppression of the γ background. Events from the $\overline{\nu}_e + p \rightarrow e^+ + n$ reaction in the plastic (200 events$\cdot$day^{-1}!) were efficiently identified by the detection of the annihilation γ rays and/or by the delayed neutron capture signal in the plastic scintillator or the NaI. These events thus only contributed very little to the background.

The detector was placed at 11.2 m from the core of the reactor operated at the time at a power of 1800 $MWth$. Since the cross section is small and the signature rather poor it was necessary to reduce the background from natural activities to a minimum by selecting clean materials for all the components. To illustrate the low level achieved let us just mention that the singles count rate, normalized to detector mass, is comparable to the one in the best Ge detectors used underground, under optimal conditions, in searches for double β decay.

Events were recorded during 64.6 days reactor on and 60.7 days reactor off. The count rates in two bins of electron recoil energy T were:

$T(MeV)$	Events/day		
	Reactor on	Reactor off	on-off
1.5-3	45.1 ± 1.0	39.2 ± 0.9	5.9 ± 1.4
3-4.5	2.4 ± 0.19	1.2 ± 0.14	1.2 ± 0.25

In spite of the low singles background the signal to noise ratio is somewhat marginal, which shows how difficult the experiment is. These data were compared with standard model predictions based on eq. 6 and on the then accepted reactor spectrum, resulting from a calculation [18]. Agreement was found and the value $sin^2\theta_W = 0.29 \pm 0.05$ was derived. This is consistent with the best value to date obtained by the CHARM collaboration [19] in $\nu_\mu e^-$ scattering, an other purely leptonic process: $sin^2\theta_W = 0.211 \pm 0.037$.

We now know however that the reactor spectrum used in this original analysis was much too hard. Vogel and Engel [7], using the presently best determination of the reactor spectrum and fixing $sin^2\theta_W$ to the presently accepted value find that the measured rates in the two energy bins given above are 1.35±0.4 and 2.0±0.5 times larger than the expected ones. Taken litterally this discrepancy indicates new physics beyond the standard model. One can reconcile the data with eq. 6 by introducing a neutrino magnetic moment $\mu_\nu = (2-4) \cdot 10^{-10}$, of the right order for the spin flip mechanism in the sun to occur! Of course, the experiment is very difficult, and the discrepancy may be due to an instrumental problem. Nevertheless it appears important to perform a new experiment to clear up the situation. With better signal to noise and a lower detection threshold one would be sensitive to values of μ_ν around 10^{-11}. One would also be able to remeasure $sin^2\theta_W$ with good sensitivity in a purely leptonic process.

How can one improve? The background level achieved in the Savannah River experiment is already impressive, but one can probably do better now. Much progress has been done these recent years in low background techniques, and cleaner materials may be available. Also one could think of improving the signature by replacing the inner plastic scintillator by a finely segmented array of scintillating fibers or capillaries filled with liquid scintillator. This would allow a better anti-Compton suppression and, hopefully, a coarse measurement of the electron scattering angle for the events with large recoil. New more efficient fibers and capillaries are in developpement for high energy experiment [20].

5 Neutrino oscillations $\overline{\nu}_e \leftrightarrow x$

Soon after the standard model was shown experimentally to work very well theorists began speculating about physics beyond it. One good place to look for new physics are the neutrino masses (see ref. [3], [21]). In particular the weak neutrino eigenstates ν_l may not be mass eigenstates ν_i with mass eigenvalue m_i but rather superpositions of them

$$\nu_l = \sum_{i=1}^{N} U_{li}\nu_i, \text{ with } M\nu_i = m_i\nu_i, \text{ and } l = e,\ \mu, \ldots \qquad (9)$$

in which case neutrino oscillations will occur. Here U_{li} is an $N \times N$ mixing matrix. In the simplest two neutrino model (N=2 in eq. 9) U_{li} depends on a single parameter, the mixing angle θ:

$$U_{li} = \begin{pmatrix} cos\theta & sin\theta \\ -sin\theta & cos\theta \end{pmatrix} \qquad (10)$$

and a neutrino created in the state ν_l with energy E_ν has a probability

$$P_e(L, \Delta m^2, E_\nu) = 1 - \frac{1}{2}sin^2 2\theta \left[1 - cos\frac{2.53\Delta m^2(eV^2)L(m)}{E_\nu(MeV)}\right], \qquad (11)$$

with

$$\Delta m^2 = m_1^2 - m_2^2,$$

of still having the original flavor l after travelling a distance L. To search for oscillations one can either look for the disappearance of neutrinos with the original flavor in a beam in function of the distance to the source, or look for the appearance of neutrinos with a different flavor. Disappearance experiments are more general since they probe oscillations into any state and not into a specific one, but they are less sensitive to small mixings than appearance experiments. It is easier to measure a small contamination in a flux than a small variation of it.

The masses, and thus Δm^2, are expected to be small, if anything, and one thus has to go to long distances L or low energies E_ν to achieve good sensitivity. Felix

Boehm was one of the very first to realize that reactors, with their high fluxes of MeV $\overline{\nu}_e's$, are ideal to perform disappearance experiments sampling the channel $\overline{\nu}_e \leftrightarrow x$. He started a collaboration which included, in the beginning, collaborators from Caltech, TU Munich and ISN Grenoble. If I am allowed a personal note I would like to add that this was, for those like me who got the chance to participate, the beginning of a very exciting time. The first experiment was performed at the 57 MW research reactor of the Institut Laue-Langevin in Grenoble [22]. The distance from the reactor core to the detector was 8.76 m. The experiment was continued at the Goesgen 2800 $MWth$ PWR in Switzerland, with SIN replacing ISN Grenoble in the collaboration [23]. The $\overline{\nu}_e$ flux was measured with an upgraded version of the detector at three distances successively: 37.8 m, 45.9 m and 64.7 m.

The best results were obtained at Goesgen. The reactor has a nearly cylindrical core with 3.2 m diameter and 3.4 m height. The detector was located outside the containment building, which provides 8 m of concrete shielding from the core. To reduce the background from the nucleonic component of the cosmics and from natural activities the detector was completely surrounded by various shielding layers, including 2 m of concrete on the sides and 4 m on top.

The detector used the same detection reaction $\overline{\nu}_e + p \rightarrow e^+ + n$ (eq. 1) as the first Savannah River experiment. A time coincidence between positron and neutron was required as the signature for a neutrino event. The detector system (fig. 6) consisted of five planes of six target cells filled with liquid scintillator (377 l total volume) serving as proton target, positron detector and neutron moderator, alternated with four ^3He wire chambers (neutron detectors). The neutrino energy was determined from the measured positron kinetic energy E_{e^+} using eq. 2. Pulse shape discrimination was used to suppress fast neutron background in the target cells. The position of an event along the length of a target cell was obtained from the time difference of the photomultiplier pulses at either end. Each ^3He chamber had 16 groups of resistive wires arranged in one plane. The position along a wire was determined by charge division. In spite of the complexity of the design it was possible to achieve a very low level of natural radioactivity. The requirement of a position correlation between target cell and neutron counter events reduced the accidental background by a factor of 7, at the modest cost of an 8% loss in detection efficiency. The entire apparatus was surrounded by a liquid scintillation detector to veto the cosmic μ's.

The measured spectra are somewhat washed out by several instrumental effects. The energy response function was calculated by Monte Carlo taking into account the energy resolution of the detector (18% $FWHM$ at 1 MeV), annihilation in flight and at rest, positron bremsstrahlung and the wall effect. The probability to detect a neutron created in a target cell was measured with a $Sb(Be)$ neutron source and was 21.7 ± 0.8 %. The total acceptance after applying all cuts was 16.6 ± 0.6 %.

At each position the reactor off spectrum was measured during the one month long annual shutdown of the reactor. After background subtraction a total of 10930 ± 220 events with $0.7\ MeV < E_{e^+} < 5.6\ MeV$ were seen at 37.9 m in 3441 h of data aquisition, 10590 ± 190 at 45.9 m in 4893 h and 8787 ± 325 at 64.7 m in 8608 h (fig. 7).

The simplest way to analyze the data is to compare the three spectra looking for differences in shape and normalization resulting from oscillations (eq. 11), after correcting for the different solid angles and the small changes ($< 2\%$) due to slightly different mix of fissile isotopes and taking into account the finite energy and spatial resolution. This analysis is free of uncertainties affecting the cross section of the detection reaction, the absolute reactor neutrino flux, and the acceptance of the detector. As it turns out, the three spectra are in good agreement and there is no evidence for oscillations of the type $\overline{\nu}_e \leftrightarrow x$. Oscillations with small mixings, long (compared with the distances) or short (compared with the core size) lengths are also allowed. But, performing a maximum likelyhood ratio test in the frame of the two neutrino model, a large area in the Δm^2 versus $sin^2 2\theta$ plane is excluded as shown in fig. 8a for $0.02 < \Delta m^2 < 2eV^2$.

As a consistency check, and to be sensitive to large values of Δm^2, the three Goesgen positron spectra were compared to the expected spectra determined from the neutrino yield and the cross section of the detection reaction. The error on the normalization of 6 % must then be taken into account. It results essentially from the uncertainties in the acceptance of the detector and the expected neutrino yield. Anyway, the expected spectra assuming no oscillations are in good agreement with the measured ones (fig. 7), in shape and normalization (the ratio of the measured to expected integrated rates is 1.04 ± 0.06). Improved limits on the parameters Δm^2 and $sin^2 2\theta$ are derived (fig. 8a). For $\Delta m^2 > 2\ eV^2$ one gets $sin^2 2\theta < 0.18$ ($90\% CL$).

Other similar experiments have been performed using the same detection reaction $\overline{\nu}_e + p \rightarrow e^+ + n$. In Bugey (France) a simplified version of the Goesgen detector, with poorer background suppression was used. The detector was located inside the containement building, in the subbasement, much closer to the core. The spectrum was measured at 13.6 m and 18.3 m. Early indications for oscillations [24] were not confirmed and withdrawn [25]. The UC Irvine group took data at two distances from the Savannah River reactor (18.2 m and and 33.8 m) and found no strong evidence for oscillations [26]. But better results were obtained in Rovno (USSR) [27]. A tank containing 240 l of scintillator loaded with Gd for neutron detection and viewed by photomultipliers on two opposite sides was set up in a lab located underneath the core of a PWR. Excellent positron spectra with high statistics were measured at 18 m and 25 m. They agree well with each other after correcting for the differences in solid angle. Although somewhat less sensitive the experiment thus confirms the Goesgen limits which are still the most stringent ones.

The Goesgen exclusion plot is compared in fig. 8b with limits from appearance experiments studying the channel $\nu_\mu \leftrightarrow \nu_e$ conducted at various accelerators. One sees that reactor experiments have a much better sensitivity for low values of Δm^2.

How can one improve the sensitivity? After all, neutrinos may oscillate, with parameters not too far beyond existing limits. The Bugey group has built a new detector with an acceptance of ~ 80 %, much larger than in the previous Goesgen or Bugey designs [28]. The detector is a stack of cells filled with NE 320 liquid scintillator and viewed by photomultipliers at two ends. NE 320 is loaded with 0.15 % ^6Li and the neutrons are detected using the reaction

$$n_{th} +^6 Li \rightarrow \alpha + t + 4.8\ MeV. \tag{12}$$

Pulse shape discrimination is used to identify the α and the t. The detector has a total volume of 600 l and is being operated underneath the reactor at a distance of 16 m. The count rate, 90 counts·hour^{-1}, is quite high. A second identical detector is being installed outside the reactor building, at a distance of 40 m. Both detectors will be operated simultaneously in the hope to reduce systematic uncertainties. The experiment will have a better sensitivity than the Goesgen one toward small mixings $sin^2 2\theta$, but not toward small Δm^2. To improve in that direction it is foreseen to set up a third detector having twice the size at a distance of 120 m.

Felix Boehm is planning an experiment which would improve the sensitivity in a much more drastic way. The aim is to have a sensitivity extending down to $\Delta m^2 \sim 10^{-4}\ eV^2$ so as to fill the gap between existing limits and the region where solar neutrino experiments become very sensitive thanks to the Mikheyev-Smirnov-Wolfenstein resonant amplification (see [3], [21]). To achieve this a distance $L \sim 10\ km$ from the source to the detector is necessary. One needs around 10000 events to have a positron spectrum with decent statistics. Assuming that one is working at a standard PWR with a power around 3000 $MWth$, that the experiment should not last more than 2 to 3 years, and that one can achieve an acceptance of $\sim$ 80 %, one can calculate that the detector must have a mass of 1 kt. This is not particularly large by today's standards, but the problem here is that one wants to see positrons with energies of a few MeV only. To have a reasonable signal to background ratio the background normalized to mass must be 10^3 to 10^4 times better than in the Goesgen experiment! This may, however, be possible.

It seems that a detector using Gd loaded liquid scintillator would work. The neutron capture in Gd is followed by the emission of γ's having a total energy around 8 MeV. This allows to set the neutron threshold reasonably high in a large volume detector, where the γ's will be contained. This will help in keeping the background low. Calculations performed in Felix Boehm's group indicate that one can achieve the desired neutron detection efficiency of 80 % in a large volume detector with a loading of 0.1 % only. With this one should be able to maintain a light transmission length of several meters required in a large size detector. Tests on various scintillators are presently underway in Felix's group. Various geometries (plain, segmented) are being investigated. To reduce the background from natural activities it will be necessary to use very clean components. To fight the cosmics background one will have to place the detector 200 to 300 m underground.

It is foreseen to first perform an experiment with a smaller, 20 to 30 t, prototype of similar design located much closer to the reactor. That prototype will already produce useful data, and will serve later as monitor.

6 Conclusion

Past experiments with reactor neutrinos have produced important results and had a profound impact on our knowledge of particle physics. Neutrino physics nevertheless still has open questions. New projects are currently being considered to adress some of these questions. I hope they will be carried out.

References

[1] F. Reines and C.L. Cowan, Phys. Rev. 92(53)830; F. Reines, in Neutrino Physics and Astrophysics, Edited by E. Fiorini, Plenum Press, p. 11

[2] C.L. Cowan et al., Science 124(56)103; F. Reines et al., Phys. Rev. 117(60)159

[3] For a general discussion of neutrino experiments see F. Boehm and P. Vogel, Physics of Massive Neutrinos, Cambridge University Press

[4] F.T. Avignone and Z.D. Greenwood, Phys. Rev. C 22(80)594

[5] P. Vogel et al., Phys. Rev. C 24(81)1543

[6] H.V. Klapdor et al., Phys. Rev. Lett. 48(82)127; Phys. Lett. B112(82)22

[7] P. Vogel and J. Engel, Phys. Rev. D 39(89)3378

[8] F. v. Feilitzsch et al., Phys. Lett. 118B(82)162

[9] K. Schreckenbach et al., Phys. Lett. 160B(85)325

[10] K. Schreckenbach et al., in Weak and Electromagnetic Interactions in Nuclei, Proc. Int. Symp., Heidelberg 1-5 July 1986, Springer, p.759

[11] M.F. James, J. Nucl. Energy 23(69)517; J.M. Paratte, Swiss Federal Institute for Reactor Research, Report No. TM-45-81-19 (unpublished)

[12] B. Kayser et al., Phys. Rev. D 20(79)87

[13] R.C. Allen et al., Phys. Rev. Lett. 55(85)2401

[14] B. Kayser, Phys. Rev. D 26(82)1662

[15] A. Mann, Proc. Workshop on Neutrino Telescopes, Venezia, November 17-19, 1988, Milla Baldo-Ceolin editor, p. 56

[16] M.B. Voloshin, M.I. Vysotskii and L.B. Okun, Sov. Phys. JETP 64(86)446

[17] F. Reines, H.S Gurr and H.W. Sobel, Phys. Rev. Lett. 37(76)315

[18] F.T. Avignone, Phys. Rev. D 2(70)2609

[19] J. Dorenbosch et al., Zeit. Phys. C 41(89)567

[20] V. Zacek, private communication

[21] S.P. Rosen, Neutrino Oscillations, these proceedings

[22] H. Kwon et al., Phys. Rev. D 24(81)1097

[23] G. Zacek et al., Phys. Rev. D 34(86)2621

[24] J.-F. Cavaignac et al. Phys. Lett. B 148(84)387

[25] H. de Kerret, in 5th Force and Neutrino Physics, O. Fackler and J. Tran Thanh Van editors, Frontières, 1988, p.125

[26] H. Sobel et al., in Neutrino 86, Proc. 12th Int. Conf. Neutrino Physics and Astrophysics, Sendai, T. Kitagaki and H. Yuta editors, World Scientific, Singapore(89)

[27] A.I. Afonin et al., Sov. Phys. JETP 67(88)213

[28] J. Collot, in 5th Force and Neutrino Physics, O. Fackler and J. Tran Thanh Van editors, Frontières, 1988, p.153

[29] L.A. Ahrens et al., Phys. Rev. D 31(85)2732

[30] C. Angelini et al., Phys. Lett. B 179(86)307

[31] L.S. Durkin et al., Phys. Rev. Lett. 61(88)1811

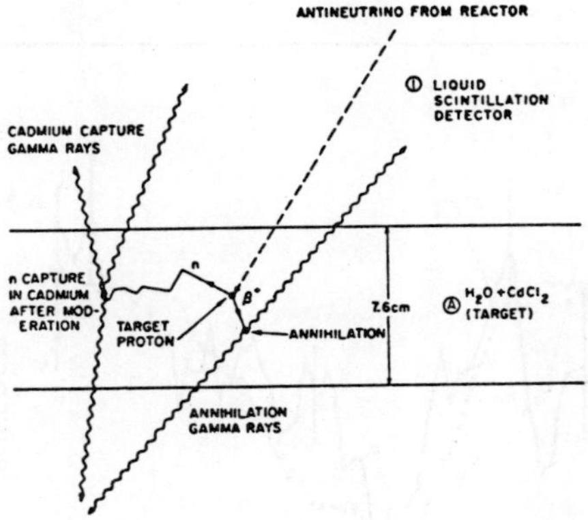

Figure 1: The principle of the first Savannah River detector.

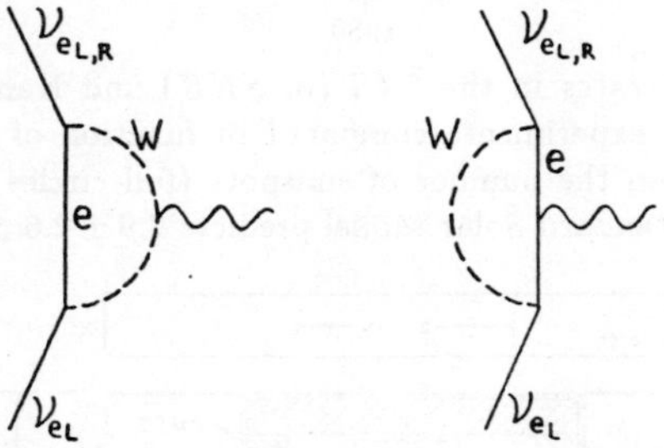

Figure 2: Two graphs contributing to the electrodynamics of the ν_e.

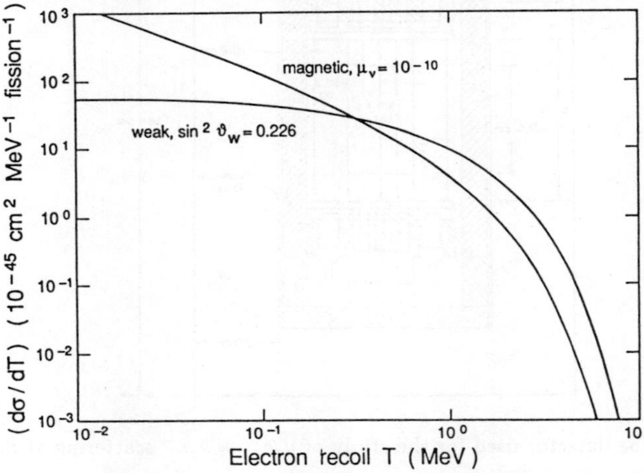

Figure 3: The expected cross section for $\overline{\nu}_e e^- \rightarrow \overline{\nu}_e e^-$ scattering averaged over the reactor $\overline{\nu}_e$ spectrum. The contributions from weak interaction alone and from a magnetic moment $\mu_\nu = 10^{-10}$ are shown [7].

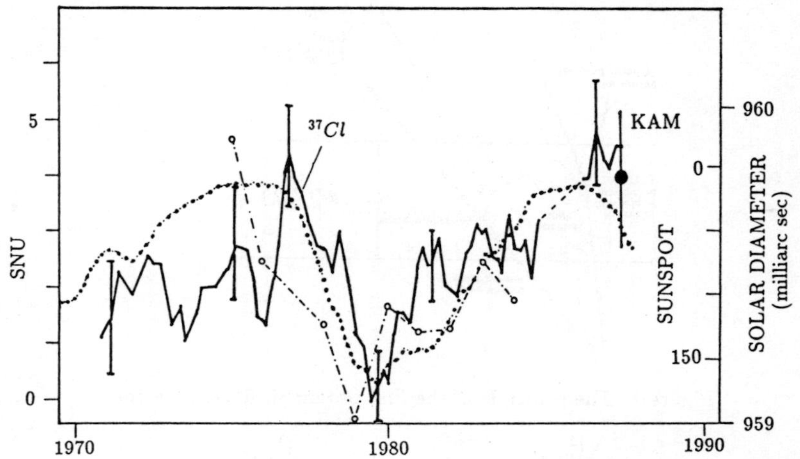

Figure 4: The count rates in the ^{37}Cl (in SNU) and Kamiokande (equivalent units) solar neutrino experiments compared in function of time with the solar activity measured from the number of sunspots (full circles) and solar diameter (open circles). The Standard Solar Model predicts 7.9 ± 2.6 SNU.

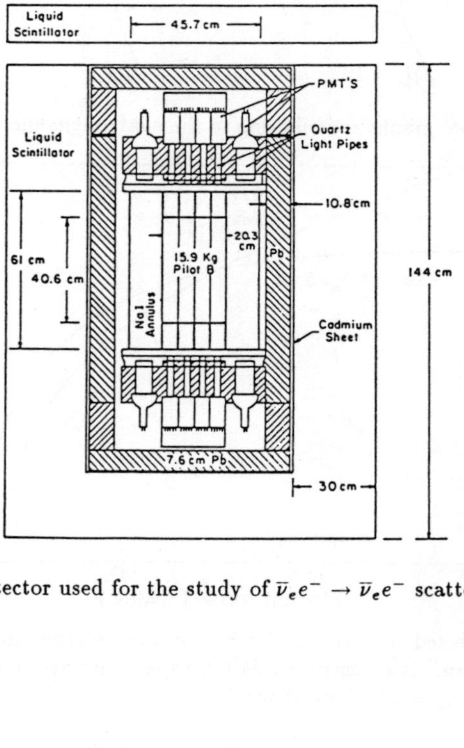

Figure 5: The detector used for the study of $\overline{\nu}_e e^- \rightarrow \overline{\nu}_e e^-$ scattering at Savannah River [17].

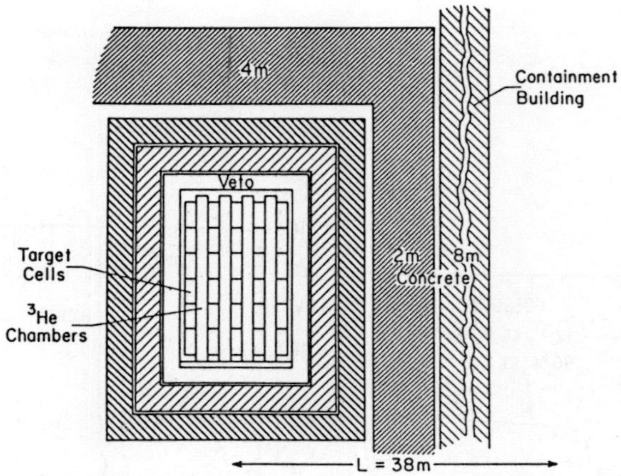

Figure 6: The detector arrangement at the Goesgen reactor.

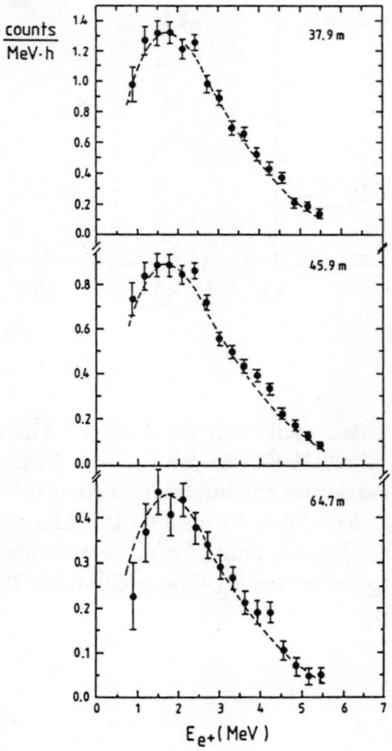

Figure 7: The positron spectra from the reaction $\overline{\nu}_e + p \rightarrow e^+ + n$ measured in Goesgen. The vertical scales have been adjusted for the differences in solid angle. The broken curves are the expected spectra assuming no oscillations.

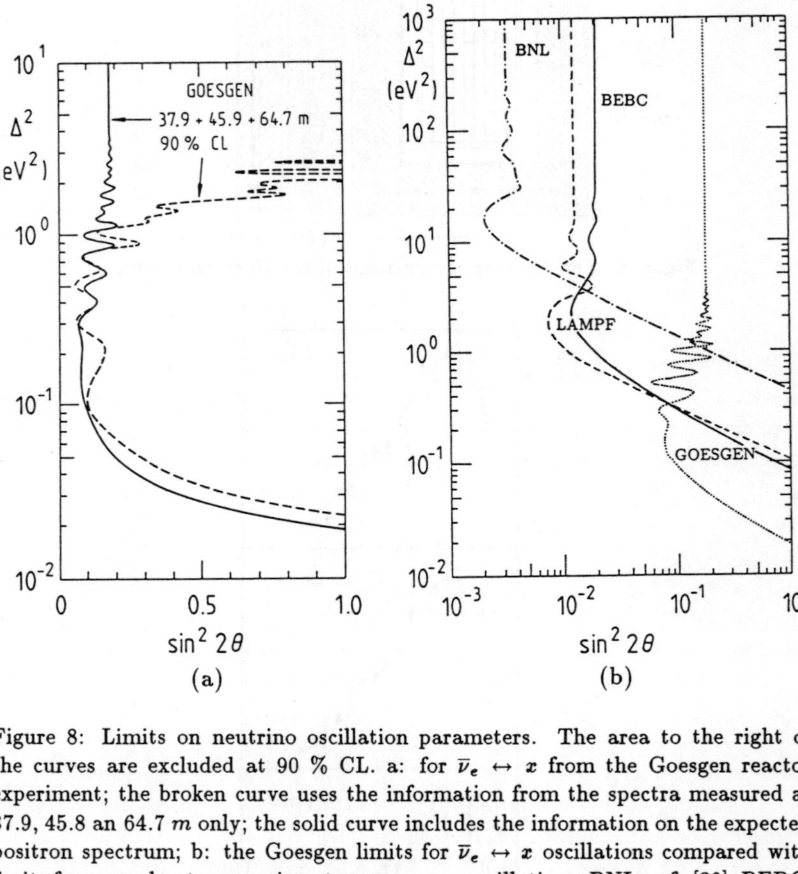

(a) (b)

Figure 8: Limits on neutrino oscillation parameters. The area to the right of the curves are excluded at 90 % CL. a: for $\overline{\nu}_e \leftrightarrow x$ from the Goesgen reactor experiment; the broken curve uses the information from the spectra measured at 37.9, 45.8 an 64.7 m only; the solid curve includes the information on the expected positron spectrum; b: the Goesgen limits for $\overline{\nu}_e \leftrightarrow x$ oscillations compared with limits from accelerator experiments on $\nu_\mu \leftrightarrow \nu_e$ oscillations; BNL: ref. [29], BEBC: ref. [30], LAMPF: ref. [31].

Reorientation in gauge space: How to measure the diabolic pair-transfer in rotating nuclei+

J. de Boer
Sektion Physik der Universität München
Am Coulombwall 1, D-8046 Garching, FRG

C.H. Dasso
The Niels Bohr Institute, University of Copenhagen
Blegdamsvej 17, DK-2100 Copenhagen Ø, Denmark

G. Pollarolo
Dipartimento di Fisica Teoretica, Università di Torino
and INFN, Via Pietro Giuria 1, I-10125 Torino, Italy

It has been predicted by Ring and coworkers[1] that the sign of the matrix elements for the transfer of a pair of neutrons from a rotational state in ^{162}Dy to a state with the same spin in ^{164}Dy, might change sign as a function of spin above a diabolic spin-value, $I_D \approx 11$. We find from our calculations that this diabolic sign-change manifests itself in the probabilities for excitation of rotational states with $I > I_D$ in ^{164}Dy when the bombarding conditions are chosen properly: i) The bombarding energy remains below the Coulomb barrier (absorption and nuclear forces are unimportant (relative orbit is hyperbolic), and ii) the strength of Coulomb excitation is such that at t = 0, the time of closest approach when the transfer is active, spin states just above and below I_D in ^{162}Dy are excited with large and roughly equal probability (destructive interference between the pair-transfer paths in the diabolic case).

The figures show the results of calculations for 1000 MeV ^{208}Pb scattered at 180 degrees from ^{162}Dy:

Figure 1: Population of the spin states in ^{162}Dy at the turning point of the relative orbit (t = 0). The states with $I \approx I_D$ are strongly populated.

Figure 2: Relative population of the spin-states in ^{164}Dy, following the transfer of a neutron pair from a state in ^{162}Dy with the same spin. All transfer formfactors were chosen to have the same value (non diabolic transfer).

Figure 3: Same as fig. 2, but using transfer formfactors with opposite sign above and below I_D (diabolic transfer).

Comparisons of the relative population of states with spin $I > I_D$ in ^{164}Dy show a marked reduction in the diabolic case, stemming from the destructive interference. Experimental observation of this effect would be easy if diabolic and non-diabolic measurements could be directly compared. In the absence of such a possibility the trend of the populations with bombarding energy and/or scattering angle must be carefully investigated.

Fig. 1
Excitation probabilities of rotational states in ^{162}Dy as a function of spin for 1000 MeV ^{208}Pb projectiles approaching the target nucleus head on up to the turning point (t = 0)

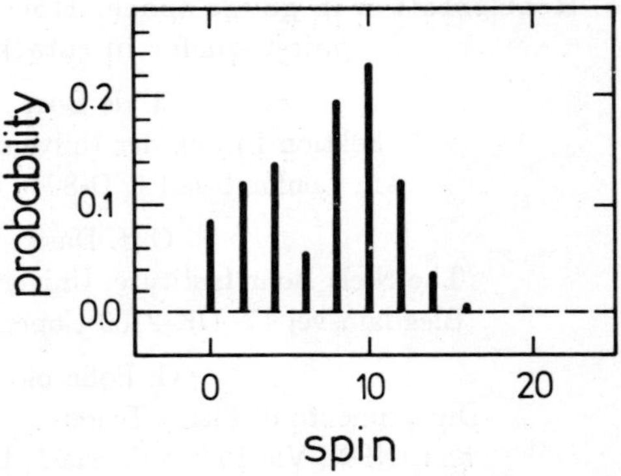

Fig. 2
Relative probabilities for the excitation of rotational states in ^{164}Dy following non-diabolic pair-transfer from ^{208}Pb onto ^{162}Dy

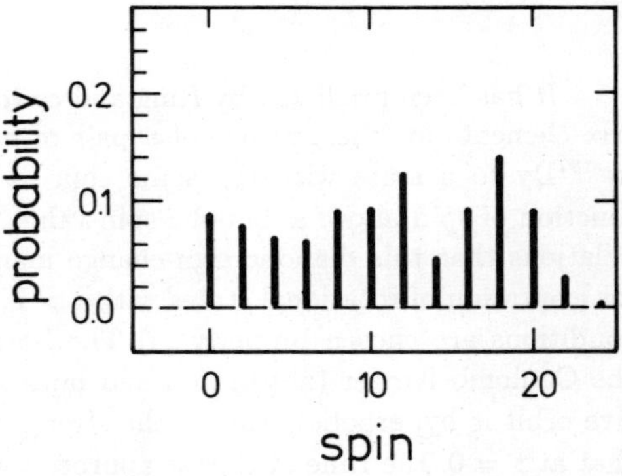

Fig. 3
Relative probabilities for the excitation of rotational states in ^{164}Dy following diabolic pair-transfer from ^{208}Pb onto ^{162}Dy

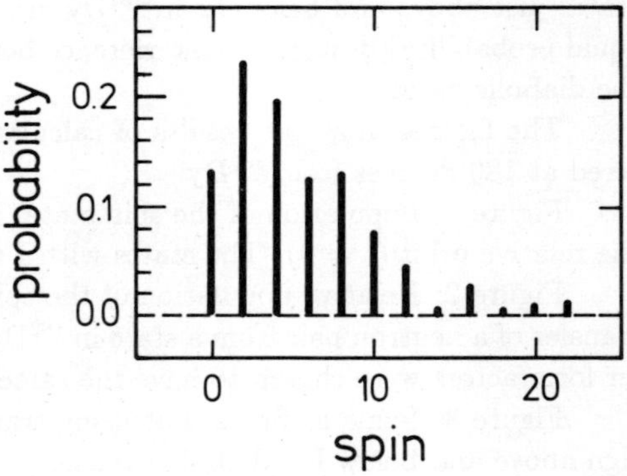

References
[1] R.S. Nikam, P. Ring and L.F. Canto, Phys. Lett. **B185** (1987) 269
+ Supported by BMFT grant 06 LM 171 II and by GSI

RECENT RESULTS ON THE NEUTRON LIFETIME AT ILL

W. Mampe
Institut Laue-Langevin
156X
38042 Grenoble Cédex, France

ABSTRACT

Two recent experiments at the ILL using stored ultra-cold neutrons have obtained for the neutorn lifetime the values 877±10 s and 887.6±3 s.

Study of the β-decay of the neutron $n \rightarrow p+e^-+\bar{\nu}_e+782$ keV allows to determine the weak interaction coupling constants g_A and g_V without any further input. One part of this information is contained in the neutron's weak decay rate:

$$\tau_n^{-1} = \frac{\ln 2}{K} \cdot f^R \cdot (g_V^2 + 3\, g_A^2)$$

With a constant K and the corrected phase space factor f^R. The neutron lifetime plays furthermore an important role in astrophysics and cosmology.

Earlier experiments measured the decay products from a given neutron beam volume. More recent experiments successfully use neutron beam pulses or trap the decay products in 4π geometry.

An alternative to beam experiments became feasible with the availability of intense ultra cold neutron (UCN) sources. UCN ($E \leq 3.10^{-7}$ eV) can be stored with loss rates much smaller than the neutron decay rate in material bottles (due to a repulsive neutron-nuclear interaction potential) or in magnetic traps (via a magnetic force on the neutrons magnetic moment). Such storage allows to obtain τ_n from a relative measurement:

$$\tau_n = \frac{\ln(N(t_1)/N(t_2))}{t_2 - t_1}$$

by repeated filling - storage (time = t_i) - counting (N(i)) cycles, under the condition that the filling density is constant and that there are no competing UCN losses - unless they are well under control.

Two experiments of this kind have been performed recently at the new very cold and ultracold neutron source of the ILL.

1) Magnetic Storage

A superconducting magnetic hexapole torus (diameter 1 m) has been designed and constructed by W. Paul's group in Bonn[1] and installed at the vertical guide of the ILL, extracting very cold neutrons and UCN from the cold source. The neutrons are forced on the circular orbit by the hexapole field acting on the magnetic moment. Betatron oscillations leading to neutron losses aditional to those due to beta decay are suppressed by a superimposed decapole field. The storage volume is filled to the steady state density, corresponding to ~1000 neutrons/cycle, by a movable guide. After variable confinement times (up to 1.5 h equivalent to 6 lifetimes), the remaining neutrons are detected by moving a ^{3}He/Ne proportional counter through the beam cross section. The described measuring cycle was repeated many times and under variation of the essential geometric and magnetic parameters as there are the beam acceptance, the magnetic field strength and the filling and detecting procedure. The resulting decay curves show, after an initial beam stabilisation period ($t < 450$s) a pure exponential decay the slope of which representing the inverse neutron lifetime. The final result obtained for the lifetime of the free neutron from all measuring cylces is $\tau_n = (877 \pm 10)$s [2]. Though there was no experimental evidence one cannot exclude completely long term intrinsic losses which might influence the result within the given error limits. Even in this most conservative view the result represents a very valuable lower limit for the neutron lifetime which no other method could produce up to now.

2) UCN Storage in a Liquid Walled Bottle

Coating a rectangular glass box with a Hydrogene free oil [3] which has low UCN reflection losses we obtained storage times longer than 700 s in a 70 l volume. The UCN were admitted into the bottle by opening a door until the steady state density is achieved. After a preselected storage time ranging from 100 s to 7200 s the remaining UCN were released towards a ^{3}He detector by opening a second door. This procedure was repeated for different storage times, temperatures, vacua etc.The non vanishing reflection loss rate $\Theta_m = \mu(v).v/\lambda$ (where $\mu(v)$ is the mean loss probability per reflection, v the neutron velocity and λ the mean free path) could be eliminated by varying λ, i.e. the bottle volume to surface ratio. For this purpose the bottle rear wall was movable. τ_n is obtained by extrapolating the λ dependence of the measured lifetimes to infinite bottle volume. A < 1% correction to the experimental result is necessary, essentially due to the effect of gravity on the UCN. Our final result is $\tau_n = (887.6 \pm 3)$s[4]. Combining this result with those reported later than 1980 we obtain (887.6 ± 2.6)s. Together with the neutron β-asymmetry result $\lambda = g_A/g_V = -1.261(4)$[5] we get both coupling constants g_A and g_V from neutron decay alone:

$g_V^n = 1.4205(42).10^{-62}$ J.m^3 and $g_A^n = -1.7921(29).10^{-62}$ J.m^3.

It seems worth noting that the improved value of the neutron lifetime allows together with the observed primordial Helium mass fraction to set a limit on the number of existing neutrino generations: $N_\nu = 2.6 \pm 0.3$[6].

REFERENCES

1) Kügler, K.J., Paul, W. and Trinks, U., Nucl. Instr. Meth. A 228 (1985) 240
2) Paul, W., Anton, F., Mampe, W., Paul, L. and Paul, S., Z. Phys. C, in print
3) Ageron, P., Mampe, W., Bates, J.C. and Pendlebury, J.M., Nucl. Instr. Meth. A 249 (1986) 261
4) Mampe, W., Ageron, P., Bates, J.C., Pendlebury, J.M. and Steyerl, A., Phys. Rev. Lett. 63 (1989) 593
5) Erozolimskii, B., in: Workshop on Fundamental Physics with Slow Neutrons, Grenoble 1989, Eds. Dubbers, D. Mampe, W. and Schreckenbach, K., Nucl. Instr. Math. A, in print
6) Schramm, D.N. and Kawano, L. in the same Workshop as ref 5).

Solid and questionable Results from Lepton Production in very Heavy Ion Reactions

E. Kankeleit
Institut für Kernphysik
Technische Hochschule, D 6100 Darmstadt, Fed. Rep. Germany

Fifteen years ago quite a number of fundamental questions [1] of production and behavior of supercritical atomic systems with the 1s state "diving into the Dirac sea", "the electron falling upon the nucleus", the sawtooth shaped positron spectra expected from a supercritical reaction and the spontanous decay of the "charged vacuum" initiated an experimental programm in our institute using the UNILAC at GSI.

The first instrument set up for the positron spectroscopy was an Orange type spectrometer [1] including a positron anihilation detector. The spectra [3] gave first experimental evidence for the Frankfurt multi channel calculations of a coulombic lepton excitation process[2]. To increase detection efficiency we built a solenoidal transport system with kinematical coincidences, with which the highest Z system, U-Cm, could be studied[3]. To improve the quality in separation of positrons from the many orders of magnitude higher yield of δ-electrons the TORI spectrometer[4] was constructed, which uses the opposite drift of positrons and electrons in an S-shaped toroidal magnetic field. This feature allows with unsurpassed efficiency simultanous detection of both electrons and positrons over a broad dynamic energy range.

From a range of united charges and distances of closest approach, including deep inelastic reactions, systematic features did become obvious from the data. It turned out that a characteristic reaction time, called $\hat{t}$, plays a decisive role. For coulombic trajectories it is half the distance between the extrema of $\dot{R}/R$, with R(t) the distance between the reaction partners, or more general for charge distributions as in fusing nuclei twice the rms charge radius of the reacting system.[5]. The lepton spectra are due to a monopole conversion process and depending on the lepton densities at the origin, a kind of Fermi function, and kinematically on the Fourier transform of $\dot{R}/R$.

After having the knowledge about atomic and nuclear lepton production we now use these processes as a tool to derive from the spectra information about $\dot{R}/R$, -in complete analogy to diffraction spectroscopy. Because high Fourier contributions are essential experiments with detection of electrons beyond 8 MeV have been performed. With a combined analysis of both lepton spectra rather detailed information about the decelaration and separation process in deep inelastic reactions for different energy loss windows were obtained[6]. There are discrepancies to theory[7], which might be attributed to the poorly understood friction.

At the beginning of this research we still were hoping for long sticking times long enough that would allow an enhancement in the positron spectra at the diving energy of the 1s state. Indeed we did find a structure at 680 keV in the positron spectra in the U+U 7.5 MeV/u reaction, but its energy being too high for this process we considered an axion decay as a possible source[8]. But after some experiments it was never seen again.

The more we were impressed by the booming lines and lepton coincidences seen by the EPOS and ORANGE groups. To find independent evidence for the decay of a

[1]fundamental: nearly thirty years ago I certainly learned a lot from working with Felix Boehm on fundamental problems, for which I like to thank him with the best wishes for his 65. birthday.

neutral particle we repeated under practically identical conditions the U-Th 5.85 MeV/u experiment [9], in which an impressive peak was seen in the coincidence plane. Because of the drift properties of the TORI back to back coincidences can be determined with better coincidence probability and this with simultanous forward correlation detection. But no lines were observed in the sum spectra and also all other scans (as is usual in this field) through the very many parameters did not reveal a window with significant structures beyond a smooth background determined by a fair method[10].

But in later EPOS experiments[12] with much improved statistics (using together with the ORANGE group a considerable part of available UNILAC beam time) the peak did also not show up. Instead new lines were found, being now part of "multiple line structures"[11].

Concerning this part of research there are obviously questionable experimental and definitly *no physical* results.

References

[1] E. Moll and E. Kankeleit, Nukleonika 7, 180 (1965)

[2] G. Soff, et al., Phys. Rev. Lett. 43 (1979) 1981;
U. Müller et al., Phys. Rev. C30 (1984) 1199.

[3] H. Backe, L. Handschug, F. Hessberger, E. Kankeleit, L. Richter, F. Weik, R. Willwater, H. Bokemeyer, P. Vincent, Y. Nakayama, and J. S. Greenberg, Phys. Rev. Lett. 40,1443 (1978)

[4] E. Kankeleit, U. Gollerthan, G. Klotz, M. Kollatz, M. Krämer, R. Krieg, U. Meyer, H. Oeschler, P. Senger, Nucl. Instr. Meth. A234 (1985) 81

[5] E. Kankeleit, Nukleonika 25 (1980) 253.
E.Kankeleit, in Nuclear Interactions, B. A. Robson, ed., Lecture Notes in Physics, Vol 92, Springer-Verlag, New York (1979), p. 306
E. Kankeleit, in: International Advanced Courses on "Physics of Strong Fields", ed. by W. Greiner, Plenum Press, (1987)

[6] R. Krieg, E. Bozek, U. Gollerthan, E. Kankeleit, G. Klotz-Engmann M. Krämer, U. Meyer, H. Oeschler, P. Senger, Phys. Rev. C34 (1986) 562.
M. Krämer, B. Blank, E. Bozek, E. Kankeleit, G. Klotz-Engmann, C. Müntz, H. Oeschler and M. Rhein, Phys. Lett. B201, 215 (1988)

[7] H. Feldmeier, in: Nuclear Structure and Heavy-Ion Dynamics, LXXXVII Corso Bologna, ed. by L. Moretto and R. Ricci,North Holland Amsterdam, (1984).

[8] H. Backe, Habilitationsschrift THD (1980), GSI proposal 217

[9] T. Cowan et al., Phys. Rev. Lett. 56, (1986) 444

[10] M. Rhein, B. Blank, E. Bozek, E. Ditzel, H. Friedmann, H. Jäger, E. Kankeleit, G. Klotz-Engmann, M. Krämer, V. Lips, C. Müntz, H. Oeschler, A. Piechaczek, I. Schall, C. Wille,Proc. of the VIIIth Moriond Workshop, Les Arcs (1989)

[11] W. Koenig, et al., Phys. Lett. B218, 12 (1989)

[12] H. Bokemeyer, P. Salabura, D. Schwalm, K. E. Stiebing Proc. of the VIIIth Moriond Workshop, Les Arcs (1989)

CP VIOLATION IN *B* DECAYS

Boris Kayser
Division of Physics
National Science Foundation
Washington, D.C. 20550

ABSTRACT

The study of *CP*-violating effects in *B* decays will be a good test of whether *CP* violation is caused by the known weak interaction. If this is its origin, then large, cleanly-predicted *CP*-violating effects are expected in certain neutral *B* decays to hadronic *CP* eigenstates. The phenomenology of *CP* violation in the *B* system is reviewed, and the genesis of these large effects is explained. It is shown that large, cleanly-predicted effects are also expected in some decays to states which are not *CP* eigenstates. The combined study of the latter decays and those to *CP* eigenstates may make it possible to obtain a statistically-significant *CP*-violating signal with fewer *B* mesons than would otherwise be required.

I. MOTIVATION

Thus far, *CP* violation has been seen only in certain *K* decays. Since these decays are weak, it is natural to imagine that *CP* violation may be produced by the known weak interaction which is described in the standard model. If this is indeed the case, then there must be large, cleanly-predictable *CP*-violating effects in certain *B* decays. (Here, by "large" we mean of order 10% or more.) By contrast, if *CP* violation comes, not from the known weak force but from physics at some very high mass scale, then, in general, one does not expect large effects in *B* decays. For, compared to the high mass scale, both the *s* quark in a *K* and the *b* quark in a *B* are massless. Thus, one does not expect the high-mass physics to distinguish between these two quarks. Hence, *CP*-violating effects in B decays should be about as large as they are seen to be in K decays; [1] that is, of order 10^{-3} or less. Moreover, if *CP* violation comes from physics beyond the standard model, then one certainly does not expect in B decays the *specific detailed pattern* of large *CP*-violating effects predicted by the standard model. Thus, the study of *CP*-violating effects in B decays will be a very good test of whether *CP* violation is due to the known weak force or not.

It should be noted that even large *CP*-violating effects will not be easy to observe in B decays. The typical effect to be sought will pertain to some particular exclusive decay mode. Accurate information on this mode will not be easy to obtain, simply because B mesons can decay in many possible ways, so that the branching ratio for any particular exclusive decay is rather small. (Branching ratios of order 10^{-3} are typical.) When experimental efficiencies, the need for tagging in some experiments, *etc.*, are taken into account, it is estimated that $\gtrsim 10^8$ B mesons will be needed to study the *CP*-violating effect in a typical promising decay mode. [2] As a result, there is considerable interest in the possibility of constructing a "B factory" which would produce the required number of B mesons.

II. *CP* VIOLATION IN THE STANDARD MODEL

In the standard model, *CP* violation comes from complex phase factors. Omitting gamma matrices, in this model the weak interaction between the W boson and the quarks has the form

$$\mathcal{H} = \frac{g}{\sqrt{2}} W^+ \sum_{i,j=1}^{3} \bar{u}_i V_{ij} d_j + \frac{g}{\sqrt{2}} W^- \sum_{i,j=1}^{3} \bar{d}_j V_{ij}^* u_i \, . \tag{1}$$

Here u_i runs over the positively-charged quarks u, c and t, while d_j runs over the negatively-charged ones d, s, and b. The quantity g is a (real) overall coupling strength, and V is a unitary matrix known as the quark mixing matrix. Now, under CP, particles go into their antiparticles. Thus,

$$(CP)\mathcal{H}(CP)^{-1} = \frac{g}{\sqrt{2}}W^- \sum_{i,j=1}^{3} \bar{d}_j V_{ij} u_i + \frac{g}{\sqrt{2}}W^+ \sum_{i,,j=1}^{3} \bar{u}_i V_{ij}^* d_j \,. \tag{2}$$

Comparing Eqs. (1) and (2), we see that if V is real, $\mathcal{H}$ is CP-invariant, but if V is not real, $\mathcal{H}$ and $(CP)\mathcal{H}(CP)^{-1}$ differ. To be sure, if we look just at the terms in $\mathcal{H}$ with a particular i and j, then the complexity of V_{ij} does not prove anything, because any particular V_{ij} can always be made real by redefining the field d_j so as to absorb the phase of V_{ij}. However, Kobayashi and Maskawa showed [3] that when V is 3×3 or larger (that is, when the number of quark generations is three or more), it is possible for V to contain complex phase factors which cannot *all simultaneously* be eliminated by rephasing of the various quark fields. The weak interaction is then genuinely CP-noninvariant.

Interestingly enough, studies of the shape of the Z^o resonance have just yielded strong evidence that the number of generations is precisely three, [4] the smallest number that allows V to contain phases which cannot be removed by rephasing of fields, and which, therefore, can lead to CP-violating effects. (As is well-known, for three generations, just one such phase is possible.)

If the weak interaction (1) acts at the vertices of some Feynman diagram, complex values of the V_{ij} will make these vertices complex. As a result, the diagram as a whole will have a complex overall phase. *If* this diagram interferes with another one, this complex phase can have consequences and physical CP-violating effects can be produced. In short, in the standard model, physical CP-violating effects always require an *interference* between different Feynman diagrams, or between different amplitudes.

III. CHARGED B DECAYS

The interferences which generate the most promising CP-violating effects in neutral B decays are of a different nature from those which operate in charged

B decays. Therefore, we shall treat these two cases separately, starting with the charged B mesons.

If CP invariance holds, then the amplitude $\langle f|R|i\rangle$ for any initial state $|i\rangle$ to decay to any final one $|f\rangle$ obeys the constraint

$$\langle f|R|i\rangle = \langle CP[f]|R|CP[i]\rangle \ . \tag{3}$$

As a result, if

$$\Gamma\left[B \to a_1 + \cdots + a_n\right] \neq \Gamma\left[\bar{B} \to \bar{a}_1 + \cdots + \bar{a}_n\right] \ , \tag{4}$$

where a_i is some particle and $\bar{a}_i$ is, of course, its antiparticle, then CP is violated. With few exceptions, all theoretical studies of CP violation in B decays have focussed on this potential CP-violating effect.

In the standard model, this effect, like all other violations of CP, would have to come from an interference. In charged B decays, the interference would be between two weak decay amplitudes. Without loss of generality, we may take one of these amplitudes, A_1, to be real, and write the second amplitude as $A_2 e^{i\delta_{WK}} e^{i\delta_{ST}}$. Here, A_2 is real, δ_{WK} (the "weak phase") is the phase of some combination of elements and complex conjugates of elements of the mixing matrix V, and δ_{ST} (the "strong phase") is a phase associated with strong interaction effects such as final state rescattering. Denoting the rates for the B decay and the CP-conjugate $\bar{B}$ decay in Eq. (4) by Γ and $\bar{\Gamma}$, respectively, we have

$$\Gamma = \left|A_1 + A_2 e^{i\delta_{WK}}\, e^{i\delta_{ST}}\right|^2 , \tag{5a}$$

and

$$\bar{\Gamma} = \left|A_1 + A_2 e^{-i\delta_{WK}}\, e^{i\delta_{ST}}\right|^2 \ . \tag{5b}$$

To see why the weak phase factors in these two decays are complex conjugates of each other, we note that if some term in the weak Hamiltonian occurs in a diagram for the B decay, the Hermitean conjugate of this term will occur in the corresponding diagram for the $\bar{B}$ decay. Thus, where V_{ij} occurs in the B decay, V_{ij}^* occurs in the $\bar{B}$ decay. However, apart from the phases in V, the standard-model weak interaction is CP-invariant, as is the strong interaction. Hence, due to the constraint (3) for the CP-invariant case, neither the sizes of the two amplitudes in Γ, nor the strong phase of the second amplitude, change when we go to $\bar{\Gamma}$.

Comparing Eqs. (5), we see that for Γ and $\bar{\Gamma}$ to differ, both of the interfering amplitudes must indeed be present, the weak phase must be present (or the B and $\bar{B}$ decay amplitudes are equal), *and* the strong phase must be present (or the B and $\bar{B}$ decay amplitudes are complex conjugates of each other). A large value of $(\Gamma - \bar{\Gamma})/(\Gamma + \bar{\Gamma})$ requires that the two interfering amplitudes be of comparable size, and that both δ_{WK} and δ_{ST} be large.

The difference (4) between rates for CP-conjugate decays is not the only possible signal of CP violation. There are other signals which involve the momenta and/or spins of the decay products. [5] Under CP, the state $|a(\vec{p}, \lambda)\rangle$ of a particle a with momentum $\vec{p}$ and helicity λ transforms according to

$$CP|a(\vec{p}, \lambda)\rangle = \eta|\bar{a}(-\vec{p}, -\lambda)\rangle , \tag{6}$$

where η is a phase factor. As a result, Eq. (3) implies that if

$$\begin{aligned} &\Gamma\left[B \to a_1(\vec{p}_1, \lambda_1) + \cdots + a_n(\vec{p}_n, \lambda_n)\right] \\ &\neq \Gamma\left[\bar{B} \to \bar{a}_1(-\vec{p}_1, -\lambda_1) + \cdots + \bar{a}_n(-\vec{p}_n, -\lambda_n)\right] , \end{aligned} \tag{7}$$

then CP is violated, even if there is no difference between these two decay rates after they are integrated over all outgoing momenta and spins.

As an example, consider the decays

$$\begin{aligned} B^{\pm} \longrightarrow\ & {K^*}^{\pm} + \psi \\ & \hookrightarrow K^{\pm} + \pi^{o} \end{aligned} , \tag{8}$$

and imagine an experiment that studies the distribution in the angle θ between the K momentum in the K^* rest frame, and the K^* momentum in the B rest frame. The details of this example are very relevant to a promising neutral B

decay to be discussed later, so let us examine them. Since B mesons are spinless, $B \to K^*\psi$ yields isotropically-distributed K^* and ψ particles whose helicities are equal. Hence, assuming the data are summed over the momenta and spins of the decay products of the ψ , the distribution in θ is given by the incoherent sum

$$\frac{d\Gamma(B^\pm)}{d(\cos\,\theta)} \propto \sum_\lambda \Gamma\Big(B^\pm \to {K^*}^\pm(\lambda)\psi(\lambda)\Big)\, \frac{d\Gamma({K^*}^\pm(\lambda) \to K^\pm\pi^o)}{d(\cos\,\theta)}\ . \tag{9}$$

Since for the spin-one K^*

$$\frac{d\Gamma\Big(K^*\,(\lambda = \pm 1) \to K\pi\Big)}{d(\cos\,\theta)} = \frac{\sin^2\theta}{2}\,, \tag{10a}$$

and

$$\frac{d\Gamma\Big(K^*\,(\lambda = 0) \to K\pi\Big)}{d(\cos\,\theta)} = \cos^2\theta = 1 - \sin^2\theta\ , \tag{10b}$$

the angular distribution of Eq. (9) has the form

$$\frac{d\Gamma(B^\pm)}{d(\cos\,\theta)} = a^\pm + b^\pm \sin^2\theta\ , \tag{11}$$

where $a^\pm$ and $b^\pm$ are constants. From Eqs. (9) and (10), it follows that any observed difference between a^+ and a^- , or between b^+ and b^- , would imply a CP-violating asymmetry of the form (7) among the rates for the decays $B^\pm \to {K^*}^\pm(\lambda)\psi(\lambda)$. Now, suppose, for example, that the rates for decay into the $K^*\psi$ states with $\lambda = 1$, 0 , and -1 are all comparable, but that only the rates for decay into the $\lambda = 0$ state involve a large CP-violating asymmetry. Then the asymmetry in the a-coefficient, $(a^+ - a^-)/(a^+ + a^-)$, would be larger than the one which remains when the angular distribution (11) is integrated over all angles. Thus, it could be useful to determine this angular distribution and extract the a coefficient. For each event, the angle θ would be measured in any case, since the K and π momenta must be determined in order to reconstruct the K^*.[6,7]

IV. NEUTRAL B DECAYS

The special feature which distinguishes CP violation in neutral B decays from that in charged B decays is the fact that neutral B mesons mix. That is, the $\overline{B^o}$, which is a $b\bar{d}$ quark bound state, and its antiparticle the B^o, which is a $\bar{b}d$ bound state, metamorphose into each other. They do this as a result of box diagrams in which, for example, $b\bar{d} \to t\bar{t}$ via W exchange, and then $t\bar{t} \to d\bar{b}$ via a second W exchange. To observe the mixing experimentally, one must, of course, be able to tell a $\overline{B^o}$ from a B^o. This one can do via the semileptonic decays of these particles. The quark process underlying the semileptonic decay of a $\overline{B^o}$ is the decay $b \to c\ell^-\bar{\nu}$ (where $\ell = e, \mu$, or τ), while that underlying the decay of a B^o is $\bar{b} \to \bar{c}\ell^+\nu$. Thus, the charge of the daughter lepton indicates whether the parent was a $\overline{B^o}$ or B^o. It is found that the probability that a neutral B born as a B^o will decay as a $\overline{B^o}$ (that is, yielding an ℓ^-), relative to the probability that it will decay as a B^o (that is, yielding an ℓ^+), is[8]

$$\frac{\Gamma\left(B^o \to \overline{B^o} \to \ell^-\right)}{\Gamma\left(B^o \to \ell^+\right)} = 0.18 \pm 0.05 \,. \tag{12}$$

We see that $B^o - \overline{B^o}$ mixing is quite large.

Diagonalization of the $B^o - \overline{B^o}$ mass matrix yields two mass eigenstates

$$|B_{1,2}\rangle = \frac{1}{N}\left[p|B^o\rangle \pm q|\overline{B^o}\rangle\right] \,, \tag{13}$$

with $N = \left(|p|^2 + |q|^2\right)^{\frac{1}{2}}$. We shall denote the mass difference of $|B_2\rangle$ and $|B_1\rangle$ by Δm, their average mass by M, and both of their widths, which are expected to be approximately equal,[9] by Γ. Due to the $B^o - \overline{B^o}$ mixing, a neutral B born at time $t = 0$ as a $|B^o\rangle$ evolves in time t into a state $|B^o(t)\rangle$ which is a linear combination of $|B^o\rangle$ and $|\overline{B^o}\rangle$ and is given by

$$|B^o(t)\rangle = e^{-\frac{\Gamma}{2}t}\, e^{-iMt}$$

$$\times \left\{ \cos\left(\frac{\Delta m}{2}t\right)|B^o\rangle + i\frac{q}{p}\sin\left(\frac{\Delta m}{2}t\right)|\overline{B^o}\rangle \right\} . \tag{14a}$$

Similarly, a B born as a $|\overline{B^o}\rangle$ evolves in time t into a state $|\overline{B^o}(t)\rangle$ given by

$$|\overline{B^o}(t)\rangle = e^{-\frac{\Gamma}{2}t}\, e^{-iMt} \times$$

$$\times \left\{ \cos\left(\frac{\Delta m}{2}t\right)|\overline{B^o}\rangle + i\frac{p}{q}\sin\left(\frac{\Delta m}{2}t\right)|B^o\rangle \right\} . \tag{14b}$$

Since $|q/p| \simeq 1$,[9] $|B^o(t)\rangle$ oscillates back and forth between pure B^o and $\overline{B^o}$ components of equal strength, apart from the exponential decay of the entire wave function.

Suppose, now, that f is a final state into which either a (non-evolved) B^o or $\overline{B^o}$ can decay. (Examples are $f = \pi^+\pi^-$, and, less obviously, $D^+\pi^-$.) Then there are two paths through which a particle born as a B^o can ultimately become the state f. First, the B^o can evolve into a B^o, and then decay into f with some amplitude w_f. Alternatively, the B^o can evolve into a $\overline{B^o}$, and then this $\overline{B^o}$ can decay into f with some amplitude $\bar{w}_f$. This situation is pictured below:

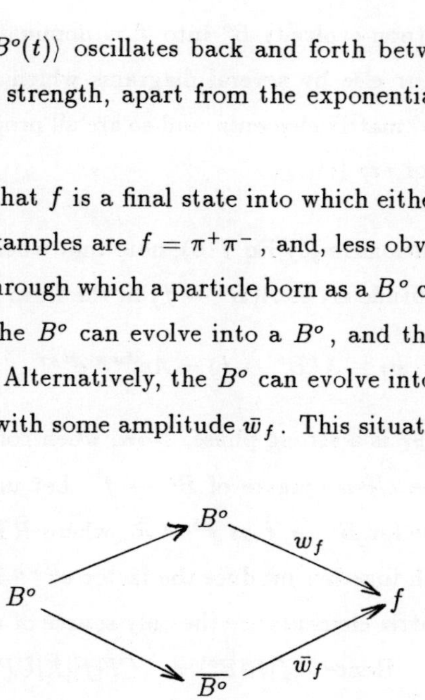

(15)

Through the *interference* between the amplitudes for the two paths from the initial B^o to f, phases in the standard-model quark mixing matrix can produce CP-violating effects.

To maximize these effects, we would like the interfering amplitudes to be of comparable size, just as in the charged B case. Now, since $|q/p| \simeq 1$, the amplitudes for $B^o \to B^o$ and $B^o \to \overline{B^o}$ are of equal size (see Eq. (14a) with, for example, t such that $\cos\left(\frac{\Delta m}{2}t\right) = \sin\left(\frac{\Delta m}{2}t\right)$). Thus, we desire that

$$\left|w_f\right| = \left|\bar{w}_f\right| . \tag{16}$$

It has been observed[10,11] that this equality will indeed hold when —

1. f is a CP eigenstate,

   and

2. the decay of a (non-evolved) B^o into f is dominated by a single weak decay diagram, or else by several diagrams which all involve the same combination of V matrix elements, and so are all proportional to the same weak phase factor $exp\ (i\delta_{WK})$.

To see why these conditions imply Eq. (16), note that when condition (2) holds, we can write the amplitude $w_f \equiv A(B^o \to f)$ in the form

$$w_f \equiv A(B^o \to f) = A e^{i\delta_{WK}} e^{i\delta_{ST}} , \tag{17a}$$

where A is real and δ_{ST} is a strong phase. Now, when condition (1) holds, the process $\overline{B^o} \to f$ is the CP-conjugate of $B^o \to f$. Let us write the transition operator R responsible for $B^o \to f$ as $e^{i\delta_{WK}}\tilde{R}$, where $\tilde{R}$ is R with all of its V matrix elements, which together produce the factor $e^{i\delta_{WK}}$ in Eq. (17a), factored out. Since these V matrix elements are the only source of CP violation in $B^o \to f$, $\tilde{R}$ is CP-invariant. Hence, $\langle f|\tilde{R}|B^o\rangle = \langle CP[f]|\tilde{R}|CP[B^o]\rangle = \eta_f\langle f|\tilde{R}|\overline{B^o}\rangle$, where η_f is the CP parity of f. From this relation and Eq. (17a), it follows that

$$\bar{w}_f \equiv A\left(\overline{B^o} \to f\right) = \eta_f\, A e^{-i\delta_{WK}} e^{i\delta_{ST}} . \tag{17b}$$

Thus, $\left|\bar{w}_f\right| = \left|w_f\right|$.

We see that decays which yield CP eigenstates and are dominated by a single weak phase factor are especially interesting. From Eqs. (14) and (16), it is easily shown that the rate at which an evolved $B^o(t)$ or $\overline{B^o}(t)$ will undergo such decays is given by

$$\Gamma\left(\overset{(-)}{B^o}(t) \to f\right) \propto |w_f|^2 e^{-\Gamma t}\left\{1 \pm Im\left(\frac{p}{q}\rho_f\right)\sin(\Delta m\, t)\right\} . \tag{18}$$

Here the upper and lower signs on the right-hand side correspond, respectively, to a parent $B^o(t)$ and a parent $\overline{B^o}(t)$. The quantity ρ_f is defined by

$$\rho_f \equiv \frac{A\left(B^o \to f\right)}{A\left(\overline{B^o} \to f\right)} . \tag{19}$$

We observe that the decay rates (18) are not purely exponential, but have exponential behavior modulated by a sinusoidal oscillation. The $\sin(\Delta m\ t)$ term results from the interference between the amplitudes for the two paths in (15). Now, it is trivial to show that when CP invariance holds, $CP|B^o(t)\rangle = |\overline{B^o}(t)\rangle$, so that according to Eq. (3), the two decay rates in Eq. (18) must be equal. Since the $\sin(\Delta m\, t)$ term contributes oppositely to these two rates, this term is clearly a CP-violating effect. Moreover, since $\left|(p/q)\,\rho_f\right| \simeq 1$, the coefficient of this effect, $Im\,[(p/q)\,\rho_f]$, can be of order 0.1, or even of order 1.0, if the CP-violating phases in $(p/q)\,\rho_f$ are large. As we have seen, this possibility of very large CP-violating sine waves in the $\overset{(-)}{B^o}(t)$ decays described by Eq. (18) arises from the fact that the amplitudes whose interference produces these sine waves are of equal size.

(Much of the analysis leading to Eq. (18) could be repeated for neutral K decays. There too one can have a CP-violating sine wave whose coefficient is the imaginary part of a quantity $[(p/q)\,\rho_f]_K$ of unit magnitude. However, in the K meson case one does not expect the phase of this quantity to be large. The reason has to do with the fact that while three generations are required for CP violation, K mesons are composed of quarks from only the first two generations, and the quarks of the third generation are coupled only very weakly to those of the first two generations through the quark mixing matrix. [12] There is no such suppression of CP violation in B decays.)

In addition to being potentially large, the coefficient $Im\left[(p/q)\,\rho_f\right]$ of the sine wave in the $\overset{(-)}{B^o}(t)$ decay rates of Eq. (18) can be cleanl y predicted in terms of quark mixing matrix elements alone, and does not depend on poorly-known hadronic matrix elements. [9,13] To see this, we note first that from the diagonalization of the $B^o - \overline{B^o}$ mass matrix, it follows that $p/q = \left[A\left(b\bar{d} \to \bar{b}d\right)/A\left(\bar{b}d \to b\bar{d}\right)\right]^{1/2}$. Here $A\left(b\bar{d} \to \bar{b}d\right)$ is the amplitude for a box diagram which involves t and $\bar{t}$ exchanges and causes $\overline{B^o} \to B^o$, and $A\left(\bar{b}d \to b\bar{d}\right)$ is the amplitude for the analogous box diagram for the inverse process. The only difference between these two diagrams is that each V_{ij} in the one is replacesd by V_{ij}^* in the other. Clearly, p/q then depends only on the elements of V. That $\rho_f \equiv A\left(B^o \to f\right)/A\left(\overline{B^o} \to f\right)$ also depends only on the elements of V (and on the known CP parity of the final state of interest) follows immediately from Eqs. (17a) and (17b). [14]

At present, the matrix V is not fully known, so the values of $Im\left[(p/q)\,\rho_f\right]$ for various decays cannot as yet be determined. Nevertheless, from what *is* known of V, interesting statements can already be made. For example, consider the decays $\overset{(-)}{B^o}(t) \to K_s^o + \psi$, where the final state is a CP eigenstate, and most analyses conclude that the decay process is strongly dominated by a single quark diagram in which $b \to c\bar{c}s$ via W exchange. [14] It is already known [15] that

$$0.08 \le \left|Im\left(\frac{p}{q}\,\rho_{K_S\psi}\right)\right| \le 1\,. \tag{20}$$

That is, a large CP-violating asymmetry is predicted. [16]

In view of the clean relation between the quantities $Im\left[(p/q)\,\rho_f\right]$ and the parameters of the quark mixing matrix, one can imagine that perhaps measurements of neutral B decays into a few CP eigenstates will be used to eliminate the remaining unknowns from this matrix, and then measurements of decays into additional CP eigenstates will be used to cleanly test the hypothesis that CP violation does indeed come from phases in this matrix.

V. DECAYS TO NEAR *CP* EIGENSTATES

Since even a large CP-violating sinusoidal oscillation will be difficult to observe because of limited statistics, it is of interest to ask whether one can identify additional decay modes which do not yield CP eigenstates but in which large, cleanly-predicted CP-violating oscillations are nevertheless expected. Some of these new modes may be particularly good candidates for study. In addition, if one can find situations where a decay to a non-CP-eigenstate is predicted to exhibit *exactly the same* oscillation as some related decay to a CP eigenstate, then perhaps the data for these related decays can be combined to yield statistically-significant signals of CP violation without the need for quite so many B mesons as would otherwise be required.

A class of decays which has been found [17] to be promising as an additional source of large, cleanly-predicted CP-violating signals is that which yields "near CP eigenstates" — states, such as $\Lambda_c\bar{\Lambda}_c$, which are CP-self-conjugate collections of particles, but are not CP eigenstates because they are mixtures of partial waves with different CP parities. In $\overset{(-)}{B^o}(t) \to \Lambda_c\bar{\Lambda}_c$, for instance, the $\Lambda_c\bar{\Lambda}_c$ pair will be in some superposition of ${}^1S_0$ and ${}^3P_0$ states. Now, individually, each of these states is indeed a CP eigenstate, but the two states have opposite CP parities, so the $\Lambda_c\bar{\Lambda}_c$ final state is not a CP eigenstate. As a result, the decay rates $\Gamma\left(\overset{(-)}{B^o}(t) \to \Lambda_c\bar{\Lambda}_c\right)$ do not have forms quite so simple as those of Eq. (18), and the CP-violating effect in these rates is very difficult to predict and possibly dynamically suppressed. [17,18] Nevertheless, it is still possible to extract a cleanly-predicted, non-suppressed CP-violating signal from the data if one measures, in addition to the partial decay rate, angular distributions of Λ_c and $\bar{\Lambda}_c$ decay products which enable one to study the B decays to ${}^1S_0$ and

${}^3P_0$ $\Lambda_c\bar{\Lambda}_c$ pairs separately. For each of these separate decay modes, the final state is a CP eigenstate, so that the rate is given by Eq. (18), with its cleanly-predicted CP-violating effect.

To see explicitly how angular distribution measurements can yield clean CP-violating signals in B decay to a near CP eigenstate, let us consider, not $B \to \Lambda_c\bar{\Lambda}_c$, which may or may not be of practical utility, but the rather promising decays [17]

$$\overset{(-)}{B^o}(t) \longrightarrow K_+^* + \psi \qquad\qquad (21)$$
$$\hookrightarrow K_S^o + \pi^o$$

Here K_+^* is the CP-even linear combination of K^{*^o} and $\overline{K^{*^o}}$. In $\overset{(-)}{K^{*^o}} \to K_s^o\pi^o$, the final state is a CP eigenstate with $CP = +1$, and in this strong decay CP is conserved. Thus, in the sequence (21) K_+^* is the only combination of K^{*^o} and $\overline{K^{*^o}}$ that can contribute. Now, just as in $\overset{(-)}{B^o}(t) \to \Lambda_c\bar{\Lambda}_c$, so too in $\overset{(-)}{B^o}(t) \to K_+^*\psi$ the final state is a CP-self-conjugate collection of particles, but not a CP eigenstate because it is a mixture of partial waves with different CP parities. It is convenient to describe the situation in terms of helicity states. Since the outgoing particles in $\overset{(-)}{B^o}(t) \to K_+^*\psi$ must have the same helicity, there are three possible $J = 0$ final states: $|K_+^*(0)\psi(0)\rangle$, $|K_+^*(+)\psi(+)\rangle$, and $|K_+^*(-)\psi(-)\rangle$, where the quantities in parentheses are the helicities. Now, CP reverses helicity, so the first of these states is a CP eigenstate, but the other two are not. Thus, we would like to find some angular distribution measurement which would select the decays into $|K_+^*(0)\psi(0)\rangle$.

From the parallel between the sequence $\overset{(-)}{B^o}(t) \to K_+^*\psi \to (K_s^o\pi^o)\psi$ and the decays $B^\pm \to K^{*^\pm}\psi \to (K^\pm\pi^o)\psi$ discussed earlier, it is obvious that measurement of the distribution in the angle θ between the K_s^o momentum in the K_+^* rest frame, and the K_+^* momentum in the $\overset{(-)}{B^o}(t)$ rest frame, will do the trick.

In view of Eqs. (9)-(11), this distribution is given by

$$\frac{d\Gamma\left[\overset{(-)}{B^o}(t)\right]}{d\left[\cos\,\theta\right]} = \Gamma\left[\overset{(-)}{B^o}(t) \to K_+^*(0)\psi(0)\right] +$$

$$+\left\{\frac{\Gamma\left[\overset{(-)}{B^o}(t) \to K_+^*(+)\psi(+)\right] + \Gamma\left[\overset{(-)}{B^o}(t) \to K_+^*(-)\psi(-)\right]}{2}\right.$$

$$\left. -\Gamma\left[\overset{(-)}{B^o}(t) \to K_+^*(0)\psi(0)\right]\right\} \sin^2\theta \,. \qquad (22)$$

Note that the only B decays which contribute to the θ-independent term in this distribution are those to the CP eigenstate $K_+^*(0)\psi(0)$.

The state $K_+^*\psi$ has the same quark content as $K_s^o\psi$. Thus, if the single quark diagram in which $b \to c\bar{c}s$ via W exchange dominates $\overset{(-)}{B^o}(t) \to K_s^o\psi$, as is generally believed, then this same single diagram also dominates $\overset{(-)}{B^o}(t) \to K_+^*\psi$. Hence, the $\overset{(-)}{B^o}(t)$ decays into the CP eigenstate $K_+^*(0)\psi(0)$ meet both of the conditions following Eq. (16). Consequently, the rates for these decays are given by Eq. (18). Furthermore, the CP-violating quantity $Im\,[(p/q)\rho_f]$ in these rates is exactly the same as in those for $\overset{(-)}{B^o}(t) \to K_s^o\psi$, apart from a minus sign reflecting the fact that (for $J=0$) $CP(K_s^o\psi) = -1$, while $CP\left(K_+^*(0)\psi(0)\right) = +1$ (see Eqs. (17)). Thus, for the θ-independent term in $d\Gamma\left[\overset{(-)}{B^o}(t)\right]/d\left[\cos\,\theta\right]$, Eq. (22), we have[19)]

$$\Gamma\left[\overset{(-)}{B^o}(t) \to K_+^*(0)\psi(0)\right] = |w_o|^2\, e^{-\Gamma t}\left\{1 \mp \alpha\, \sin(\Delta m\, t)\right\}\,, \qquad (23)$$

where $w_o \equiv A\left(B^o \to K_+^*(0)\psi(0)\right)$, and

$$\alpha \equiv Im\left(\frac{p}{q}\rho_{K_s\psi}\right) . \tag{24}$$

To obtain the *θ-dependent* term in $d\Gamma\left[\overset{(-)}{B^o}(t)\right]/d[\cos\theta]$, we may form from $|K_+^*(\pm)\psi(\pm)\rangle$ the CP eigenstates

$$|(K_+^*\psi)_{CP=\pm}\rangle \equiv \frac{1}{\sqrt{2}}\left[|K_+^*(+)\psi(+)\rangle \pm |K_+^*(-)\psi(-)\rangle\right] , \tag{25}$$

and define the amplitudes $w_\pm \equiv A\left(B^o \to \left(K_+^*\psi\right)_{CP=\pm}\right)$. The rates for decay into these CP eigenstates are given by

$$\Gamma\left[\overset{(-)}{B^o}(t) \to \left(K_+^*\psi\right)_{CP=+}\right] = |w_+|^2 e^{-\Gamma t}\left\{1 \mp \alpha\sin(\Delta m\, t)\right\} \tag{26a}$$

and

$$\Gamma\left[\overset{(-)}{B^o}(t) \to \left(K_+^*\psi\right)_{CP=-}\right] = |w_-|^2 e^{-\Gamma t}\left\{1 \pm \alpha\sin(\Delta m\, t)\right\} . \tag{26b}$$

Note that because of the factor η_f in Eq. (17b), the signs in front of α in the rates for decay into the CP-odd state $\left(K_+^*\psi\right)_{CP=-}$ are opposite to those in the rates for decay into the CP-even states $\left(K_+^*\psi\right)_{CP=+}$ and $K_+^*(0)\psi(0)$. From Eqs. (22), (23), (25), and (26), we find that

$$\frac{d\Gamma\left[\overset{(-)}{B^o}(t)\right]}{d[\cos\theta]} = |w_o|^2 e^{-\Gamma t}\left\{1 \mp \alpha\sin(\Delta m\, t)\right\} +$$

$$+\frac{|w_+|^2 + |w_-|^2 - 2|w_o|^2}{2} e^{-\Gamma t}\left\{1 \mp \frac{|w_+|^2 - |w_-|^2 - 2|w_o|^2}{|w_+|^2 + |w_-|^2 - 2|w_o|^2}\alpha\sin(\Delta m\, t)\right\}\sin^2\theta . \tag{27}$$

We see that while the CP-violating sine wave in the θ-independent part of this distribution is cleanly-predicted, that in the $\sin^2\theta$ term has a strength which depends on a ratio of amplitudes which is difficult to calculate, and which may be either large or small.

Since the interesting part of the θ distribution is the θ-independent term which corresponds to $B \to K_+^*(0)\psi(0)$, the determination of this distribution will not be very useful if the rate for decay into $K_+^*(0)\psi(0)$ is tiny compared to the rates for decay into $K_+^*(\pm 1)\psi(\pm 1)$. However, simple-minded quark-model arguments suggest that the decays to K^* and ψ of zero helicity should be a very sizeable fraction of all the decays to $K^*\psi$.

Neutral B decay into any one of the states $K_s^o\psi$, $K_+^*(0)\psi(0)$, $K_s^o\psi'$, $K_+^*(0)\psi'(0)$, $K_s^o\,\chi_J$ (where χ_J is the ${}^3P_J$ state of charmonium), or $K_+^*(0)\chi_J(0)$ yields a CP eigenstate with exactly the same quark content. Decay into any of these states is expected to be dominated at the quark level by $b \to c\bar{c}s$ via W exchange. Thus, from Eqs. (17)-(19), if f is any one of these states,

$$\Gamma_f \equiv \Gamma\left(B^o(t) \to f\right) = |w_f|^2 e^{-\Gamma t}\left\{1 - \eta_f \alpha \sin(\Delta m\, t)\right\} , \qquad (28a)$$

and

$$\overline{\Gamma}_f \equiv \Gamma\left(\overline{B^o}(t) \to f\right) = |w_f|^2 e^{-\Gamma t}\left\{1 + \eta_f \alpha \sin(\Delta m\, t)\right\} . \qquad (28b)$$

Here $\alpha \equiv Im\left[(p/q)\,\rho_{K_s\psi}\right]$ as before, and η_f is the CP parity of f. Now, since the decays into all these final states involve exactly the same CP-violating quantity α, it may be advantageous to combine the measured decay rates [20] to form the quantity

$$\frac{\sum_f \eta_f\left(\overline{\Gamma}_f - \Gamma_f\right)}{\sum_f\left(\overline{\Gamma}_f + \Gamma_f\right)} = \alpha \sin\left(\Delta m\, t\right) . \qquad (29)$$

This procedure can perhaps reduce the total number of B mesons needed to detect and measure α.

VI. SUMMARY

The hypothesis that complex phases in the quark mixing matrix associated with the known weak interaction are the source of CP violation can be incisively tested in B decays. In particular, if this hypothesis is correct, then there must be large, cleanly-predicted CP-violating effects in certain neutral B decays to CP eigenstates. Such effects will also occur in decays to near CP eigenstates such as $K^*_+\psi$, where they can be extracted with the aid of angular distribution measurements. Combining the data on related CP-eigenstate and near-CP-eigenstate final states may reduce the total number of B mesons needed to study CP violation.

ACKNOWLEDGMENTS

It is a pleasure to thank Lincoln Wolfenstein for several enlightening conversations, and Wick Haxton and Ernest Henley for the hospitality of the Summer Institute in Theoretical Physics at the University of Washington, where a part of this work was done.

REFERENCES

1. Gilman, F., in *SLAC Heavy Flavors 1987* (Proceedings of the International Symposium on the Production and Decay of Heavy Flavors, Stanford, CA, 1987), p. 211.

2. Foley, K., *et al.*, in *Experiments, Detectors, and Experimental Areas for the Supercollider*, edited by Donaldson, R. and Gilchriese, M.G.D. (World Scientific, Singapore, 1988), p. 701; Dib, C., Dunietz, I., Gilman, F., and Nir, Y., Stanford Linear Accelerator Center preprint SLAC-PUB-5109.

3. Kobayashi, M. and Maskawa, T., Prog. of Theo. Phys. *49*, 652 (1973).

4. Abrams, G., *et al.*, Phys. Rev. Lett. **63**, 2173 (1989); Decamp, D., *et al.* (The ALEPH Collaboration), Phys. Lett. **B231**, 519 (1989); Aarnio, P. *et al.* (The DELPHI Collaboration), Phys. Lett. **B231**, 539 (1989); Adeva, B., *et al.* (The L3 Collaboration), Phys. Lett. **B231**, 509 (1989);

Akrawy, M. *et al.* (The OPAL Collaboration), Phys. Lett. **B231**, 530 (1989).

5. An analysis of such kinematically nontrivial *CP*-violating effects has been carried out by Kuroda, M., Peccei, R., Sanda, A., and the author. See Kayser, B., National Science Foundation preprint, to appear in the Proceedings of the International Workshop on Weak Interactions and Neutrinos, Ginosar, Israel, 1989, and Bélanger, G., *et al.*, in *Proceedings of the Summer Study on High Energy Physics in the 1990's*, edited by Jensen, S. (World Scientific, Singapore, 1989), p. 339.

6. For early work on *B* *CP*-violating asymmetries which involve outgoing momenta or spins, see Barshay, S. and Geris, J., Phys. Lett. **84B**, 319 (1979).

7. For a general analysis of certain kinematically nontrivial *CP*-violating effects in *B* decays to pairs of vector particles, see Valencia, G., Phys. Rev. D**39**, 3339 (1989).

8. Kreinick, D., reporting on an analysis of CLEO and ARGUS data, in a talk presented at the XIV International Symposium on Lepton and Photon Interactions, Stanford, 1989. Danilov, M., reported a similar value at the same Symposium.

9. Bigi, I., Khoze, V., Uraltsev, N., and Sanda, A., in *CP Violation*, edited by Jarlskog, C. (World Scientific, Singapore, 1989), p. 175.

10. Carter, A., and Sanda, A., Phys. Rev. Lett. **45**, 952 (1980).

11. Dunietz, I., and Rosner, J., Phys. Rev. D **34**, 1404 (1986).

12. The author thanks Lincoln Wolfenstein for pointing this out to him.

13. Foley, K., *et al.*, Ref. 2.

14. Of course, when $A(B^o \to f)$ is *not* dominated by a single weak phase factor and Eqs. (17) do not hold, ρ_f does not depend on V matrix elements alone. The degree to which a single weak phase factor does dominate various decays is considered by London, D., and Peccei, R., Phys. Lett. **B 223**, 257 (1989), by Gronau, M., Phys. Rev. Lett. **63**, 1451 (1989), and by Grinstein, B., Phys. Lett. **B 229**, 280 (1989).

15. Dib, C., *et al.*, Ref. 2.

16. That CP-violating asymmetries in B decays could well be large was pointed out by Wolfenstein, L., Nuc. Phys. **B 246**, 45 (1984).

17. Kayser, B., Kuroda, M., Peccei, R., and Sanda, A., to be published.

18. Kayser, B., Ref. 5.

19. For an interesting illustration of other CP-violation information which can be obtained by study of angular distributions, see the discussion of $\overset{(-)}{B^o}(t) \to \psi\pi^+\pi^-$ by Bjorken, J., Fermi National Accelerator Laboratory preprint 88-0736, presented at the Fourth Workshop on Recent Developments in High Energy Physics, Chania-Crete, Greece, 1988.

20. This was suggested by Schmidt-Parzefall, W., at the Workshop on B Factories and Related Physics Issues, Blois, France, 1989.

Search for the Electric Dipole Moment of the Neutron

W. Mampe
Insitut Laue-Langevin
156X
38042 Grenoble Cédex, France

ABSTRACT

The situation of the experimental search for the electric dipole moment of the neutron is reviewed. The most recent results from the only two experiments in progress are reported.

INTRODUCTION

Until today experimental evidence for CP violation comes exclusively from K° meson decays[1,2] and the baryon asymmetry of the universe. The origin of CP violation is not known up to date.

Grand unification theories incorporate CP violating terms and adjust their relevant parameters to the experimental data from the Kaon system. The present baryon asymmetry is inaccessible to direct verification in the laboratory and needs physics beyond the Standard Model. The fundamental question of the correct theory emerging from one of the numerous models could be hopefully answered with the help of further experimental input.

Since CPT symmetry is expected to remain inviolate in the most general terms, CP and T violation are connected. Thus the observation of T violation would be of the same outstanding importance. One possiblility of a direct manifestation of T violation would be the existence of a permanent electric dipole moment (EDM) of elementary particles[3]. The most promising candidate for this search has been since the early 50ies the neutron[4] due to the advantage of its simple structure as neutral elementary particle and the fact that long lived free neutrons are available form reactors in large numbers. For completeness we mention that during the last years comparable sensitivities have been achieved in EDM investigations with more complex systems, namely atoms.

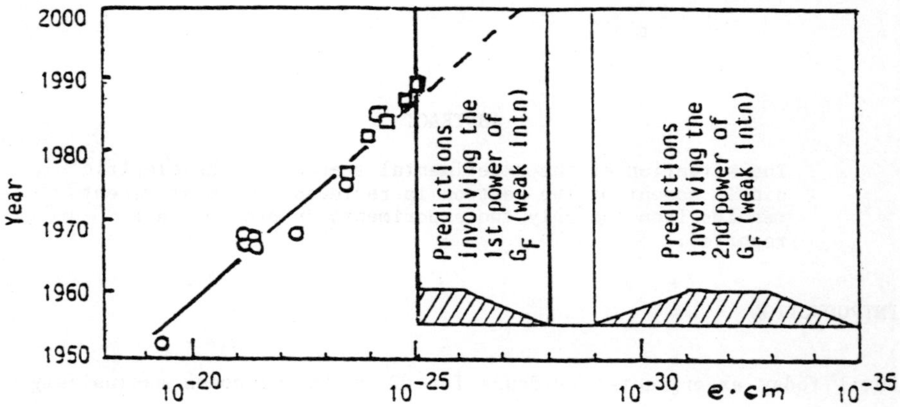

Fig. 1: Upper limits from experiments on the neutron electric dipole moment measured with neutron beams-circles, and with UCN-squares.

A large number of theoretical papers has been devoted to predictions for the possible seize of the neutron's EDM. The intense theoretical interest continues today (after elimination of some earlier models the results of which have been excluded by the experimental progress covering 6 orders of magnitude over 35 years) and produces values for the EDM which can roughly be devided into two groups: one category with neutron EDM's roughly between 10^{-25} and 10^{-27} e.cm and another one with values between 10^{-31} and 10^{-33} e.cm (Fig. 1), depending on whether the leading contribution to a finite EDM is generated by first or second order terms in the weak interaction [5]. It is thus evident that in this situation even experimental limits are extremely valuable since they can discriminate between theories which can or cannot accommodate the limit. The experimental sensitivity achieved is just at the level of 10^{-25} e.cm and has thus reached a decisive step.

Experimental Method

Both experiments at present in progress - one at the ILL in France, the other one in Gatchina in the USSR - use basically the same technic for detecting a possible EDM, i.e. a magnetic resonance spectrometer in connection with spatially confined ultra cold neutrons. Therefore, we concentrate mainly on one experiment, namely the one at ILL.

The EDM Experiment at ILL

The experiment situated at the ILL is run in a common effort by scientists from the following institutions: Harvard, ILL, Munich, Sussex, RAL, Seattle. The leading scientists involved since a long time in the experiment are: J.M. Pendlebury, K. Smith, N.F. Ramsey, K. Green and B. Heckel. The necessary ultra-cold neutrons (UCN) are produced by a recent installation[6] at ILL. The increased UCN density (about a factor of 40 compared to the original ILL UCN source) offers the main advantage over the last generation of the experiment [7]. The installation is schematically shown on Fig. 2: cold neutrons are extracted vertically by a high quality totally reflecting curved guide from the integrated liquid Deuterium (25 K) cold soruce. The guide curvature (r = 13 m) cuts the neutron spectrum to higher energies. The outcoming spectrum with a maximum at v = 50 $m.s^{-1}$ enters a neutron turbine which Doppler shifts the neutron velocities v down to the UCN region (v_{UCN} < 8 ms^{-1}). The UCN density measured is 60 per cm^3. Ultracold neutrons which are totally reflected by the container walls can be confined for several hundred seconds have replaced the earlier neutron beam experiments for two reasons: The magnetic resonance linewidth can be made 4 orders of magnitude narrower with UCN due to the long observation time. Secondly, vanishes the v x E effect (leading to an important experimental uncertainty) due to the fact that $\langle v \rangle \sim 0$ for stored ultra cold neutrons.

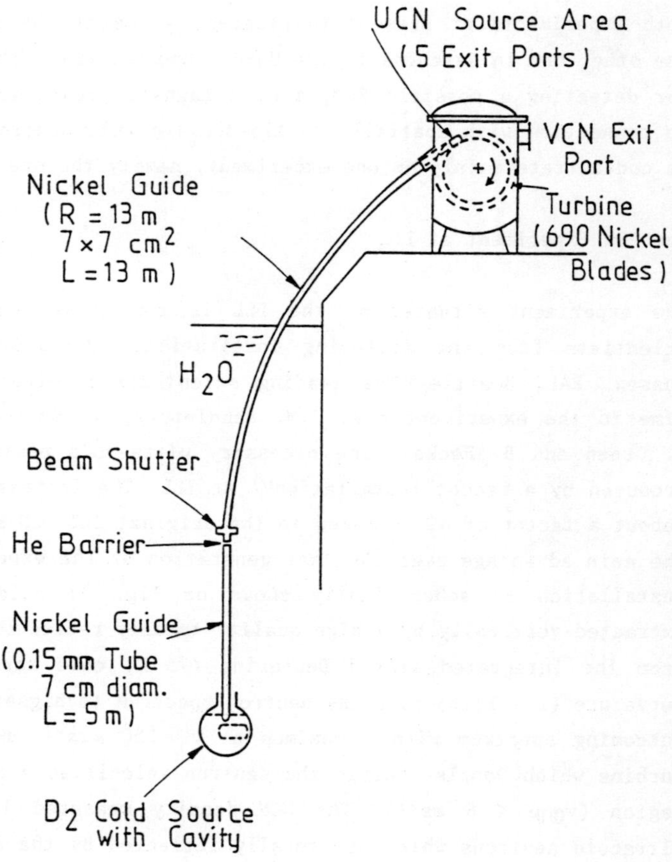

Fig. 2: Schematic view of the very cold and ultra-cold neutron source at ILL. The cold source sits in the D 2 moderator tank in a thermal flux of 4.5×10^{14} cm^{-2}.s^{-1}. The turbine is installed about 15 m above the level of the cold source.

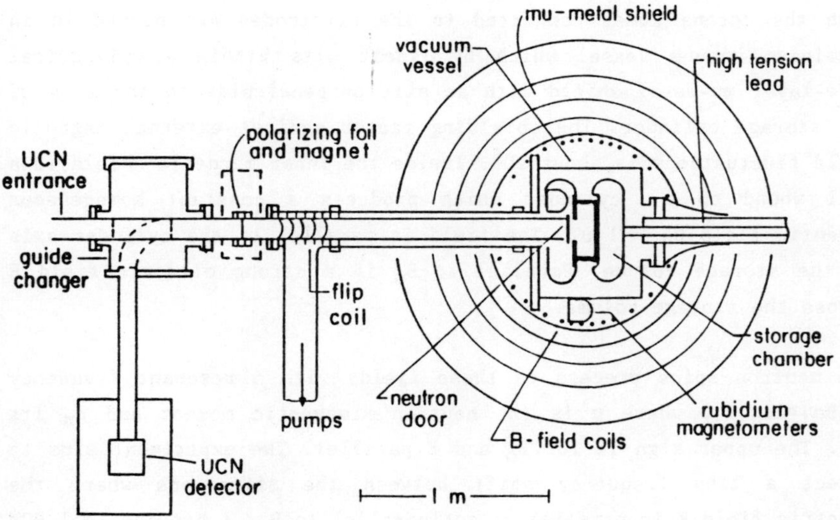

Fig. 3: Magnetic resonance spectrometer used by the collabortion a ILL for the search of the electric dipole moment of the neutron. Fig. from Ref. 8.

Fig. 3 shows the main experimetnal apparatus. The UCN of one spin component pass on their way from the turbine to the storage vessel a magnetically saturated 1 μm iron cobalt foil (polarizer). The UCN are admitted into the vessel by a 7 cm diameter hole which is closed by a sliding door after steady state density is reached (10 s).

The storage vessel is a cylinder of electrically insulating Beryllium oxide closed by Beryllium base plates which serve as electrodes for the necessary electric field. Beryllium is used as bottle wall material because of its high Fermi potential. UCN with velocities up to 6.9 $m.s^{-1}$ can be stored. The base plates are 25 cm in diamter and the cylinder is 10 cm high. The cylinder axis is horizontal and parallel to the feeding neutron guide. This storage volume together with the corona domes connected to the electrodes are housed in an aluminium vacuum vessel which by itself sits within a cylindrical five-layer mu-metal shield with an axis perpendicular to the axis of the storage cylinder. The shielding factor against external magentic field fluctuations is about 10^5. Inside the inner magnetic shield is a coil wound on a cylinder which produces a constant homogeneous magentic field B_o = 1 μT. The field is parallel to the cylinder axis of the storage volume. Parallel to B_o is a strong electric field E across the storage volume.

The neutron spins precess in these fields with a resonant frequency $-\mu|Bo|\mp 2d_n|E|$, where μ is the neutron's magentic moment and d_n its EDM. The upper sign is for B_o and E parallel. The experiment aims to detect a tiny frequency shift between the situations where the electric field E is parallel or antiparallel to B_o. A hypothetical EDM of 10^{-25} e.cm would experience less than one revolution per week in a field E = 10 $KV.cm^{-1}$. The sensitivity necessary to detect such a small effect is achieved by applying the Ramsey time separated magentic resonance technique: The UCN enter the storage volume with their spins parallel to B_o. After the bottle door is closed a 4 s pulse of an oscillating magentic field turns the spins perpendicular to B_o. The subsequent free spin precession continues for about 70 s until a second 4 s pulse from the same oscillator turns the spins coherently by another 90°. The bottle door is opened and the UCN are guided through the same FeCo foil, where their spin direction is analyzed, via a guide switch into a ^3He detector. After counting for 10 s the neutrons with the spin component for which the FeCo foil is transparenet a spin flip coil is energized and turns the remaining UCN with the opposite spin-state by 180° so that they can pass the spin analyzer and be counted also for 10 s. In this way 16 000 UCN reach the detector per cycle. A complete cycle lasts 114 s. The number of UCN detected is a function of their spin direction at the position of the spin analyser and hence of the precession frequency. The frequency scan on Fig. 4 demonstrates the full interference pattern for one spin component. During a run the oscillator is tuned to the center of one slope of the central resonance line. Reversal of the electric field should manifest in a change in neutron counts due to $d\nu = 4\ d_n.E$.

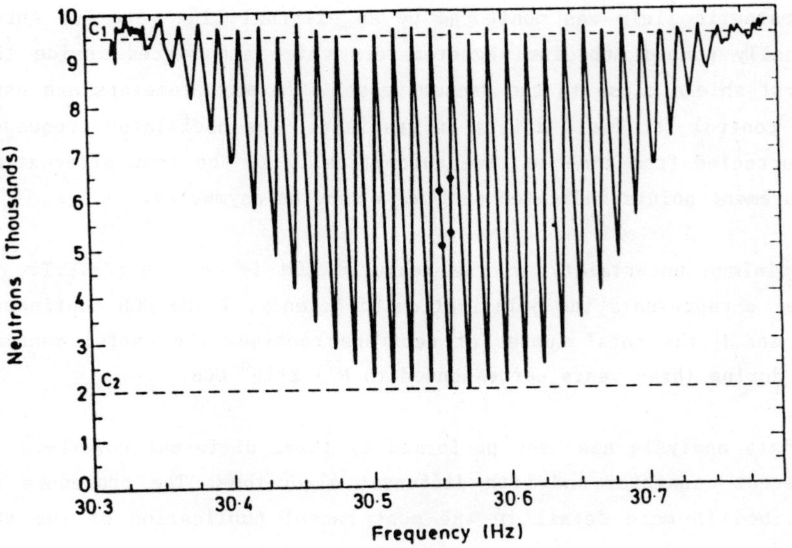

Fig. 4: Neutron magnetic resonance curve for UCN stored for 68s in a 1 μT field. The UCN counts are shown as a function of the frequency of the applied oscillating magnetic field. EDM data are taken at the four frequencies marked by crosses. Fig. from Ref. 8.

8 cycles with electric field (up to 1.6 KV cm^{-1}) parallel to B_o are followed by 4 cycles with E = 0 and subsequent 8 cycles with E antiparallel to B_o. The leakage current across the bottle was monitored and kept below 30 nA by limiting the voltage. For most of the data the leakage current was less than 5 nA. 10^{-4} torr of Nitrogen or Helium in the vacuum system helped to make sparks very rare under normal operating conditions.

The magnetic field was monitored by an external flux gate and three optically pumped Rubidium magnetometers which are placed inside the innerst shield close to the vacuum vessel. The magnetometers are used as a control for field drifts or gradients. The oscillator frequency is corrected from time to time automatically if the four alternating measurement points indicated on Fig. 4 show an asymmetry.

The minimum uncertainty in the measured EDM is $\sigma = \hbar/(2\alpha.E.T.\ \sqrt{N})$ where α represents the polarisation efficiency, T the UCN confinement time and N the total number of neutrons counted. The useful running time during three years corresponded to $N > 2*10^9$ UCN.

The data analysis has been performed by three different coworkers in different countries and with different algorithms. The procedure is described in more detail in the most recent publication of the EDM group[8]. Combining the result from the neutron resonances $-(1.9\pm2.2)*10^{-26}$ e.cm (with $\chi^2 = 3.1$ per degree of freedom) with the spurious EDMs obtained from the magnetometers (3±3), (1±1) and $(1\pm2)\text{x}10^{-26}$ e.cm which contain information on the magentic field stability, the authors conclude that the best estimate of the EDM of the neutron is:

$$d_n = -(3\pm5)\ \text{x}\ 10^{-25}\ \text{e.cm.}$$

This value is interpreted as an upper limit:

$$|d_n| < 1.2 * 10^{-25}\ \text{e.cm with 95\% confidence level.}$$

Since counting statistics contribute the minor part to the total error, special care had to be taken to check the effect of possible spurious error sources. An artificial EDM can be generated by magnetic fields correlated with the direction of the electric field. Those perturbations can be induced either by fields external to the magnetic shield or by internal fields created for instance by leakage currents. The data with E = 0 were analysed separately as well as the influence of the periodically reversed Bo field, of jumps in the external magnetic field and of sparks. The correlations between neutron and magnetometer data was studied. A detectable EDM constribution due to a vxE interaction has been excluded in the existing configuration. The detailed procedure and conclusions are presented in Ref. 8. The present experiment has reached its limits. An improved apparatus is under construction. The main modifications are a 10 times larger UCN storage volume to obtain better statistics and probably a longer storage time and most of all the replacement of the localized magnetometers by a new version which averages the real field inside the whole storage volume in the same way the neutrons do (except for a gravitational correction). The method is based on work by Lamoraux et al.[9] and uses polarized ^{199}Hg atoms. It is hoped to increase the overall sensitivity of the experiment by a factor of 5 within the next two years.

The Experiment at LNPI

The experiment at the Leningrad Nuclear Physics Institute in Gatchina has achieved its last, intermediate result in 1986 and is since under reconstruction for improvements. Various problems connected with the safety and operation of the reactor and the integrated cold source made in addition the continuation of the experimental program impossible. For these reasons we describe the LNPI experiment only shortly, with the intention to show the main differences between the two experiments as they appeared in 1986.

The project is performed with UCN at a medium power research reactor. A cold source, 1 liter of liquid hydrogen situated in the center of the reactor core, produces a measured UCN flux of 6×10^3 $cm^{-2}s^{-1}$ in the velocity range up to 7.8 $m.s^{-1}$. This flux is roughly 5 times lower than the one available at the ILL in the same velocity range.

A previous version of the experiment followed a flow through method where polarized UCN entered and left the active volume through holes without shutters. The average residence time in the magnetic and electric field region was 5 s. The result $d_n = -(2 \pm 1) \times 10^{-25}$ e.cm or $|d_n| < 4 \times 10^{-25}$ e.cm at 95% confidence level was obtained[10].

The more recent version of the experiment [11] profitted not only from a new UCN source but also from the neutron storage mode leading to a narrower resonance linewidth. A schematic view of the experiment is presented on Fig. 5. The storage cycle consisted of a 30 s filling period of polarized UCN, (50-55)s storage and a 30 s UCN emptying-counting period. 1.2 s long oscillatory field pulses were applied at the end of the filling and confinement times in order to start and stop the neutron spin precession. The direction of the superposed electric field (E = 12-15 $KV.cm^{-1}$) is changed every 40 s. Leakage currents are kept below 30 nA.

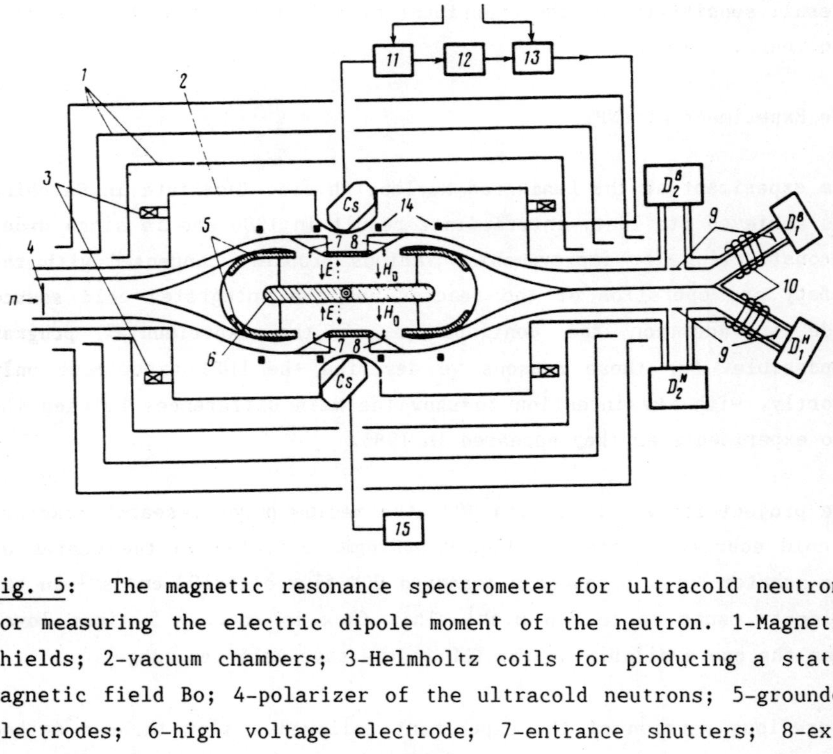

Fig. 5: The magnetic resonance spectrometer for ultracold neutrons for measuring the electric dipole moment of the neutron. 1-Magnetic shields; 2-vacuum chambers; 3-Helmholtz coils for producing a static magnetic field Bo; 4-polarizer of the ultracold neutrons; 5-grounded electrodes; 6-high voltage electrode; 7-entrance shutters; 8-exit shutters; 9-analysers; 10-rf flipper; 11-computer-controlled Cs magnetometer; 12-frequency divider; 13-computer-controlled unit for producing the rf pulses; 14-coils for producing an oscillating field; 15-Cs magnetometer; D-ultracold neutron detectors. Fig. from Ref. 11.

The main difference of the described experiment and the ILL set up is a larger, double confinement chamber for elimination of field drifts to a first approximation. The advantage of this feature is compensated in some sense by the fact that external fields are shielded less perfectly by only 3 layers of mu metal. The set up is shown on Fig. 4. The high voltage electrode is connected to the central plate separating the two confinement volumes. The outer electrodes are at ground potential. This way the electric field has opposite directions, whereas the magntetic field is the same in both chambers except for higher order terms. The bottle wall material is BeO and Be_3N_2 deposited on quartz. The total number of neutrons counted per cycle is about the same as for the ILL experiment.

Optically pumped Cesium magnetometers are used to control the magnetic field close to the confinement volume. A feed back system servos a coil around the storage volume thus stabilizing the magnetic field in the storage chambers to better than 2×10^{-12}T within 10 min.

Four detectors count simultaneously the UCN from both storage chambers and both spin components which are separated by individual spin analysers and spin flipping coils. Various combinations of these individual countrates permit some control of possible systematic errors or asymmetries of the arrangement.

The published result of this experiment including the earlier measurements is [11]:

$$d_n = -(1.4 \pm 0.6) \times 10^{-25} \text{ e.cm}$$

which is interpreted as an upper limit at 95% C.L.

$$|d_n| < 2.6 \times 10^{-25} \text{ e.cm.}$$

As far as systematic errors are concerned the same general arguments are true for both experiments with the exception that in the LNPI experiment the two layer storage chambers might see different field gradients leading to an error as one substracts the signal from both chambers.

The LNPI experiment should to our knowledge now be in a position to reach higher sensitivity as soon as the reactor is operational.

References

1 Christensen J.H., Cronin J.W., Fitch V.L. and Turlay R.: Phys. Rev. Lett. 13 (1964) 138

2 Burkhard H. et al., Phys. Lett. B206 (1988) 169

3 Purcell E.M. and Ramsey N.F.: Phys. Rev. 78 (1950) 807

4 Smith J.H., Purcell E.M. and Ramsey N.F., Phys. Rev. 108 (1957) 120 and Ramsey N.F., Ann. Rev. Nucl. Part. Sci. 32 (1982) 211

5 Pendlebury J.M., Proceedings of the 9th Workshop on Grand Unification (1988) 69, Ed. Barloutaud A., World Scientific

6 Steyerl A. et al., Phys. Lett. A16A (1986) 347

7 Pendlebury J.M. et al., Phys. Lett. 136B (1984) 202

8 Smith K.F. et al., Phys. Lett. A, in print

9 Lamoreaux S.K. et al., Phys. Rev. Lett. 59 (1987) 2275

10 Altarev I.S. et al., Phys. Lett. 102B (1981) 269

11 Altarev I.S. et al., JETP Lett. 44 (1986) 460

CHIRAL SYMMETRY AND CHARGE SYMMETRY

Ernest M. Henley
Institute for Nuclear Theory, Department of Physics, FM-15
University of Washington, Seattle, Washington 98195

ABSTRACT

The interconnection of charge symmetry and chiral symmetry breaking are discussed. Some consequences of a partial restoration of chiral symmetry in nuclei is brought out by means of the Nambu-Jona-Lasinio model.

It is well known that both charge independence and the weaker charge symmetry are approximately respected in hadronic interactions. Their breakdown is connected to both the electromagnetic interaction and to QCD. The QCD spontaneous violation of the symmetries is due to the finite mass and mass difference of the up and down quarks.

Whereas charge independence reflects full rotational symmetry in isospin space, charge symmetry only requires invariance under reflection about the plane perpendicular to the charge axis.[1] There follow various symmetries which can and have been tested. One of the most sensitive ones is the comparison of low energy scattering parameters for nn and pp because of a quasi-resonance near zero energy. For a comparison of the scattering lengths, long-range electromagnetic corrections (Coulomb and magnetic) must be applied. The best measurement of the scattering length for the nn system has been obtained from the analysis of the reaction

$$\pi^- \, d \to nn\gamma \,. \tag{1}$$

It gives[2]

$$a_{nn} = -18.6 \pm 0.4 \text{ fm} \tag{2a}$$

versus[3]

$$\text{``}a_{pp}\text{''} = -17.3 \pm 0.3 \text{ fm} \tag{2b}$$

after charge and magnetic effects have been "removed." The uncertainty in (2b) is primarily theoretical due to our lack of knowledge of the short-range nuclear forces. Other measurements of a_{nn} are possible. An example is[4]

$$\bar{\nu}\, d \rightarrow nn\, \mu^{\dagger}\ . \tag{3}$$

The results given in Eqs. (2) indicate a small breakdown of charge symmetry, with a somewhat more attractive *nn* than *pp* force. Effects of charge symmetry breaking can also be measured in the *np* system, where they cause mixing of I=0 and 1 states.[1] This mixing breaks the symmetries[1][5-6]

$$P_n(\theta) = P_p\,(\pi - \theta) \tag{4}$$

for the polarization (P) or asymmetry (a) in elastic *n-p* scattering; another test is[1][7]

$$\frac{d\sigma}{d\Omega}\,(\theta) = \frac{d\sigma}{d\Omega}\,(\pi - \theta) \tag{5}$$

in $np \rightarrow d\pi^0$, where θ is the angle of emission of the π^0. The first of these symmetries has been shown to be broken at TRIUMF at 477 MeV, with[8] $\frac{a_{nn}-a_{pp}}{a_{nn}+a_{pp}} = (3.7 \pm 2.4) \times 10^{-3}$, by determining the difference of the angle at which the asymmetry from polarized protons and neutrons changes sign. The difference in analyzing power as a function of angle is now being measured at 183 MeV at the Indiana University Cyclotron.[9] At present these are the only measurements that show non-vanishing charge-symmetry breaking in the *n-p* system.

Charge symmetry is also tested by comparing the binding energies of mirror nuclei after correcting for long-range (Coulomb and magnetic) electromagnetic effects. A well-known example is the comparison of ^{3}He and ^{3}H, where the binding energy difference[10] is 764 keV, whereas the long-range electromagnetic effects account for only[11] $\sim$ 680 keV. The remaining difference indicates a somewhat more attractive *nn* than *pp* force, in accord with Eqs. (2). The discrepancy between the calculated and measured mass differences grows with A and is known as the Nolen-Schiffer effect.[12]

Theoretically, the breaking of charge symmetry in nuclear forces occurs primarily from the *n-p* mass difference and ρ-ω, π-η-η' mixing.[1] Secondary effects occur due to meson-photon and two meson exchanges as well as in coupling constant differences.[1][13] The comparison of the *nn* and *pp* systems is sensitive to class III forces, proportional to $[\tau_3(1) + \tau_3(2)]$, and charge symmetry-breaking in the *n-p* system depends on class IV forces, proportional to $[\tau_3(1) - \tau_3(2)]$ or $[\vec{\tau}(1) \times \vec{\tau}(2)]_3$. The n-p mass difference and the meson-mixing are known to be largely due to the mass difference between up and down quarks. This mass difference for current quarks is assumed to be caused by a Higgs mechanism, since the QCD Lagrangian is chirally symmetric with zero mass quarks.

Chiral symmetry implies invariance under left-right (γ_5) transformations. I will restrict myself to SU(2) and up (u) and down (d) quarks. For a two-component Fermion field ψ, the invariance holds for the general phase transformation

$$\psi \to e^{i\vec{\alpha}\cdot\vec{\tau}\gamma_5}\psi \,, \qquad \bar{\psi} \to \bar{\psi}\, e^{i\vec{\alpha}\cdot\vec{\tau}\gamma_5} \,, \tag{6}$$

where α_i is an arbitrary phase angle and τ_i is a Pauli operator in isospin space.

For low energy nuclear physics, chiral symmetry is only an approximate symmetry. It is the interplay of charge symmetry and chiral symmetry that I want to focus on in this talk. Much of it was carried out with G. Krein[14] and some with H. Muether[15] and R-K. Su.[16]

Calculations of QCD on the lattice and other approximations suggest that chiral symmetry is restored at high temperatures and high densities.[17] Since nuclei, even in their ground states, are at finite densities, a partial restoration of chiral symmetry (PRCS) can be expected. We have studied some of these effects and I will concentrate on one of them here, related to charge symmetry and the Nolen-Schiffer effect.

The model developed by Nambu and Jona-Lasinio (NJL)[18] is a useful tractable one for examining chiral symmetry, its breaking, and its restoration. It is based on an analogy to superconductivity.[19] The energy gap is interpreted as giving rise to a finite mass to initially massless particles. In analogy to superconductivity, collective modes of excitation occur. The model thus has some features of QCD.

The Lagrangian density of the NJL model is

$$\mathcal{L} = \bar{\psi}(i\nabla\!\!\!/ - m)\psi + G\left[(\bar{\psi}\psi)^2 + (\bar{\psi}\, i\gamma_5\tau\psi)^2\right]. \tag{7}$$

It is chirally invariant if the $2\ x\ 2$ mass matrix, m

$$m = \begin{pmatrix} m_u & 0 \\ 0 & m_d \end{pmatrix} \tag{8}$$

vanishes. The interaction term in Eq. (7) can be Fierz transformed to include vector and axial vector coupling[15)18)20)], but I will restrict myself to Eq. (7).

The Lagrangian density of Eq. (7) leads to divergent results; thus a cut-off, Λ, is needed. This cut-off can be introduced covariantly or only in the three-momentum. Since the concept of density is not covariant, we use the non-covariant method because it allows one to gain additional insights more readily. In some sense, the cut-off can be viewed as representing the boundary between non-perturbative QCD and asymptotic freedom; i.e. $G \to 0$ for momenta $p > \Lambda$.

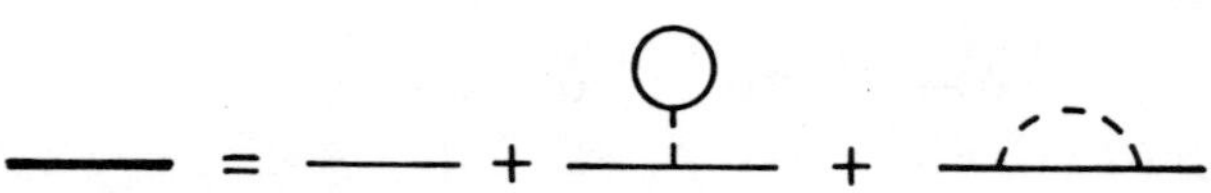

Fig. 1. Hartree-Fock approximation.

The Lagrangian density has a trivial solution and a non-trivial one computed in a self-consistent Hartree-Fock or Dyson approximation, represented by Fig. 1,

$$S_F^{-1} = (\not p - m + i\varepsilon) \to (\not p - m - \Sigma + i\varepsilon) = (\not p - M + i\varepsilon)\,, \tag{9}$$

with $M = m + \Sigma$,

$$\Sigma = -\frac{8i\,G\,n_f\,n_c}{(2\pi)^4}\int \frac{M}{p^2 - M^2 + i\varepsilon} d^4p\,, \tag{10}$$

$$M - m = -\frac{8i\,G\,M}{(2\pi)^4} n_f\,n_c \int \frac{d^4p}{p^2 - M^2 + i\varepsilon} = \frac{2G\,n_f\,n_c\,M}{\pi^2}\int_0^\Lambda \frac{p^2\,dp}{E}\,. \tag{11}$$

In Eqs. (10) and (11) n_f is the number of flavors, $n_f = 2$, and n_c the number of colors, $n_c = 3$. Eq. (11) is a nonlinear equation; it has both a trivial solution and a nontrivial one for G larger than a critical value For $m = 0$, the solution is sketched in Fig. 2 and is given by[18)]

$$\frac{\pi^2}{2G\,n_f n_c \Lambda^2} = \sqrt{\frac{M^2}{\Lambda^2} + 1} - \frac{M^2}{\Lambda^2}\ln\left(\sqrt{1 + \frac{\Lambda^2}{M^2}} + \frac{\Lambda}{M}\right).$$

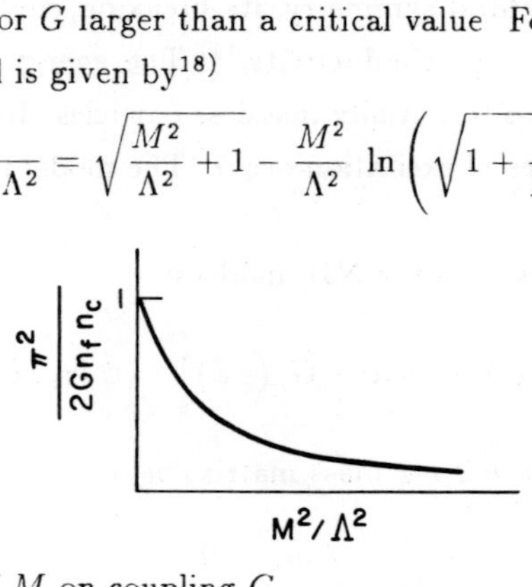

Fig. 2. Dependence of M on coupling G.

As for superconductivity, collective states are also to be expected. If $m = 0$, one of these states will be the Goldstone boson, identified here as the pion, and another one will be a massive scalar boson, here called a σ meson. These states are solutions of the Bethe-Salpeter equation in the ladder approximation, sketched in Fig. 3,

$$\Phi = <0|T(\psi(x)\bar{\psi}(y)|B> = S_F(p+\frac{1}{2}k)\,\Gamma(p+\frac{1}{2}k,p-\frac{1}{2}k)S_F(p-\frac{1}{2}k)\,. \qquad (12)$$

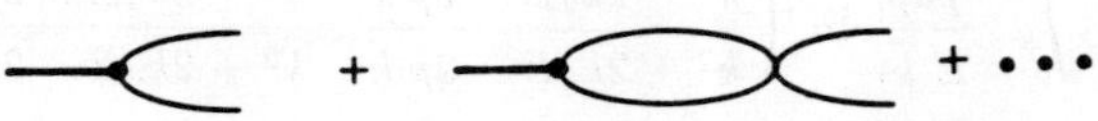

Fig. 3. Bethe-Salpeter approximation for collective state.

For a scalar Γ = constant, we have

$$_{out}<p_1'p_2'|p_1p_2>_{in} \equiv \Gamma_s = -2iG\,\bar{u}(p_1')u(p_1)\frac{1}{1-J_S(k^2)}\bar{u}(p_2')u(p_2)\,, \qquad (13)$$

where J_S is the basic quark loop,

$$J_S = 2iGn_fn_c\int\frac{d^4p}{(2\pi)^4}\,Tr[S_F(p+k/2)\,1\,S_F(p-k/2)]\,, \qquad (14)$$

with

$$S_F = (\not{p} - M + i\varepsilon)^{-1}\,. \qquad (15)$$

There is a pole in Eq. (13) when $J_s(k^2) = 1$; this pole can be viewed as the exchange of a scalar σ meson, as shown in Fig. 4

$$J_S - 1 = -\frac{m}{M} + (k^2 - 4M^2)\,I(k^2)\,. \qquad (16)$$

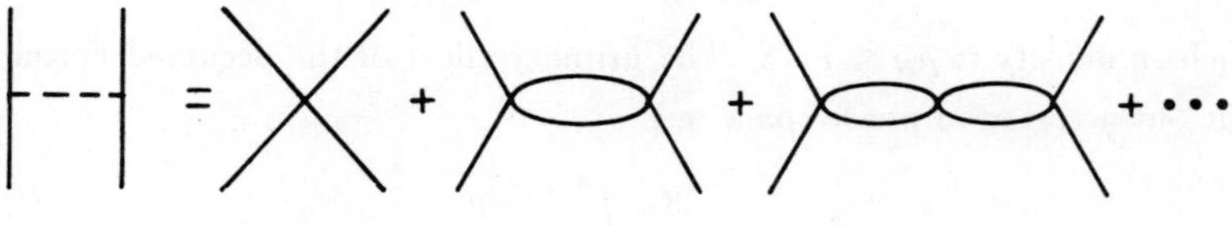

Fig. 4. Relationship of pole of renormalized propagator and meson exchange.

The pole occurs at $k = 2M$ if $m = 0$ and at a slightly higher mass when $m \neq 0$. For the pseudoscalar J_P, we have

$$J_P = 2i\, G\, n_f n_c \int \frac{d^4p}{(2\pi)^4}\, Tr\left[i\gamma_5\, \vec{\tau} S_F(p + \frac{1}{2}k) i\gamma_5\, \vec{\tau}\, S_F(p - \frac{1}{2}k)\right] , \tag{17}$$

$$J_P - 1 = -\frac{m}{M} + k^2\, I(k^2)\, . \tag{18}$$

The pole of J_P occurs at $k^2 = 0$ if $m = 0$. The integral $I(k^2)$ is given by

$$I = -\frac{G n_f n_c}{4\pi^2} \int_0^\Lambda \frac{pdp}{E|\vec{k}|} \ln\left(\frac{k^2 - 2k_0E + 2p|\vec{k}|}{k^2 - 2k_0E - 2p|\vec{k}|} \cdot \frac{k^2 + 2k_0E + 2p|\vec{k}|}{k^2 + 2k_0E - 2p|\vec{k}|}\right) . \tag{19}$$

For the applications I want to consider in this talk, finite current quark masses (presumably from a Higgs mechanism) are taken and are determined from strange meson mass differences and baryon decays,[21]

$$m_u = 4 \text{ MeV}, \;\; m_d = 10.4 \text{ MeV}\; . \tag{20}$$

First, we examine the effect of this mass difference on the constituent quark masses and the neutron-proton mass difference in free space and in a nuclear medium. Solutions of the NJL equation couple M_u and M_d,

$$M_u = m_u + 3\, M_u\, I(M_u) + \frac{7}{2}\, M_d\, I(M_d)\, .$$

$$M_d = m_d + 3\, M_d\, I(M_d) + \frac{7}{2}\, M_u\, I(M_u)\, , \tag{21}$$

$$M = \overline{M} + \frac{1}{2}\, \delta M\, \tau_3\, ,\;\; \overline{M} = \frac{1}{2}\,(M_u + M_d)\, ,\;\; \delta M = M_u - M_d\, , \tag{22}$$

$$I(M) = \frac{2G}{\pi^2}\int_0^\Lambda \frac{p^2}{E} dp\, . \tag{23}$$

At finite density ρ (or temperature T), PRCS is anticipated and occurs. We assume a Fermi gas with momentum p_F given by

$$\rho_q = \frac{1}{3\pi^2}\, n_c n_f k_F^3 = \frac{2k_F^3}{\pi^2}\; . \tag{24}$$

The nucleon density is $\rho_N \approx \rho_q/3$. The primary effect of the occupied Fermi gas is to cut out a region of phase space e.g .

$$I(M) = \frac{2G}{\pi^2}\int_{k_F}^\Lambda \frac{p^2 dp}{E}\, . \tag{25}$$

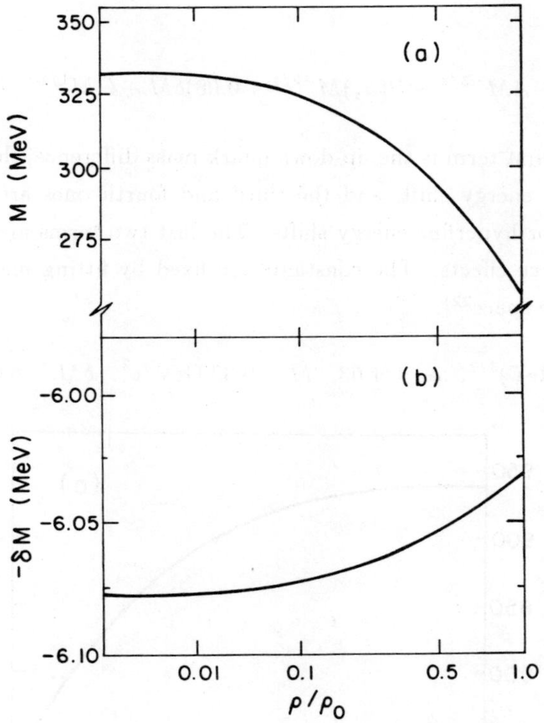

Fig. 5. Dependence of average constituent quark mass $\overline{M}$ and $\delta M = M_u - M_d$ on density ρ.

This equation makes it clear that as $k_F \rightarrow \Lambda$, chiral symmetry is restored in that $M \rightarrow m$. Fig. 5 shows the dependence of $\overline{M}$ and δM on ρ. It is of particular note that $\delta M \approx \delta m$ and hardly changes with ρ, whereas $\overline{M}(\rho_0)/\overline{M}(0) \approx 0.75$, where ρ_0 corresponds to nuclear matter density ≈ 0.17 nucleons/ fm^3. Of course, the constituent quark mass cannot be measured; however, there are numerous successful constituent quark models of the nucleon. We can therefore investigate the PRCS by means of the dependence of the average nucleon mass and the n-p mass difference on ρ. We use the model of Isgur and Karl[22)], which has been successful in fitting numerous baryon and meson masses and mass differences on the basis of constituent quarks interacting via a confining potential and with a color hyperfine interaction. In this model, the quark mass effects enter directly, in the kinetic energy, zero point motion. color hyperfine energy, and electromagnetic energies,

$$\Delta M_{pn} = (-1 + AM^{-3/2} + B(\alpha_s)M^{-9/4} + 0.08)\delta M + CM^{1/4} - DM^{-5/4}. \quad (26)$$

In Eq. (26), the first term is the up-down quark mass difference, the second term is the zero-point energy shift, and the third and fourth ones are the first and second order color hyperfine energy shifts. The last two terms are the Coulomb and magnetic force effects. The constants are fixed by fitting masses and mass differences in free space[22],

$$A = (199.4\ \text{MeV})^{3/2},\ \ \alpha_s = 1.63,\ \ M = 0.33\ \text{GeV}/\text{c}^2,\ \ \delta M = 6.6\ \text{MeV}\ . \quad (27)$$

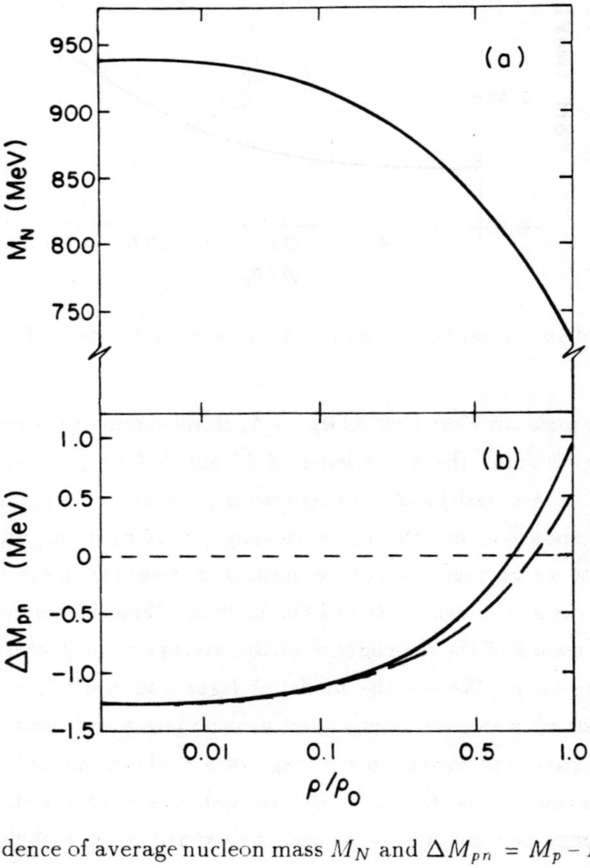

Fig. 6. Dependence of average nucleon mass M_N and $\Delta M_{pn} = M_p - M_n$ on density ρ.

Fig. 6 shows M_N and ΔM_{pn} vs ρ_N. There is a significant effect of PRCS on both M_N and ΔM_{pn}. The solid curve is for a constant α_s with ρ. The long-dashed curve shows the effect of a linear decrease of $\alpha_s = \alpha_s(0)(1 - \rho/5\rho_0)$ with density; such a decrease could occur due to asymptotic freedom. It is interesting to note that ΔM_{pn} changes sign with increasing density. The mass difference dependence ΔM_{pn} is in the right direction and larger than needed to explain the Nolen-Schiffer anomaly

$$\Delta M_{mirror} \equiv M(Z,N) - M(Z' = N, N' = Z) = \Delta M_{\text{“}em\text{''}} + \Delta M_{pn} \ . \tag{28}$$

The anomaly is that ΔM_{mirror} (*theor.*) $<$ ΔM_{mirror} (*exper.*) for all nuclei, e.g. ^{3}He, ^{3}H. As ΔM_{pn} increases, the Nolen-Schiffer discrepancy decreases. The NJL corrections for the mirror nuclei are shown below, together with the average nuclear density.

Nuclei	ρ_{av}(fm^{-3})	Discrep (MeV)	ΔM_{pn} at $\bar{\rho}$	Discrep with NJL
^{17}F-^{17}O	$\sim$ 0.09	0.3	-0.4	-0.6
^{41}Sc-^{41}Ca	$\sim$ 0.10	0.7	-0.12	-0.5

The relationship between ρ_q and ρ_N and our use of $\bar{\rho}$ are only approximate. The overestimate is probably due to these approximations, our neglect of kinetic energy effects, but, perhaps, above all, the lack of confinement in the NJL model. I will argue later that the inclusion of confinement slows the PRCS.

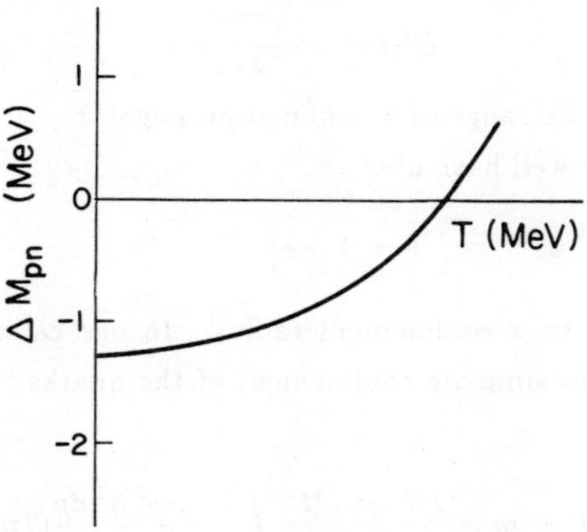

Fig. 7. Dependence of ΔM_{pn} on temperature T at zero density.

The charge symmetry-breaking effects have been examined as a function of temperature with Ru-keng Su. The effects parallel those with increasing ρ and are of interest for astrophysical applications; we are looking into them. The temperature effects on the average nucleonic mass and on ΔM_{pn} are shown in Fig. 7.

Even though the NJL model incorporates some of the effects of the long range QCD interaction, the quarks in it are not confined. There is a crude way of showing that confinement is likely to slow the restoration of chiral symmetry and thus bring the calculated Nolen-Schiffer correction into better agreement with experiment.

Consider the long range propagator of a gluon

$$\begin{aligned} G(q^2) &= \frac{1}{q^2+i\varepsilon} \to -\frac{1}{\vec{q}^{\,2}-i\varepsilon} \quad \text{if } q_0 = 0 \\ G(r) &= -1/4\pi r \ . \end{aligned} \tag{29}$$

If we introduce a soft cut-off at small momenta, e.g.,

$$G \to G' = G\,F \ , \tag{30}$$

with

$$F(q) = 1 - \frac{1}{q^2 r_0^2 + 1} \ ,$$

$$F(0) = 0 \ , \tag{31a}$$

$$F(\infty) = 1 \ , \tag{31b}$$

then

$$G'(r) = -\frac{e^{-r/r_0}}{4\pi r} \ , \tag{32}$$

corresponding to a finite range or a confined propagator.

We could equally well have used

$$F = 1 - e^{-\alpha^2 q^2} \ , \tag{33}$$

where α corresponds to a confinement radius. In our calculation, we use this low momentum c.o. to simulate confinement of the quarks.[15] The change in the equation for M is

$$M \to M' = m + \frac{2Gn_f n_c M'}{\pi^2} \int_0^{\Lambda} \frac{F(p^2)p^2\,dp}{E} \theta(p - k_F) \ . \tag{34}$$

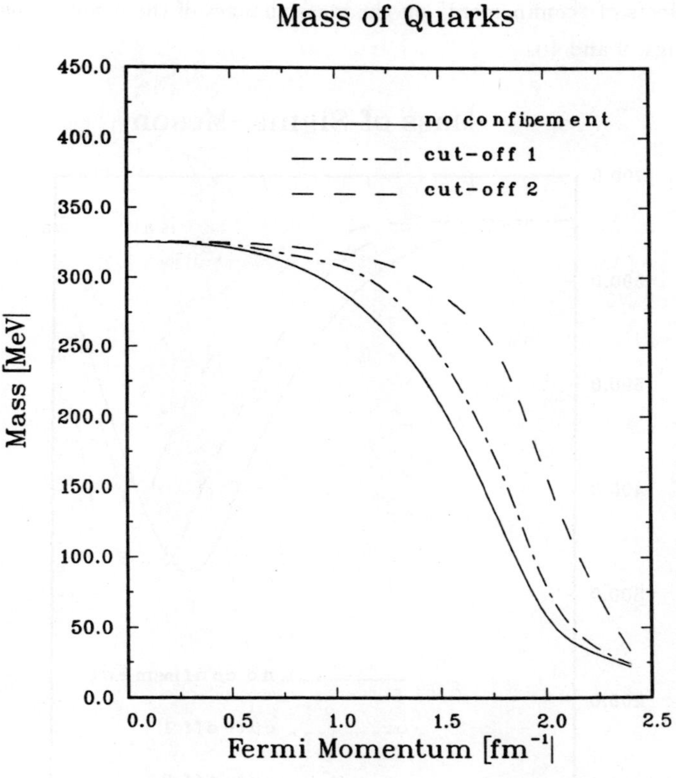

Fig. 8. Dependence of constituent quark mass on density for various confinement radii or regions. Nuclear matter density corresponds to $k_F \approx 1.35\ \mathrm{fm}^{-1}$.

Similar modifications occur for J_S and J_P, where we also simply introduce a low momentum c.o. . The density effect in a medium is small if $k_F \lesssim \alpha^{-1}$ because the lower limit of the integral is controlled by confinement rather than the Fermi momentum. It is only at high density when $k_F \gtrsim \alpha^{-1}$ that the PRCS becomes important. The net effect is therefore to slow down the PRCS with increasing density as shown in Fig. 8. For instance, we have

$$\begin{aligned}
\text{no confinement} : M^*/M &= 0.76 \text{ for } \alpha^2 = \infty \quad (r_0 = \infty) \\
\text{cut off 1} : M^*/M &= 0.85 \text{ for } \alpha^2 = 2 \times 10^{-5} (\mathrm{MeV/c})^2 \quad (r_0 \approx 1.8\ \mathrm{fm}) \\
\text{cut off 2} : M^*/M &= 0.94 \text{ for } \alpha^2 = 0.67 \times 10^{-6} (\mathrm{MeV/c})^{-2} \quad (r_0 \approx 0.32\ \mathrm{fm})\,.
\end{aligned} \tag{35}$$

The effects of "confinement" on the mass changes of the σ and π mesons are shown in Figs. 9 and 10.

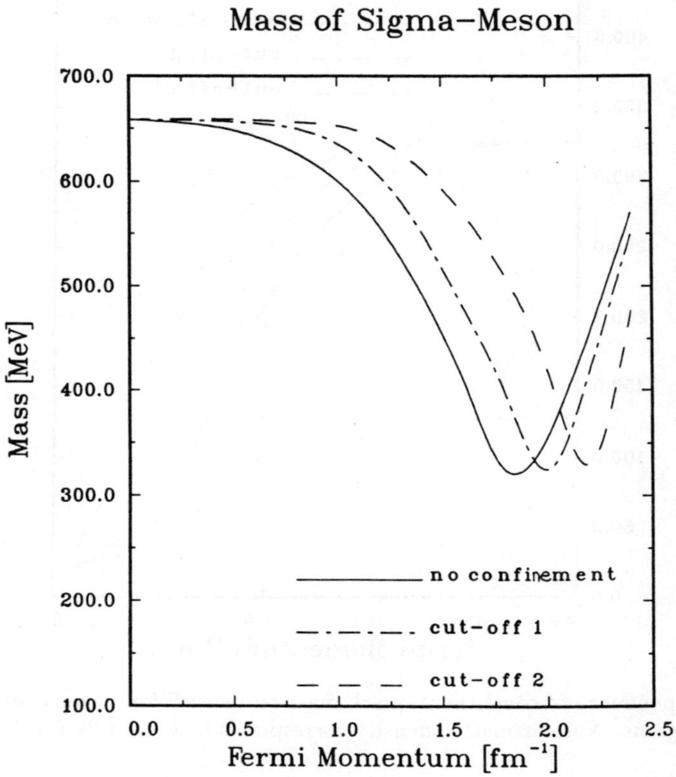

Fig. 9. Dependence of σ mass on density for various confinement radii (see Eq. (35).

I believe that the main effect of confinement is represented by this crude model. The restoration of complete chiral symmetry therefore occurs at a higher density than for an unconfined system. This slowing brings the NJL model effects into closer agreement with the measured Coulomb energy difference between mirror nuclei.

In conclusion, QCD connects chiral and charge symmetries. The NJL model allows us to investigate this connection in some detail. The PRCS in nuclei is in the right direction to help explain the Nolen-Schiffer anomaly. Although the NJL model may not be correct quantitatively, it is likely that QCD will lead to similar qualitative conclusions.

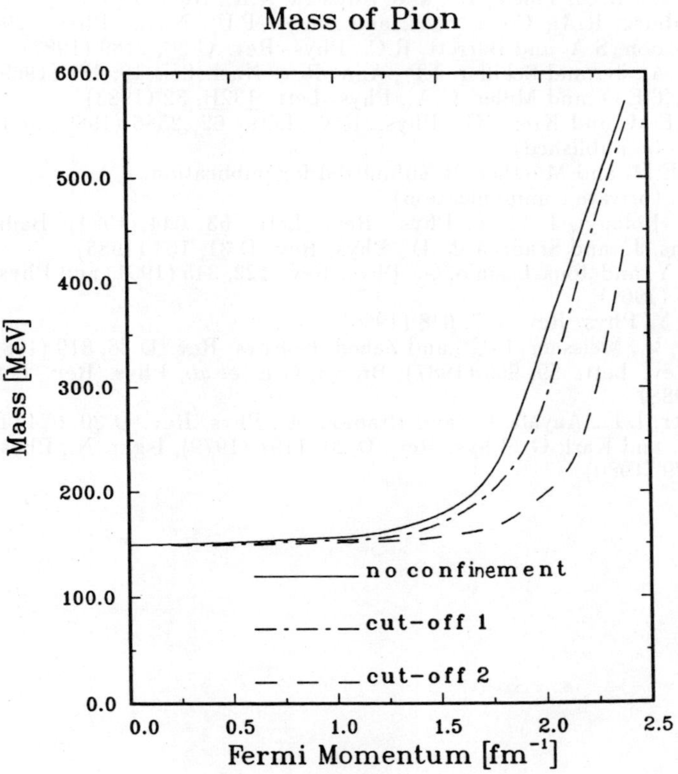

Fig. 10. Same as Fig. 9 for pion mass.

REFERENCES

1. Henley, E.M., in *Isospin in Nuclear Physics*, ed. M Rho and D.H. Wilkinson (North Holland, Amsterdam, 1969),p 15.
2. Schori, O. *et al.*, Phys. Rev. C 35, 2252 (1987).
3. Miller, G.A., Nefkens, B.M.K and Slaus, I., (to be published in Phys. Rpts.
4. Ying, S., Haxton, W., and Henley, E.M., Phys. Rev. D (to be published).
5. Cheung, C.Y., Henley, E.M., and Miller, G.A., Nucl. Phys. A305, 342 (1978); A348, 365 (1980); Gersten, A., Phys. Rev. C 24, 2174 (1981).
6. Williams, A.G., Thomas, A.W., and Miller, G.A., Phys. Rev. C 34, 756 (1987); Holzenkamp, B.H., Holinde, K., and Thomas, A.W., Phys. Lett. B195, 121 (1987).
7. Cheung, C.Y., Henley, E.M., and Miller, G.A., Nucl. Phys. A348, 365 (1980).
8. Abegg, R., *et al.*, Phys. Rev. D 39, 2464 (1989); Phys. Rev. Lett. 56, 2571 (1986).

9. Knutson, L., in *Few Body XII*, Vancouver, Canada (July 2-8, 1989) (to be published.
10. Mattauch, J.H.E., Thiele, W., and Wapstra, A.H., Nucl. Phys. 67, 1 (1965).
11. Brandenburg, R.A., Coon, S.A., and Sauer, P.U., Nucl. Phys. 294, 305 (1978); Coon, S.A. and Barrett, R.C., Phys. Rev. C 36, 2189 (1987).
12. Nolen, J.A., Jr. and Schiffer, J.P., Ann. Rev. Nucl. Sci. 19, 471 (1969).
13. Ericson, T.E.O. and Miller, G.A., Phys. Lett. 132B, 32 (1983).
14. Henley, E.M. and Krein, G., Phys. Rev. Lett. 62, 2586 (1989) and Phys. Lett. (to be published).
15. Henley, E.M. and Muether, H. submitted for publication.
16. Su, R-K. (private communication).
17. See e.g. Polonyi, J. *et al.*, Phys. Rev. Lett. 53, 644 (1984); Bailin, D., Cleymans, J., and Scadron, M.D., Phys. Rev. D 31, 164 (1985).
18. Nambu, Y. and Jona-Lasinio, G., Phys. Rev. 122, 345 (1961) and Phys. Rev. 124, 246 (1961).
19. Nambu, Y. Phys. Rev. 117, 648 (1960).
20. Bernard, V., Meissner, U-G., and Zahed, I., Phys. Rev. D 36, 819 (1987) and Phys. Rev. Lett. 59, 966 (1987); Brown, G.E. *et al.*, Phys. Rev. Lett. 60, 2723 (1988).
21. Ameteller, L.E., Aayala, C., and Bramon, A., Phys. Rev. D 30, 674 (1984).
22. Isgur, N. and Karl, G., Phys. Rev. D 20, 1191 (1979); Isgur, N., Phys. Rev. D 21, 779 (1980).

DISCRETE SYMMETRIES IN PARTICLE PHYSICS

L. Wolfenstein
Carnegie Mellon University
Pittsburgh, PA 15213

ABSTRACT

The role of the discrete symmetries P, T, and C in particle physics is reviewed. The present experimental and theoretical status of CP violation is summarized.

Three discrete symmetries have played a major role in particle physics in the last forty years

P Space reflection, or parity

T Time reversal

C Particle - antiparticle conjugation

These symmetries arose not as theoretical postulates nor as experimental discoveries but rather as theoretical discoveries. Thus all three of these symmetries were discovered to hold for quantum electrodynamics (QED), the quantum version of the Maxwell-Einstein theory of electrons and electromagnetic radiation.

As an example I will discuss my own introduction to the C symmetry. It had to do with the positronium atom whose ground states are ^1So and $^3S_1$. As you know, the ^1So states decays to two gammas but this decay is forbidden by rotational invariances for $^3S_1$, which decays into 3γ. In 1952 my colleague Sergio de Benedetti and his student Bob Siegel were experimentally studying this 3γ decay. A student, not mine, was asked to calculate the rate for the ^1So to decay into 3γ. He diligently calculated all the lowest order graphs, summed them, and after some weeks returned with the answer zero. This led me and Geoff Ravenhall[1] to look for a reason and we soon uncovered that this was an absolute selection rule following from the C invariance of QED. The ^1So state has C eigenvalue of +1 while each photon is C odd. While the concept of C was well known[2] the idea of using it as an invariance principle satisfied by the QED Hamiltonian was not. The point I want to emphasize is that the dynamical theory, QED, was the primary postulate; the symmetries were discovered later as being there.

There is one other interesting feature of the paper I wrote with Ravenhall. At the end of the paper we discuss "extension to neutral fields." We consider hypothetical neutral bosons with strong interactions and discuss the extension of the C symmetry to them. For example, if the neutral boson B^o is even under C then the decay $B^o \rightarrow \pi^o + \pi^o + \gamma$ is forbidden. We do not say "let us postulate that the strong and electromagnetic interactions of hadrons are invariant under C." Rather we say assume that we know the interaction Hamiltonian between B^o and some charged particles then we can figure out whether B^o is even or odd. We assume only one interaction term (enough for a new hypothetical particle) and so get a definite answer. The dynamics came first, the symmetry was then derived.

Today the search for decays like $^1So \rightarrow 3\gamma$ and C violation in $\eta \rightarrow \pi + \pi + \gamma$ are high on the list of "Tests of Discrete Symmetries" in the Particle Data Group listings.[3]

The change in attitude dates, of course, from the parity revolution of 1957. Before 1957 the issue in weak interaction physics was to choose the right combination of V, A, S, and T, a vast set of possibilities. A crucial experiment, which proved very difficult, was the measurement of the electron-neutrino correlation. The 1956 paper of Lee and Yang[4] had two main aspects. The first, of course, was that it pointed out that in spite of more than 25 years of beta-decay experiments there was no evidence relative to parity violation and that large parity violation was perfectly possible. The second was that, if parity were violated, a whole new set of experiments became possible, many of which proved easier than the electron-neutrino correlation. Within a few months the new experiments pointed to the universal V-A theory, put forward by Marshak and Sudarshan, Feynman and Gell-Mann and others. There were experiments, however, that did not agree with the V-A theory. Marshak and Sudarshan report[5] that they hesitated to present their theory until they visited Felix Boehm at Cal Tech, who assured them that those experiments were not to be relied on. Felix's own experiments[6] were in fine agreement with V-A.

The V-A theory involved maximum violation of P and also of C, but CP and T remained as good symmetries. Then in 1964, Fitch, Cronin, Christenson, and Turlay[7] discovered a small amount of CP violation in the decay $K^o \rightarrow 2\pi$. Amazingly, today, CP and T violation have been found only in the K^o system and the only clearcut measure of CP violation is that measured in the original experiment 25 years ago.

Let me briefly review the present status of CP Violation.[8] Five numbers describing CP violation in K^o decay have been measured: $| \eta_{+-} |$, $| \eta_{oo} |$, φ_{+-}, φ_{oo}, and δ where

$$| \eta_{+-} | e^{i\varphi_{+-}} = \frac{A(K_L \rightarrow \pi^+\pi^-)}{A(K_s \rightarrow \pi^+\pi^-)}$$

$$| \eta_{oo} | e^{i\varphi_{oo}} = \frac{A(K_L \rightarrow \pi^o\pi^o)}{A(K_s \rightarrow \pi^o\pi^o)}$$

$$\delta = \frac{\Gamma(K_L \rightarrow \pi^-\ell^+\nu) - \Gamma(K_L \rightarrow \pi^+\ell^-\bar{\nu})}{Sum}$$

Under some very general assumptions (CPT invariance, unitarity, and the

$\Delta Q = \Delta S$ rule) it is possible to reduce these to a good approximation to two real numbers $|\epsilon|$ and (ϵ'/ϵ). One writes

$$\eta_{+-} = \epsilon + \epsilon'$$

$$\eta_{oo} = \epsilon - 2\epsilon'$$

$$\delta = 2Re\ \epsilon$$

with the phases of the complex parameters ϵ and ϵ' given by

$$\phi(\epsilon) \simeq tan^{-1}\left(\frac{2\Delta M}{\Delta\Gamma}\right) = 43.67 \pm 0.14^o$$

$$\phi(\epsilon') = \delta_2 - \delta_o + \pi/2 = 48 \pm 8^o$$

where δ_I are the final state pion-pion phase shifts. Because ϵ'/ϵ is very small the original 1964 experiment already determined the value of $|\epsilon|$ to be 2.3×10^{-3}. Accepting the theoretical result that $\cos[\varphi(\epsilon') - \varphi(\epsilon)]$ is very close to unity the measurement of $|\eta_{oo}/\eta_{+-}|$ determines a real quantity ϵ'/ϵ. We recently have been given two precise experimental results

$$(CERN\ NA\ 31)^9\ \epsilon'/\epsilon = (33 \pm 11) \times 10^{-4}$$

$$(Fermilab\ E731)^{10} \epsilon'/\epsilon = (-5 \pm 15) \times 10^{-4}$$

These are remarkably precise experiments; when final results are available each should have an error $\pm 7 \times 10^{-4}$ corresponding to an error in ϵ' itself of less than 2×10^{-6}. Yet unless the results change we will still have only one confirmed non-zero number measuring CP violation. Given the importance of ϵ'/ϵ further experiments will still be needed.

After the discovery of CP violation a great variety of models of CP violation were discussed. They could be catalogued as millistrong, milliweak, and superweak according to their strength and symmetry properties (Table 1). Millistrong models imply a small CP-violating piece in the strong interactions of hadrons so C and T violation show up in all hadronic processes. CP violation in K^o decay occurs as a hadronic correction to a weak process. The best limits on such models[11] come from photon correlation experiments and from limits on the neutron and other electric dipole moments. Such models are unpopular, in any case, because QCD is the accepted theory of strong interactions and, as discussed by Herczeg and collaborators[12], it is extremely hard to construct such models. At the other extreme, superweak models are easy to construct and lead to the prediction $\epsilon' = 0$. Thus, an important reason to confirm the CERN result for ϵ' would be rule out such models as the sole origin of CP violation. The most popular models are milliweak models, in which there is a CP-violating piece of the usual weak interactions, which has a small ($\sim 10^{-3}$) effect in the K^o system. The Kobayashi-Maskawa (KM) model is the main candidate since it involves no assumption beyond the minimum particle content needed in the $SU(2) \times U(1)$ standard electroweak gauge theory. Its most clearcut prediction is a sizeable value for CP violation in suitable B^o decay experiments.[13]

I have also listed in Table 1 a fourth class of CP violation, irrelevant for K^o decay but important for electric dipole moments of hadrons. This is characterized by P violation but is limited to $\Delta S = 0$. The prototype is the "strong CP violation"[14] which will occur in the Lagrangean of QCD once there is any CP violation in the theory. The present limits on the electric dipole moment of the neutron assure us that this interaction is really very weak. However, if a non-zero value were to be observed it could be blamed on this type of interaction without affecting the discussion of CP violation in the K^o system.

The contrast between the consequences of the discovery of P violation and that of CP violation is striking:

1. P violation was a large effect, CP violation was small.
2. Countless experiments measuring P violation have followed; no really new measures of CP violation have been confirmed.
3. P violation quickly led to a clarification of the theory of weak interactions; CP violation has not affected the development of weak interaction theory.

In 1967 I gave a summary talk on weak interactions at the Heidelberg conference. I pointed out that the V-A theory was consistent with all experiments except that showing CP violation. Thus, it might seem that CP violation would be the key for the next development. In fact the next great step in weak interaction theory was taking place just at the time with the publication of Weinberg's letter on the SU(2) × U(1) theory. Its motivation had nothing to do with the CP problem. In fact the original paper involved only leptons and when it was extended to include two generations of quarks it still was CP-invariant. Weinberg's motivation was the hope that by using a gauge theory there might be a solution to the divergence problems similar to the renormalization schemes for QED. While I mentioned the divergence problems in my 1967 talk there were no pressing experimental results requiring higher-order calculations. The divergence problem was a problem for theorists and its solution led to the great experimental discoveries of 1973-1983.

The SU(2) × U(1) theory seemed to bring us back in a way to where we were when we looked at QED and discovered its symmetries. Once again we had a theory like QED with its interactions prescribed by the gauge principle. While P violation was built in because of the chiral choice for the SU(2), the gauge interaction was automatically CP and T invariant because the coupling constants could always be chosen as real. But, of course, there is more than the gauge interaction; the Lagrangean of the SU(2) × U(1) theory and of any of its extensions has the general form

$$L = L_G(W, \psi,\ \varphi) + H(\varphi) + Y(\psi, \varphi)$$

where W represent the gauge bosons, ψ the fermion fields, and φ the scalar bosons. L_G represents the free field Lagrangian plus the interactions prescribed by the gauge principle. It is CP invariant. The two additional terms involving the Higgs boson fields are required so that the gauge symmetry can be broken. H is the Higgs potential involving only the Higgs fields and Y is the Yukawa interaction between the Higgs fields and the fermion fields, required to give the fermion masses. CP violation can only arise from H and Y. There are three possible ways to do this

1. Complex coefficients in H

2. Complex coefficients in Y

3. L itself is CP invariant but the vacuum is not; that is, there are relative complex factors in the vacuum expectation values of the φ's; this is called spontaneous CP violation.

Thus CP violation is tied to the Higgs sector of the theory, the part of the standard model which we understand and trust the least. One of the reasons CP violation is so important from this point of view is because it is so tied in with the aspect of gauge theories that we are striving to really understand. This may also be the reason progress on this subject has been so slow.

In the standard KM model CP violation enters originally as complex coefficients in Y. After symmetry breaking Y determines the quark mass matrixes and CP violation shows up as the one unavoidable phase in the 3×3 mixing matrix. We can understand the value of $| \epsilon |$ and the results on ϵ'/ϵ (either the positive CERN result or the Fermilab result) using the KM model, particularly since the calculations are of limited accuracy because of the difficulties of calculating hadronic matrix elements.

It is often asked whether the KM model is correct. I would rather put it this way. Since we know CP is violated, unless there is some new symmetry, we expect a CP-violating phase in the quark mixing matrix. We do not know if this phase is large enough to completely explain the values of ϵ and ϵ'. In any case we believe there exists new physics beyond the standard model. CP-violating effects which are small in the standard model may be good places to look for new physics.

What experiments are good ones to do? Relatively large CP violation effects in certain B^o decays are expected, given the large observed $B^o - \bar{B}^o$ mixing, if the KM model explains the observed value of ϵ. These are important but very difficult experiments. On the other hand the KM mechanism by itself predicts a very small value (10^{-31} *to* $10^{-32} e - cm$) for the electric dipole moment of the neutron. Therefore, this experiment is an excellent place to look for new physics.

To be honest, theorists have not provided great insights into the origin of CP and T violation. Experimentalists should not be too persuaded by theoreticians' prejudices or intuition. The next development will probably come where we least expect it.

This work has been supported in part by the Department of Energy under contract number DE-AC02-76ER03066.

Table 1

Model	**P**	**C**	$\mid \Delta S \mid$	**CP Violating Processes**
Millistrong	$\surd$	x	0	All Involving hadrons
Milliweak	$\surd$	x	0,1	Weak
	x	$\surd$		
Superweak	$\surd$	x	2	$K^o - \overline{K^o}$ mixing
Very weak "strong CP"	x	$\surd$	0	Neutron Dipole Moment

References

1. L. Wolfenstein and D. G. Ravenhall, Phys. Rev. 88, 279 (1952).
2. W. Pauli, Rev. Mod. Phys. 13, 203 (1941).
3. Particle Data Group, Phys. Lett. B204, 46 (1988).
4. T. D. Lee and C. N. Yang, Phys. Rev. 104, 254 (1956).
5. G. Sudarshan and R. Marshak, in Fifty Years of Weak Interactions (eds., D. Cline and G. Riedasch) U. of Wisconsin (1984).
6. F. Boehm and A. H. Wapstra, Phys. Rev. 109, 456 (1958); F. Boehm et al, Phys. Rev. 108, 1497 (1957).
7. J. H. Christenson, J. H. Cronin, V. L. Fitch, and R. Turlay, Phys. Rev. Lett. 13, 138 (1964).
8. For a Review see L. Wolfenstein, Ann. Rev. Nuc. Part. Sci. 36 , 137 (1986).
9. H. Burkhardt et al, Phys. Lett. B206, 169 (1988).
10. B. Winstein, Talk at the Lepton-Photon Conference at SLAC, August, 1989.
11. For a Review see E. M. Henley, Int. Symposium on WEIN 89, Montreal, May 1989.
12. P. Herczeg, J. Kambor, M. Simonius, and D. Wyler, to be published.
13. See the talk by B. Kayser.
14. For a review see H. Y. Cheng, Phys. Reports 158, 3 (1988).

THE TRIUMF 222 MeV $\vec{P}+P$ PARITY VIOLATION EXPERIMENT

J. Birchall[1], J.D. Bowman[2], G.H. Coombes[3], C.A. Davis[4], N.E. Davison[1], H. Fielding[3], P.W. Green[3], D.C. Healey[4], E.Korkmaz[3], C.A. Miller[4], S.A. Page[1], W.D. Ramsay[1], G. Roy[3], P.W. Schmor[4], A. Sekulovich[1], J.Soukup[3], G.M. Stinson[3], W.T.H. van Oers[1]

1. University of Manitoba, Winnipeg, Manitoba R3T 2N2
2. Los Alamos National Laboratory, Los Alamos, New Mexico 87545
3. University of Alberta, Edmonton, Alberta T6G 2N5
4. TRIUMF, 4004 Wesbrook Mall, Vancouver, B.C. V6T 2A3

Abstract

An experiment is underway at TRIUMF to measure the parity violating longitudinal analyzing power A_z in $\vec{p}p$ scattering to an accuracy of 2 x 10^{-8}. The beam energy of 222 MeV and the geometry of the detectors is chosen to isolate that portion of the weak nucleon-nucleon interaction arising from ρ-meson exchange. The high value of the parity allowed analyzing power at 222 MeV makes the experiment extremely sensitive to systematic errors. Instrumentation which provides very accurate measurement of beam position and shape is now being tested and a precision polarization profile monitor is under development.

1. INTRODUCTION

In this experiment a longitudinally polarized proton beam is incident on a liquid hydrogen target. The experiment is designed to measure the small ($\sim 10^{-7}$) parity violating longitudinal analyzing power $A_z = \frac{\sigma^+ - \sigma^-}{\sigma^+ + \sigma^-}$, where the σ^+ and σ^- refer to the cross sections for scattering in the two helicity states.

$A_z(\theta)$ may be expanded as a sum of parity mixed partial waves, $A_z(\theta)=A_z(\theta)_{SP}$ $+A_z(\theta)_{PD}+\cdots$. The shape of $A_z(\theta)$ is determined by the strong interaction while the magnitude is set by the weak interaction. Monte carlo simulations using realistic assumptions about the detector geometry indicate the SP term integrates to zero at an energy of 222 MeV. Detection geometry has been chosen to give cancellation of the SP term in both a scattering and a transmission experiment.

The weak interaction between nucleons may be parameterized by a set of 6 weak meson-nucleon coupling constants.[1)] The SP term in $A_z(\theta)$ is proportional in roughly equal parts to the constants h_ρ and h_ω (corresponding to ρ and ω meson exchange), whereas the PD term arises almost entirely from h_ρ. The TRIUMF experiment is thus designed to isolate h_ρ.

2. INSTRUMENTATION

Random changes in beam properties such as position and intensity will average out and result only in a longer required run time. Coherent changes, on the other hand, can give rise to a false parity violating signal. Corrections must be made for these effects. If the experiment is to reach a precision of $\pm 2 \times 10^{-8}$, it is felt no correction should be larger than about 10^{-7}, and that the error in any given correction must be less than $\pm 5 \times 10^{-9}$. The main corrections concern beam intensity, position and transverse polarization.

2.1 Intensity The beam current before and after the liquid hydrogen target is measured with hydrogen filled transverse field ionization chambers and the scattered beam is measured with an axial field chamber. Preliminary measurements with the optically pumped polarized ion source being developed at TRIUMF have set an upper limit of 0.1% on coherent intensity modulation. To keep the resultant false signal below 10^{-7}, assuming 2% scattered in the target, the transverse field chamber gains would have to track well enough that the gain difference between the two chambers is no more than $(\frac{dg}{g}) = 4 \times 10^{-6}(\frac{dI}{I})$. Tests of the linearity of the chambers show that for the designed beam current of 500nA, the chambers will have to be operated at a hydrogen pressure of $\frac{1}{5}$ atmosphere and an electric field of 1 kV/cm.

2.2 Polarization Profile One unfortunate feature of the 222 MeV measurement is the rather large parity allowed analyzing power, $A_y \simeq 0.3$, which is at least an order of magnitude larger than that experienced in lower energy experiments at 15 MeV and 45 Mev.[2,3,4)] This gives a great sensitivity to first moments of transverse polarization. For proper control of systematic errors xP_y and yP_x must be measured to an accuracy of 0.12 mm in one second. A scanning polarimeter designed to meet these specifications is to be tested in the coming year.

2.3 Position To reduce sensitivity to transverse polarization components the position of the beam will be fine tuned using a purely transverse polarization. Once the axis of zero sensitivity is found, the beam must be held very precisely in that position. We have developed a fast beam position servo based on split plate position monitors which will hold the beam position fixed to within a few μm, well within our specifications.

2.4 Intensity Profile The split plate of the beam position system only controls the *median* of the beam. To measure and correct for any variation in the true centroid of the beam, A low mass intensity profile monitor is being developed. The design calls for a series of 1 mm foil strips in vacuum. The secondary electron current is amplified by a preamplifier on each wire. A 32 channel device has been built and is now being tested.

References

1. Donoghue, J.F. et al., Ann. Phys.(N.Y.) 124, 449 (1980)
2. Kistryn, S. et al., Phys. Rev. Lett. 58, 1616 (1987)
3. Nagle, D.E. et al., AIP Conf. Proc. 51, 224 (1978)
4. Eversheim, P.D., Proceedings of the Symposium/Workshop on Spin and Symmetries, TRIUMF report (1989)

WHAT WE LEARNED FROM WEAK DECAYS*

Berthold Stech

Institut für Theoretische Physik
Universität Heidelberg
Philosophenweg 16

Abstract

A short review of recent progress in understanding weak decays is presented. Using decay constants and current matrix elements as manifestations of non-perturbative quark-quark and quark-antiquark correlations a coherent picture of dominant effects in K, D and B-decays emerges. In particular, the magnitude of the $|\Delta I| = 1/2$ enhancement observed in hyperon and K-decays comes out correctly as a consequence of quark-quark correlations in scalar states.

Historical Reminiscences

1956-1957 with Felix Boehm in Pasadena

Before describing recent progress in understanding weak decays, I will shortly go back to 1956-57 where I had the privilege to be a post-doc at Caltech. Here I met Felix Boehm whom we honour at this meeting. We soon became friends, and I profited much from his friendship and like to thank him on this occasion. I came, together with Hans Jensen, to Caltech in Dick Feynman's car from Seattle where Lee and Yang suggested the breakdown of mirror symmetry in weak decays. It was exciting for me to hear Feynman's views. My interest was particularly intense since in 1955 Jensen and I had published two papers on β-decay and μ-decay postulating γ_5-symmetry for weak interactions[1]. But we had "cleverly" restricted this symmetry to respect parity. At Caltech Murray Gell-Mann and Dick Feynman did not particularly like our theory because of the non-invariance of all mass

* Supported in part by the Deutsche Forschungsgemeinschaft

terms under γ_5-transformations. Spontaneous symmetry breaking had not been thought of in this connection and not for a long time. Felix was very eager to measure parity non-conserving effects and was among the first to do so. I remember well the lively discussions Kurt Alder, Aage Winther and I had with him about possible experiments which could probe the form of weak interactions. The simple generalization of the Heidelberg γ_5-symmetry requirement for each fermion field and thus to a parity non-conserving interaction selected the correct $V - A$ form of the weak interaction. But at that time experiments on He^6 favoured the tensor interaction apparently ruling out this possibility. Even after Wu, Telegdi, Frauenfelder, Boehm, and others had firmly established maximal parity non-conservation in several processes, it was not completely certain whether or not Fermi transitions (V or S) lead to the same electron helicity as Gamow-Teller transitions (A or T). Boehm, Novey, and Barnes performed the corresponding experiment[2)]. I am also on this paper, but my main contribution was to do the running with the short-lived radioactive source from Kellogg to Felix' laboratory. As it is well known, Feynman and Gell-Mann and Sudershan and Marshak then had the courage to distrust and disregard the result of the He^6 recoil experiment - Felix had made critical remarks about it - and postulated the $V - A$ interaction. Later on γ_5-symmetry - chiral symmetry - was also successfully applied to strong interactions. This time it really had to respect parity conservation, of course. Coming back from Caltech to Heidelberg I initiated this idea in 1959, but again in an incomplete way[3)]. Chiral symmetry of strong interactions was then further developed by Gürsey and finally brought into a correct form in 1967 by Cronin and Wess and Zumino[4)]. During the time at Caltech I was often invited to Felix' and Ruth's home. Together with Volker Soergel we also played music together. Felix was kind enough not to throw me out although my playing the violin must have been a cruel shock to his sensitive ears. The good contact to eminent scienctists at Caltech and particularly the close friendship with Felix meant and still means very much to me.

1. Weak Decays and QCD

In the last few years the Standard Model of electroweak and strong interactions was subjected to many tests. The main conclusions are:
i) The validity of Standard Model is remarkable, no deviation could be observed (but the Higgs sector is still untested). We know most parameters of the model with reasonable accuracy: the gauge coupling constants and the quark and lepton masses (except m_t).
ii) We can have confidence in the relevance of the Kobayashi-Maskawa matrix for quark mixing and even for CP-symmetry violation, if ϵ'/ϵ is indeed of order 10^{-3} and $|V_{ub}/V_{cb}| \approx 0.1$, as obtained from the upper part of the inclusive lepton momentum spectrum in B-decays.

In the old days β-decay was used advantageously to uncover the properties of nuclei. Today, weak decays are a perfect instrument to study QCD, in particular non-perturbative long-range QCD. The reasons are:

- The initial state consists of a single isolated particle.
- The transition operator is known (at the scale of the W-boson mass) and has simple Lorentz and flavour quantum numbers.
- Perturbative QCD effects can reliably be calculated.

However, as long as one was restricted to the physics of u, d, and s quarks, a coherent picture about the interplay of weak and strong interaction could not be obtained. In this regime only few fundamentally different channels are open. With the discovery of charmed and bottom quarks the situation changed drastically. Now one can study a large variety of decays with different particle masses, spins, and flavours. Thanks to the great efforts of several experimental groups much experimental material is already available[5]. In this talk I will have to leave out many relevant points and details. The interested reader is referred to survey articles[5][6] and to the original literature where also many more references can be found.

Typical non-perturbative quantities are the meson decay constants, the most well-known of it is the π-meson decay constant f_π:

$$\langle 0|(\bar{u}\gamma_\mu\gamma_5 d)|\pi^-\rangle = if_\pi p_\mu^\pi$$
$$f_\pi \simeq 0.132\ GeV\ . \tag{1}$$

Decay constants f_π, f_η, f_ρ... are manifestations of strong $q\bar{q}$ correlations in colour singlet states. They cannot be calculated in perturbative theory. Most of them, f_η, f_{D^*}, ..., can be measured only indirectly. Calculations rely on valence quark models for mesons, on QCD sum rules or on lattice gauge theories. At present, none of these methods has an accuracy better than $\approx 20\%$. A comparison between theory and experiment appears to be of vital importance for learning more about long-range QCD.

Exclusive semileptonic decays involve the matrix elements of quark currents. The Lorentz structure of these matrix elements is generally known from mass, spin, and parity determinations, the remaining dynamics rests in the formfactors. Fortunately, the q^2-dependence of these formfactors can often be approximated by pole formulas. These poles correspond to particles or resonances which carry the quantum numbers of the quark currents, i. e. the relevant flavour quantum numbers and spin 1 for the transversal part and spin zero for the longitudinal part of the currents. Within this approximation it remains to calculate these formfactors at a single value of q^2 only. An example is the process $D^0 \to K^- e^+ \bar{\nu}$. Only the transversal formfactor, which is dominated by the D_s^* meson pole, is needed here. But the longitudinal formfactor is also of interest, because it contributes to nonleptonic decays. Most calculations of formfactors have been performed using valence quark wave functions for the initial and final mesons. Perhaps surprisingly, it turned out that the valence quark picture works reasonably well. Simple wave functions give already good semiquantitative results.

Calculations with relativistic wave functions in the infinite momentum frame[7)] are presumably trustable for formfactor values near $q^2 = 0$, a region which gives important contributions to the upper end of the lepton spectrum. Some of these formfactor values have also been estimated by QCD sum rule calculations with very similar results[8]. Non-relativistic wave functions with relativistic corrections[9] are mainly applicable to formfactors near $q^2 = (m_i - m_f)^2$, where the initial and final mesons are both at rest. In spite of the successes mentioned, none of these approaches are particularly satisfying. But one can hope that more precise experimental data will reveal the deficiencies of the models and will stimulate new and better theoretical attempts. In particular, transition matrix elements between a very heavy and a very light particle will provide important information on the wave functions in the confinement region. The most drastic manifestations of non-perturbative QCD occur, however, in non-leptonic decays to which I will turn now.

2. Non-leptonic Decays

The transition operators for non-leptonic decays consist of products of 4 quark field operators practically at the same space time points. The strong interaction of these quarks causes dramatic effects. Unfortunately, however, only the hard gluon corrections to the non-leptonic Hamiltonian can be calculated in a straightforward way. The strong additional influence of non-perturbative QCD is evident from the empirically observed drastic $|\Delta I| = 1/2$ enhancement over $|\Delta I| = 3/2$ transitions in strange particle decays. This enhancement is much larger than the modest enhancement expected from short-distance QCD effects. Also the observed lifetime difference of D^0 and D^+ mesons must be due to the non-perturbative properties of QCD.

Recently, a semiquantitative understanding of energetic exclusive decays as they occur in D and B-decays has been achieved[10]. The starting point is the effective Hamiltonian which already incorporates the perturbative QCD effects.

For the special case of D-decays it reads:

$$H_{eff} = \frac{G}{\sqrt{2}} \Big\{ c_1(\mu)(\bar{u}d')_{V-A}(\bar{s}'c)_{V-A} + c_2(\mu)(\bar{s}'d')_{V-A}(\bar{u}c)_{V-A} + \text{Penguins} \Big\} + h.c. \tag{2}$$

In this expression G denotes the Fermi constant and d', s' are the Cabibbo-rotated down and strange quark field operators, respectively.* $c_1(\mu)$ and $c_2(\mu)$ are scale-dependent Wilson coefficients. The matrix elements of the quark fields have to be evaluated at the same scale to give a scale-independent result. For a scale $\mu \geq 1\ GeV$ the coefficients of the Penguin operators are small. Thus, the $V - A$ structure of the uncorrected original interaction remains practically intact. Very important, however, $c_1(\mu)$ is different from one and a neutral current product not present in the original Hamiltonian is induced with coefficient $c_2(\mu)$. At the scale of the charmed quark one has

$$c_1(m_c) \simeq 1.24 \qquad c_2(m_c) \simeq -0.47\,. \tag{3}$$

At the scale of the bottom quark the calculation gives

$$c_1(m_b) \simeq 1.11 \qquad c_2(m_b) \simeq -0.25\,. \tag{4}$$

For the process $D^0 \to \rho^0 \bar{K}^0$, for example, one can now make use of the fact that the $\bar{K}^0$ meson can directly be generated by the $(\bar{s}d)$ current in (2). Since vector and axial vector matrix elements increase with particle momenta, this particular contribution will be the more important, the higher the available energy is. Thus, the nonleptonic matrix element of our example factorizes into the vacuum-$\bar{K}$-amplitude and the $D^0 \to \rho^0$ current matrix element. The coefficient in front is here c_2. We call this procedure[10)] "new factorization", as opposed to "old factorization", in which c_2 is replaced by $c_2 + \frac{1}{3}c_1$ (or c_1 by $c_1 + \frac{1}{3}c_2$ in a transition

* The modification due to quark mixing with the third family can safely be neglected in (2).

with a current-generated charged meson). The idea is to stick to the Hamiltonian in the form (2) when saturating with colour singlet states. Contributions with the colour factor 1/3, which are obtained by factorizing the Fierz-transformed form of (2) go always together with contributions from colour octet intermediate states which we do not know how to handle. It seems better to leave these parts out altogether. Although there is no real justification for this procedure, only some support from the $1/N_c$ expansion method[11)] and from numerical QCD sum rule results[12)], this simple recipe works well:

- New factorization gives a successful semiquantitative description of many exclusive decays[5)10)].

In D^+-decays the two amplitudes proportional to c_1 and c_2 both contribute and interfere destructively because of the negative sign of c_2/c_1. This phenomenon explains the reduced width of most D^+-decays and gives also an understanding of the observed lifetime difference between D^+ and D^0 [10)20)]. The surprising success of new factorization does not only hold for D-decays, but equally well for B-decays, where the coefficients c_1 and c_2 have different values. In table 1 theoretical and experimental B-decay branching ratios are shown. The theoretical values are obtained using $\tau_B = 1.14 \cdot 10^{-12}$ *sec.*, $|V_{cb}| = 0.047$, $c_1 = 1.11$ and $c_2 = -0.25$.

Of course, there are still many problems: The valence quark models used are very simple and should be improved. Their range of validity is unknown. Final state interaction effects are also not known. They cause rescattering processes and channel mixing. Thus, quantitative predictions cannot be made. But much can be learned about these effects, about Penguin transitions, rare decays, and CP-violation in the B-system in dedicated experimental and theoretical studies.

3. The $|\Delta I| = 1/2$ Rule in Strange Particle Decays

For more than 30 years the empirical $|\Delta I| = 1/2$ rule was a complete mystery. Do we now understand this rule better? The answer is yes! Firstly, we understand

Table 1 B-decay branching ratios (%)

Decay	Experiment[13)]	Theory[10)]
$\bar{B}^0 \to D^+\pi^-$	$0.27 \pm 0.08 \pm 0.05$	0.50
$\bar{B}^0 \to D^*\pi^-$	$0.33 \pm 0.09 \pm 0.06$	0.38
$\bar{B}^0 \to D^*\rho^-$	$1.9 \pm 0.9 \pm 1.3$	1.2
$\bar{B}^0 \to D^*a_1^-$	$1.8 \pm 0.5 \pm 1.2$	1.7
$\bar{B}^0 \to \bar{K}^0 J/\psi$	$0.06 \pm 0.03 \pm 0.02$	0.05
$\bar{B}^0 \to \bar{K}^* J/\psi$	$0.11 \pm 0.05 \pm 0.03$	0.23
$\bar{B}^0 \to D^+D_s^-$	2.1 ± 1.1	$0.76(f_{D_s}/f_K)^2$
$\bar{B}^0 \to D^*D_s^-$	3.2 ± 1.7	$0.31(f_{D_s}/f_K)^2$
$B^0 \to D^0\pi^-$	$0.30 \pm 0.06 \pm 0.04$	0.34
$B^- \to K^- J/\psi$	$0.08 \pm 0.02 \pm 0.02$	0.05
$B^- \to D^0 D_s^-$	3.7 ± 1.9	$0.76(f_{D_s}/f_k)^2$

why the factorization contribution gives too small an effect. In s-quark decays the direct generation of a π-meson does not give a dominant contribution because of the low energy release in this process: $f_\pi p_\pi$ is too small. Also the inclusion of Penguin diagrams, which involve the large matrix elements of (pseudo)scalar currents, e.g.

$$\langle 0|i\bar{u}_i\gamma_5 d_i|\pi^-\rangle = g_\pi, \quad g_\pi(1GeV) \simeq (0.17 \pm 0.03)GeV^2 \tag{5}$$

does not significantly improve this result. The corresponding Wilson coefficients are too small.

Secondly, we observe that the short distance QCD effects are largest in the colour 3^* channel of the $s+u \rightarrow d+u$ reaction, a fact pointing to large quark-quark correlations. Already from the very existence of baryons it is evident that there are strong long-range binding forces present in this channel.

Thirdly, it is possible to rewrite the effective weak Hamiltonian in terms of local scalar and pseudoscalar currents just composed of these (us) and (du) quark pairs[14)]

$$H_{eff}^{|\Delta S|=1} = \frac{G_F}{\sqrt{2}} V_{us} V_{ud}^* c_- \{\epsilon_{kij}(\bar{u}_i^c(1-\gamma_5)d_j)^\dagger \epsilon_{klm}(\bar{s}_l^c(1-\gamma_5)u_m) + h.c.\} \qquad (6)$$
$$+ \text{ product of colour sextet currents } + \text{ penguin terms.}$$

Here $c_- = c_1 - c_2$ is a scale-dependent Wilson coefficient, and u^c, s^c denote charge conjugate fields. Evidently, the weak Hamiltonian can annihilate and create correlated pairs of quarks (diquarks). We note that the product of colour triplet currents shown explicitly in (6) transforms like an $I = \frac{1}{2}$ operator, since the scalar (ud) diquark current has necessarily $I = 0$. Correlated $0^\pm$ diquark states $(ds), (su)$, and (ud) — as they exist inside baryons — form the components of an $SU(3)$ flavour antitriplet. "Diquark decay constants" can be defined in analogy to (5), e.g. for the third component

$$\langle 0|\epsilon_{ijk}\bar{d}_i^c\gamma_5 u_j|(ud)_l\rangle = \sqrt{\frac{2}{3}}\delta_{kl}g_3^D, \qquad (7)$$

where $\sqrt{\frac{2}{3}}$ is a colour factor. The couplings g^D are of importance if they are large. The weak Hamiltonian then annihilates and creates colour antitriplet diquark states in a very effective way. From (1) and (7) we see that this decay mechanism will be dominant compared to the direct generation of a final meson as long as $g^D \gg f_\pi m_q$, where m_q is the mass of the decaying quark. This relation is well fulfilled in strange particle decays, as we will see. The product of colour sextet currents in (6) will behave quite differently. Since the short-distance force in this case is repulsive, a bound colour sextet diquark seems very unlikely. To handle this term, it should therefore be rewritten in the usual form.

Numerical values for the diquark decay constants have been obtained in a careful QCD sum rule analysis[15),16)] and turned indeed out to be large, similar in magnitude to g_π. In particular, one finds

$$c_- g_2^D g_3^D = (0.09 \pm 0.03)\ GeV^4 \tag{8}$$

for the combination which enters our calculations of decay amplitudes. It is important to note that this combination is almost scale-independent. The perturbatively calculated anomalous dimension factors cancel[16)]. The numerical value (8) is valid for a scalar (ud) diquark mass of 0.5 GeV and increases slightly for larger mass values. The strong $q-q$ correlation in scalar states can be used to construct a model in which these correlations are simulated by diquark fields[17)]. The part of the effective Hamiltonian shown explicitly in (6) takes then the simple form

$$H_{eff}^D = W \phi_L^+ \lambda_6 \phi_L \tag{9}$$

$$\text{with} \quad W = \frac{G}{\sqrt{2}} V_{ud}^* V_{us} \frac{2}{3} c_- g_2^D g_3^D \quad \text{and} \quad \phi_L = \phi_+ - \phi_-$$

where $\phi_\pm$ denotes normalized diquark fields with even (+) and odd (-) intrinsic parities.

Remarkably, the interaction (9) together with the central value of the numerical constant (8) explains and relates a great variety of processes:

i) By describing the proton as a quark-diquark composite, application of eq. (9) leads immediately to the correct value of the proton decay constant as it is known from QCD sum rules for baryons[15)].

ii) The Hamiltonian (9) gives a good description of the enhanced $|\Delta I| = 1/2$ amplitudes in hyperon decays. In fact, if the contributions from new factorization are added to these amplitudes, all hyperon decays, S- and P-wave amplitudes, are in good agreement with the experimental ones[18)].

iii) With the same Hamiltonian also the $K \to 2\pi$ processes can be calculated[17),19)]. Contrary to the case of baryons, where correlated $q-q$ states are already present, diquarks can appear here in virtual states only. In spite of the complexity of the corresponding diagrams shown in Fig. 1, quantitative calculations can be performed if one uses for the diquark-meson interaction the stringent requirements of chiral symmetry[17)19)]. The result depends on the diquark masses, i.e. the effective masses of the correlated $q-q$ state, but not severely. The loop integrals are ultraviolet convergent (or logarithmically divergent if the mass of the 0^- diquark is pushed to infinity). Taking the $0^+(ud)$ diquark mass in the region of 500 MeV and adding the contribution from factorization, the experimental amplitude is well reproduced. One can even eliminate the explicit dependence on effective diquark masses by relating the amplitude to $f_k - f_\pi$. One finds for the virtual diquark contribution to the decay amplitude the simple formula

$$\frac{1}{i} A_D(\bar{K}^0 \to \pi^+\pi^-) \simeq W \, \frac{f_K - f_\pi}{f_\pi^2} \tag{10}$$

with W defined as before (9). The conclusion is that quark-quark correlations in scalar states — simulated here by local fields — are the long-sought source for the $|\Delta I| = 1/2$ enhancement in hyperon decays and in K-decays!

Quark-quark correlations do play a role also in D- and B-decays. However, since the scalar matrix element (7) does not increase with energy, it cannot compete anymore with vector and axial vector matrix elements for high energy transitions. Still, in D-decays 20 to 30 % of the inclusive non-leptonic decay rates could be due to decay modes where initially correlated $q-q$ and $\bar{q}-\bar{q}$ states are emitted which then rearrange and form mesons[20)]. In B-decays the formation of diquarks can play only a minor role since almost all decays are very energetic. The weak currents create mesons and excited mesons. However, B-decays to baryons are an important exception. The creation of diquarks appears to be the first step in the formation of baryons[20)]. Several, but not yet tested, selection rules follow from this assumption.

4. The $K_L - K_S$ Mass Difference

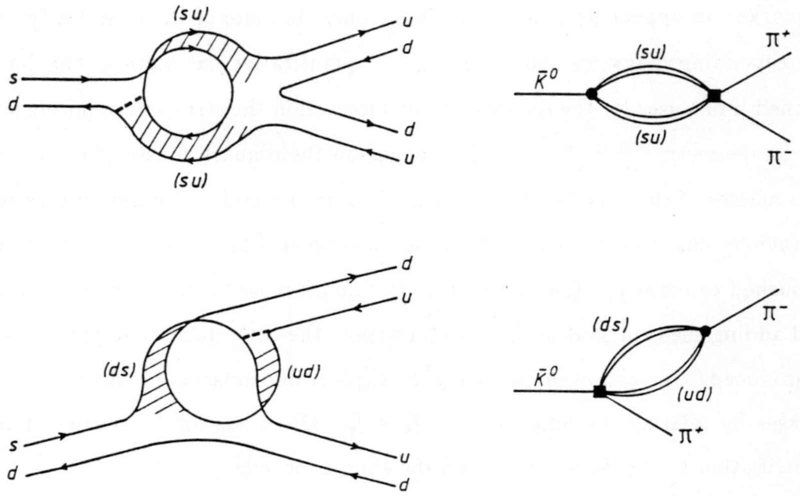

Fig. 1: Examples of quark-quark correlations in the decay $\bar{K}^0 \rightarrow \pi^+\pi^-$. The W exchange is indicated by dashed lines. In the corresponding Feynman diagrams of the diquark model (see text) the weak and strong vertices are indicated by full circles and squares, respectively.

As an additional application of the diquark picture the mass difference between the long- and short-lived K^0-mesons has been studied[19)]. The usual box diagram calculation of the corresponding $|\Delta S| = 2$ transition (Fig 2a) supplemented by short-distance QCD corrections gives a much too small value. Long-distance effects from single meson poles which represent the major part of $q - \bar{q}$ correlations in colour singlet states (Fig. 2b) gives a negative contribution to the mass difference $\Delta m = m_L - m_S$. It essentially cancels the box diagram result. The effect of $q - q$ correlations in scalar states (Fig. 2c), on the other hand, is large. It has been calculated using (9) and the chiral model for the diquark-meson interaction applied before. The result is very encouraging: The diquark loop integral converges (even if the 0^--diquark masses are pushed to infinity) and gives the large positive

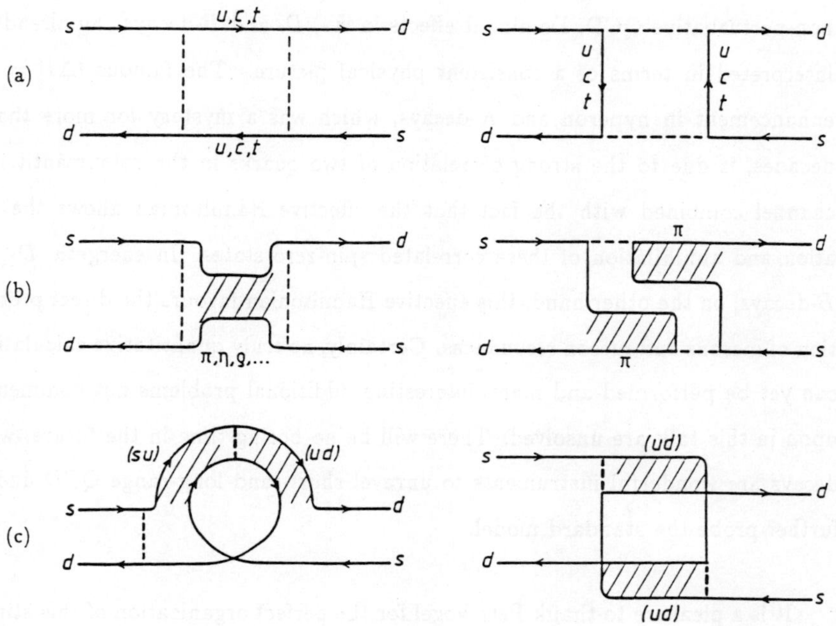

Fig. 2: Different contributions to $K^0 - \bar{K}^0$ mixing: (a) short-distance contributions from box diagrams, (b) intermediate particle states due to quark-antiquark correlations, (c) intermediate diquark states due to quark-quark correlations.

value[19)]

$$\Delta m_K^D \approx 3.6 \cdot 10^{-6}\ eV \tag{11}$$

which is very close to the experimental one. But because of the dependence of the theoretical estimate on the 0^+ diquark mass — which we took again to be 500 MeV — and on the precise value of W which enters here quadratically, the theoretical uncertainties are larger than in the examples given above.

5. Summary

The study of weak decays in the past provided for most important insights into the properties and symmetries of particle interactions. Today, the experimental data on D- and B-decays allow us to investigate the unknown region of non-perturbative QCD. Dominant effects in K-, D- and B-decays can already be interpreted in terms of a consistent physical picture. The famous $|\Delta I| = 1/2$ enhancement in hyperon and K-decays, which was a mystery for more than 3 decades, is due to the strong correlation of two quarks in the colour antitriplet channel combined with the fact that the effective Hamiltonian allows the creation and annihilation of these correlated spin-zero states. In energetic D- and B-decays, on the other hand, this effective Hamiltonian favours the direct production of mesons and meson resonances. Certainly, no truly quantitative calculations can yet be performed and many interesting additional problems not commented upon in this talk are unsolved. There will be no boring time in the future, weak decays are wonderful instruments to unravel short- and long-range QCD und to further probe the standard model.

It is a pleasure to thank Petr Vogel for the perfect organization of this stimulating Symposium. I also like to thank G. Dosch, M. Jamin, M. Neubert, and Q. P. Xu for collaborating in part of the work reported here.

REFERENCES

1) Stech, B., and Jensen, J.H.D., Z. Phys. 141, 175, 403 (1955).

2) Boehm, F., Novey, T.B., Barnes, C.A., and Stech, B., Phys. Rev. 108, 1497 (1957).

3) Stech, B., Phys. Rev. Lett. 2, 514 (1959);
Kramer, G., Rollnik, H., and Stech, B., Z. Phys. 154, 564 (1959).

4) Gürsey, F., N. Cim. 16, 230 (1960);
Cronin, J.A., Phys. Rev. 161, 1483 (1967);
Wess, J., and Zumino, B., Phys. Rev. 163, 1727 (1967).

5) Hitlin, D., Nucl. Phys. B (Proc. Suppl.) 3, 179 (1988);
R. Morrison, R., and Witherell, W., UCSB-HEP 89-01, Annual Rev. of Nucl. and Particle Science V39 .

6) Rückl, R., Habilitation Thesis, Universität München 1983, unpublished;
Bigi, I.I., Nucl. Phys. B (Proc. Suppl.) 7A, 318 (1989);
Wirbel, M., DOTH 88/2, Progress in Particle and Nuclear Physics, in print.

7) Wirbel, M., Stech, B., and Bauer, M., Z. Phys. C 29, 637 (1985);
Körner, J.G., and Schuler, G.A., Z. Phys. C 38, 511 (1988).

8) Dominguez, C.A., and Paver, N., Phys. Lett. B207, 499 (1988); Z. Phys. C 41, 217 (1988).

9) Grinstein, B., Isgur, N., and Wise, M., Phys. Rev. Lett. 56, 298 (1988) ;
Altomari, T., and Wolfenstein, L., Phys. Rev. Lett. 58, 1583 (1987);
Isgur, N., et al., Phys. Rev. 39, 799 (1989).

10) Bauer, M., and Stech, B., Phys. Lett. 152B, 380 (1985);
Bauer, M., Stech, B., and Wirbel, M., Z. Phys. C 34, 103 (1987).

11) Buras, A.J., Gerard, M., and Rückl, R., Nucl. Phys. B268, 16 (1986).

12) Blok, B. Y., and Shifman, M.A., Yad. Fiz. 45, 221, 478, 841 (1987);
Shifman, M.A., Nucl. Phys. B (Proc. Suppl.) 3, 289 (1988).

13) Stone, S., CLEO CBX-89-48;
Schroeder, H., ARGUS, private communication.

14) Stech, B., Phys. Rev. D36, 975, (1987).

15) Dosch, H.G., Jamin, M., and Stech, B., Z. Phys. C42, 167 (1989).

16) Jamin, M., and Neubert, M., Heidelberg preprint HD-THEP-89-27.

17) Stech, B., Nucl. Phys. B (Proc. Suppl.) 7A, 106 (1989);
Neubert, M., and Stech, B., HD-THEP-89-11, Ann. N. Y. Acad. Sci..

18) Stech, B., and Xu, Q.P., in preparation.

19) Neubert, M., and Stech, B., HD-THEP-89-20; Phys. Lett. in print.

20) Stech, B. in: *CP Violation, Adv. Ser. on Directions in High Energy Physics*, Vol. 3, ed. by C. Jarlskog (World Scientific, Singapore 1989), p. 680;
Stech, B., Nucl. Phys. B (Proc. Suppl.) 1B, 17 (1989);
Stech, B., *Diquarks*, ed. Anselmino and Predazzi, World Sci. (1989) 277.

PRIMORDIAL NUCLEOSYNTHESIS: THE NEXT SIGNIFICANT FIGURE

GARY STEIGMAN

Department of Physics and Astronomy

The Ohio State University

Columbus, Ohio 43210, USA

ABSTRACT

We review the current status of the comparison between the predictions of primordial nucleosynthesis in the "standard" hot big bang model and the abundances of the light elements derived from observational data. Our goals here include testing the consistency of the standard model and probing – to two significant figures – the constraints on cosmology and particle physics which follow from the agreements between theory and observation. Current data are consistent with the standard hot big bang model and demand that baryons contribute between 2 and 12 percent of the critical density of the Universe. Thus, although baryons can not "close" the Universe, most of the baryons in the Universe are "dark". We constrain the number of (equivalent) light neutrino flavors to $N_\nu \leq 3.4$, in excellent agreement with very recent results from $e^\pm$ colliders.

I. Introduction

The synthesis of the light elements in the first thousand seconds of the evolution of the Universe provides a unique probe of cosmology and particle physics. Primordial nucleosynthesis offers the only observationally accessible test of the hot big bang model at early epochs and, consistency between the predicted abundances of D, ^{3}He, ^{4}He and ^{7}Li and those inferred from observations constrains the baryon density of the Universe[1] as well as the number of species of light particles[2]. Detailed comparisons between theory and observation have revealed that the simplest, "standard", hot big bang model is in accord with the abundances of the light elements; that the present density of nucleons is such that the Universe cannot be "closed" by nucleons but, some – perhaps most – of the nucleons in the Universe are "dark"; that the number of (equivalent) light neutrino species is no larger than four[3]. For a review of big bang nucleosynthesis – theory and observation – see Boesgaard and Steigman[4] and, especially, the original literature cited therein.

In recent years both the theoretical and observational data have become more accurate and this justifies a reevaluation of the predictions emerging from primordial nucleosynthesis with a goal of extracting more accurate – and constraining – bounds to cosmology and particle physics. At stake in such an investigation is the overall consistency of the standard hot big bang model. Such a study has been begun[5] and is continuing[6]; here I will provide an overview of the current status. Since it is my purpose here to test the "standard" hot big bang model, I will limit my discussion to primordial nucleosynthesis in the context of that model. To provide a setting for the comparison between the predictions and the observations, I will begin with a brief overview of the physics of big bang nucleosynthesis (for further details and references see ref. 1). Next, I will comment, again briefly, on those aspects of the nuclear physics which have changed and, how these changes modify earlier conclusions (for further details see refs. 5&6). Then, with some reference to more recent data, I will provide an overview of the observational situation. The predictions and observations will be compared and the inferred constraints on cosmology and particle physics presented and compared with relevant observational and experimental data.

II. The Physics of Primordial Nucleosynthesis

In the context of the (simplest) standard hot big bang model, it is assumed that the early Universe is homogeneous, is expanding isotropically, is "radiation dominated" (the total energy density is dominated by the contributions from relativistic particles), is "non-degenerate" (the neutrino (or antineutrino) asymmetry is not nine, or more, orders of magnitude larger than the baryon asymmetry). Ordinary – Einstein – General Relativity is assumed. With these minimal assumptions (i.e., fewest adjustable parameters), it may be shown that leptons (neutrinos, electrons, etc.), baryons (neutrons and protons) and photons were in thermodynamic equilibrium when the Universe was hotter than T $\sim$ few MeV, at times earlier than $t \sim 0.1$ sec. It is worth emphasizing that, as a result of thermodynamic equilibrium at early epochs, we may begin our discussion of primordial nucleosynthesis at (relatively) late epochs (t $\gtrsim$ 0.1 sec) when the Universe is (relatively) cool (T $\sim$ few MeV) and dilute (the mean separation of nucleons is $\ell_N \sim 5 \times 10^5 T_{MeV}^{-1}$ fermis). Therefore, no exotic cosmology or high energy physics is required to follow the evolution of the abundances of the elements synthesized in the first few minutes of the evolution of the Universe.

The relevant nuclear physics can be (and has been) studied in the laboratory. Since two-body reactions are responsible for the production (and destruction) of the relevant nuclei, the predicted abundances depend on (only) one adjustable parameter: the nucleon density at the epoch of nucleosynthesis. However, as the Universe expands, the density of all particles decreases. Therefore, it is more convenient to adopt as the free parameter, a quantity that is independent of epoch. The "comoving" number of nucleons would be such a parameter. A comoving volume may be defined by the number of photons (actually, the entropy) it contains. The predicted abundances of D, ^{3}He, ^{4}He and ^{7}Li will, then, depend on the quantity $\eta \equiv N/\gamma$; for convenience we will often use $\eta_{10} \equiv 10^{10}\eta_{10}$. Since the present photon temperature is known[7] ($T_\gamma = 2.78 \pm 0.01K$), so too is the present density of photons and, in terms of η_{10}, the nucleon density as well. If we compare the nucleon density to the "critical" density ($\Omega_N \equiv \rho_N/\rho_c$, where $\rho_c = 2.6h_{50}^2 keVcm^{-3}$, and h_{50} is the present value of the Hubble parameter in units of $50kms^{-1}Mpc^{-1}$),

$$\Omega_N h_{50}^2 = 1.54 \times 10^{-2} T_{2.78}^3 \eta_{10}, \tag{1}$$

where $T_\gamma = 2.78T_{2.78}K$. As a preview of results to come, note that for $2.76 \leq T_\gamma \leq 2.80K$, $0.8 \leq h_{50} \leq 2$ and, $3 \leq \eta_{10} \leq 10$, $0.01 \lesssim \Omega_N \lesssim 0.25$. It is one of our goals here to tighten these bounds – and, we will! (However, note that the largest uncertainty is that due to the lack of a precise value for the Hubble parameter.)

At early times deuterons are produced rapidly via $n + p \to D + \gamma$. However, at these early times, the Universe is very hot and, there are many high energy gamma rays present to photodissociate the deuteron: $\gamma + D \to n + p$. Since $n_\gamma \sim 10^{9-10} n_N$, the deuteron is destroyed before it can capture another nucleon. As a result there is no platform from which the heavier nuclei can be synthesized. No nucleosynthesis of any significance can occur until the Universe cools to $T \lesssim T_{D\gamma} \approx 0.1$MeV ($t \gtrsim t_{D\gamma} \approx 100$ sec.) when nucleon capture competes favorably with photodissociation of the deuteron. For $T \lesssim T_{D\gamma}$, ^{3}H, ^{3}He and ^{4}He are synthesized very quickly. There is, however, a gap at mass-5 (no stable nucleus) and, as a result, it is very difficult to synthesize nuclei heavier than ^{4}He; small amounts of mass-7 are produced via $^3H(\alpha,\gamma)^7$Li and ^{3}He$(\alpha,\gamma)^7Be$. Because of the gap at mass-5, and since ^{4}He is the most tightly bound of the light nuclei, almost all the neutrons present when nucleosynthesis begins at $T_{D\gamma}$ are incorporated in ^{4}He. Therefore, whereas the predicted abundances of D, ^{3}He (^{3}H decays to ^{3}He) and ^{7}Li (^{7}Be captures an electron and decays to ^{7}Li) are sensitive to the nucleon abundance (η), the predicted abundance of ^{4}He is relatively insensitive to η but does depend on the weak interaction rates which determine the neutron abundance. Any changes which leave more neutrons yield more ^{4}He.

Primordial nucleosynthesis, which begins only when $t \geq 100$ sec, abruptly ends at $t \sim 10^3$ sec when the thermal energies have become so low that further nuclear reactions are quenched by the Coulomb barriers.

Since D and ^{3}He are burned to ^{4}He, the predicted abundances of D and ^{3}He decrease with increasing nucleon abundance. For $1 \leq \eta_{10} \leq 10$, a good fit to the detailed nucleosynthesis calculations[5,6] is

$$10^5 y_{2P} \equiv 10^5 (D/H)_P = 46\eta_{10}^{-5/3}. \tag{2}$$

Over the same range, $y_{3P} \equiv (^3He/H)_P$ decreases very slowly from $y_{3P} = 3 \times 10^{-5}$ at $\eta_{10} = 1$ to $y_{3P} = 0.9 \times 10^{-5}$ at $\eta_{10} = 10$.

Mass-7 (^{7}Li and ^{7}Be) can be synthesized via ^{3}H$(\alpha,\gamma)^7$Li and ^{3}He$(\alpha,\gamma)^7$Be and, as a result, the predicted abundance has a minimum ($y_{7P} \approx 0.9 \times 10^{-10}$) for $\eta_{10} \approx 3$ and increases for smaller and larger values of the nucleon abundance; $y_{7P} \lesssim 3 \times 10^{-10}$, for $1.3 \lesssim \eta_{10} \lesssim 5.6$.

Although the predicted ^{4}He mass fraction Y_P is insensitive to η_{10}, Y_P is sensitive to the competition between the weak interaction rates and the universal expansion rate since this competition determines the number of neutrons available to be incorporated in ^{4}He. Here, we are referring to the charged-current weak interactions $p + e^- \leftrightarrow n + \nu_e$ and $n + e^+ \leftrightarrow p + \bar{\nu}_e$ (beta-decay, $n \leftrightarrow p + e^- + \bar{\nu}_e$, is relatively unimportant since nucleosynthesis starts at $t \approx 10^2$ sec. and the neutron mean-life is $\tau_n \sim 10^3$ sec), whose rates may be written as

$$\Gamma_{wk} = n_e < \sigma_{cc} v > \quad \propto \quad T^5/\tau_n. \tag{3}$$

In (3) we have used the facts that $n_e \propto T^3$, $\sigma_{cc} \propto E^2 \propto T^2$ and, we have "normalized" σ_{cc} through the neutron mean-life[8] (from ref. 5: $\tau_n = 889.8 \pm 4.4$ sec; this corresponds to a half-life: $\tau_{1/2} = 10.28 \pm 0.05$ min.). The weak rate, Γ_{wk}, should be compared to the expansion rate of the Universe H (the Hubble parameter) which, for these epochs is,

$$H \quad \propto \quad \rho_{TOT}^{1/2} \quad \propto \quad g_{eff} T^2, \tag{4}$$

where $\rho_{TOT} = (g_{eff}/2)\rho_\gamma$ and $\rho_\gamma = aT^4$. g_{eff} counts the effective number of relativistic degrees of freedom which contribute to the total density. In the standard model we count photons ($g_\gamma = 2$), electrons and positrons ($g_e = 4 \times 7/8$; relativistic fermions contribute 7/8 as much as relativistic bosons[9]) and three families of left handed neutrinos ($g_\nu = 3 \times 2 \times 7/8$) so that $g_{eff} = 43/4$. For additional, two-component, relativistic particles (neutrino families or their equivalent),

$$g_{eff} = \frac{43}{4}\left[1 + \frac{7}{43}(N_\nu - 3)\right]. \tag{5}$$

In comparing (3) and (4) we see that for $T < T_*$, where $\Gamma_{wk}(T_*) = H(T_*)$, the weak rates are slow compared to the expansion rate and, approximately, the neutrons "freeze-out" at their equilibrium value at $T_*(\propto g_{eff}^{1/6}\tau_n^{1/3})$,

$$(n/p)_* = exp(-\Delta m/T_*), \tag{6}$$

where Δm is the neutron-proton mass difference. The expected primordial mass fraction of ^{4}He is then,

$$Y_P = \frac{2n}{n+p} \lesssim \frac{2n_*}{n_* + p_*}. \tag{7}$$

It follows then that, if g_{eff} and/or τ_n increase, then T_* increases and the neutron abundance freezes-out earlier, when it is larger. From a fit (accurate to $\pm$ 0.001 or better) to the calculated ^{4}He abundance for $2.5 \leq \eta_{10} \leq 10$, $2 \leq N_\nu \leq 4$ and $10.1 \leq \tau_{1/2} \leq 10.5$ min., we find[5,6]

$$Y_P = 0.228 + 0.010\ \ln\eta_{10} + 0.012(N_\nu - 3) + 0.018(\tau_{1/2} - 10.28). \tag{8}$$

III. Primordial Abundances

The "heavy" elements (carbon and heavier) – "metals" in the astronomers' jargon – are synthesized in processes associated with stars. The intermediate mass nuclei – Li, Be, B – are believed to be synthesized in collisions between cosmic ray nuclei (mainly protons and alphas) and interstellar gas nuclei (protons, alphas and C,N,0 nuclei)[10,11]. However, in such cosmic ray "spallation" reactions roughly equal amounts of ^{6}Li and ^{7}Li are produced, in contrast to the observed solar system ratio[12] of ^{7}Li /^{6}Li ~ 12. If ^{7}Li were solely the consequence of cosmic ray spallation, ^{6}Li, ^{9}Be, ^{10,11}B would be overabundant. Conversely, if spallation is the origin of the observed ^{9}Be, ^{10,11}B, then ^{6}Li is accounted for but ^{7}Li must have another source. Indeed, the low abundance of lithium observed in the oldest – PopII – stars (y_7(PopII)$\approx 1 - 2 \times 10^{-10}$), when compared with the higher abundance observed in PopI stars and in the meteorites[12] (y_7(PopI)$\approx 1 - 2 \times 10^{-9}$) suggests, strongly, that $y_{7P} \approx y_7(PopII)$ and that lithium has been produced in the course of galactic chemical evolution[13]. The absence of observed lithium in the interstellar gas in the LMC in front of SN87A argues in favor of a low primordial abundance of lithium[14]. Following the summary of the PopII data and the analysis of Deliyannis etal[15], I will adopt the conservative bound,

$$(Li/H)_P \equiv y_{7P} \lesssim 2.5 \times 10^{-10}. \tag{9}$$

Setting ^{4}He aside for the moment, consider next deuterium. Being so weakly bound and with such a small coulomb barrier, any deuterium incorporated in stars is quickly burned away. For the same reason, no astrophysical site has ever been found where deuterium can be synthesized in astrophysically interesting amounts (i.e., comparable to the abundances observed). This leads to the conclusion that the D/H ratio has decreased from its primordial value ($y_{2P} \gtrsim y_{2obs}$). The observational situation, summarized in Boesgaard and Steigman[4], hasn't changed. From data on the abundances of deuterated molecules in the atmospheres of the giant planets, as well as observations of absorption due to D in the interstellar gas in front of (relatively nearby) hot stars,

$$(D/H)_P \equiv y_{2P} \gtrsim y_{2obs} \gtrsim 2 \times 10^{-5}. \tag{10}$$

As we will see shortly, it is very important to bound y_{2P} from above. This may be done as follows. Primordial deuterium can be reduced significantly if a large fraction of the Galaxy has been cycled through stars. However, in stars, D is burned to ^{3}He and, some ^{3}He survives stellar processing[15]. Indeed, cool (low mass) stars are net producers of ^{3}He[16]. Too much destruction of deuterium would, therefore, result in the production of too much ^{3}He. From observations of ^{3}He in different meteorites, the sum of the presolar abundances of D plus ^{3}He may be bounded[5],

$$3.1 \lesssim 10^5 y_{23\odot} \lesssim 5.1, \tag{11}$$

where, $y_{23} \equiv (D + {}^3\mathrm{He})/\mathrm{H}$. The solar system bounds may then be used to infer an upper bound to the sum of the primordial abundances of D and ^{3}He[1,5],

$$10^5 y_{23P} \lesssim 11. \tag{12}$$

Last, we turn to ^{4}He, the second most abundant element (after hydrogen) in the Universe. Since ^{4}He is so abundant, it can be observed throughout the Universe via its emission lines in regions of hot, ionized gas (HII Regions). As a corollary, the abundance of ^{4}He may be determined with a much higher accuracy than that of the other, less abundant, light elements. This fact is crucial since Y_P is so insensitive to the nucleon abundance (see eq. 8). Great care must be taken to extract, from current observations, an accurate estimate of the primordial mass fraction Y_P (e.g., see refs 4-6, 17, 18). The ^{4}He emission lines observed from HII regions result from recombinations ($He^+ \rightarrow He^\circ, He^{++} \rightarrow He^+$) and, therefore, leave unprobed the neutral helium content (in regions where hydrogen is totally ionized, some helium may still be neutral). The neutral helium fraction must be extracted from model fits to the observations or, guessed at; the most common assumption is that there is no neutral helium. This introduces some – difficult to assess quantitatively – uncertainty in the derived helium abundances. Further, ^{4}He is produced as stars burn hydrogen. Some of the observed ^{4}He is the debris of earlier generations of stars. To extract the primordial abundance it is necessary to correct for stellar and galactic evolution. Following Peimbert and Torres-Peimbert,[19] this is usually attempted by correlating the observed ^{4}He abundance with that of some element (usually oxygen[19]; more recently nitrogen[17] and carbon[20]) produced in stars. Comparing recent ^{4}He determinations in high excitation, low metallicity, extragalactic HII regions[17,20] with the corresponding abundances of C, N, O, Olive etal[5] have determined,

$$Y_P = 0.23 \pm 0.01. \tag{13}$$

Because the uncertainties are dominated by unknown systematic errors, it is difficult to assess the statistical significance of these bounds. For comparison between theory and observation, I will adopt: $Y_P \lesssim 0.240$ (but, be forewarned that the third significant figure is uncertain; Y_P^{MAX} may be 0.235 or 0.245).

IV. Confrontation of Theory With Observations

We are now in a position to compare the predictions of big bang nucleosynthesis in the standard model with the observational data. To assess the consistency of the standard model we must determine if there exists a value (or range of values) of the nucleon-to-photon ration η for which the predicted and observed abundances (within their uncertainties) are in agreement.

We have noted earlier (eq. 10) that $y_{2P} \gtrsim 2 \times 10^{-5}$ and, that a good fit to the predicted deuterium abundance is (eq. 2) $y_{2P} = 4.6 \times 10^{-4} \eta_{10}^{-5/3}$ so that,

$$(D) \quad \eta_{10} \lesssim 6.6. \tag{14}$$

However, the PopII abundance of lithium provides a tighter constraint (eq. 9); $y_{7P} \lesssim 2.5 \times 10^{-10}$ requires[5],

$$(Li) \quad \eta_{10} \lesssim 5.0. \tag{15}$$

To provide a lower bound to η, we employ the constraint (eq. 12) from the sum of the abundances of deuterium and helium-3, $y_{23P} \lesssim 11 \times 10^{-5}$, which requires[5]

$$(D +^3 \text{He}) \quad \eta_{10} \gtrsim 2.6\,. \tag{16}$$

Thus, although the standard model is, in principle, falsifiable (it could have emerged that the upper bound to η was smaller than the lower bound), there is a consistent range for the universal abundance of nucleons,

$$2.6 \lesssim \eta_{10} \lesssim 5.0. \tag{17}$$

What of the ^{4}He abundance? For the standard case of three families of light ($\lesssim$MeV), left-handed neutrinos ($N_\nu = 3$), we predict (for $\tau_n = 889.8 \pm 8.8$ sec and η_{10} constrained by eq. 17) that the primordial helium mass fraction should lie in the range[5] (see, also, eq. 8),

$$0.235 \leq Y_P(\text{pred.}) \leq 0.246\,. \tag{18}$$

If (13) and (18) are compared (and, the warning after eq. 13 is heeded), it is clear that consistency extends to helium-4 as well.

As Steigman, Schramm and Gunn[2] noted, the helium abundance provides a bound on the number of "equivalent" light neutrinos (N_ν; see eqs. 5&8). From eq. 8 we may write,

$$N_\nu = 4.0 + 20\left(\frac{Y_P - 0.24}{0.24}\right) - 0.8\ln\eta_{10} - 1.5(\tau_{1/2} - 10.28). \tag{19}$$

For $\eta_{10} \geq 2.6$, $\tau_{1/2} \geq 10.18$ (95% C.L. lower limit) and $Y_P \leq 0.24$, we predict

$$N_\nu \leq 3.4. \tag{20}$$

(Note that if we had chosen $Y_P^{MAX} = 0.245$, we would have predicted $N_\nu \leq 3.8$; the bounds from the current data exclude an additional family of light neutrinos.)

V. Summary and Conclusions

We have reviewed the synthesis of the light elements in the context of the standard hot big bang cosmology and have compared the predicted abundances of D, ^{3}He, ^{4}He and ^{7}Li with the current observational data. The standard model is revealed to be consistent with the existing data. It has been one of our goals here to attempt to place a "second significant figure" on the constraints emerging from the comparison of theory and observation. For the range of nucleon abundances for which the predicted and observed light element abundances are in accord ($2.6 \leq \eta_{10} \leq 5.0$), we find (see eq. 1; we adopt 95% C.L. limits $T_\gamma = 2.78 \pm 0.02K$),

$$0.040 \leq \Omega_N h_{50}^2 \leq 0.077. \tag{21}$$

For a Hubble parameter in excess of [21] $40\text{kms}^{-1}Mpc^{-1}$,

$$\Omega_N \leq 0.120. \tag{22}$$

By more than a factor of eight, the Universe fails to be "closed" by nucleons.

For a Hubble parameter less than $100 kms^{-1} Mpc^{-1}$, the lower bound to the present nucleon density is $\Omega_N \geq 0.010$. However, such a Universe ($h_{50} = 2$) would be very young. If we assume that $\Omega_{TOT} = 1$, the present age of the Universe is $t_0 = 13 h_{50}^{-1}$ Gyr. It follows from (21) that,

$$\Omega_N \geq 0.040 \left(\frac{t_0}{13Gyr} \right)^2 . \tag{23}$$

If, for example, $t_0 \gtrsim 10 - 13$Gyr, then $\Omega_N \gtrsim 0.024 - 0.040$.

If the mass to light ratio inferred from the luminous matter in galaxies[22,23] were typical of the matter throughout the Universe,

$$\Omega_{LUM} \leq 0.006 h_{50}^{-1} . \tag{24}$$

Comparing (21) and (24) we see that most of the nucleons in the Universe must be dark.

$$\Omega_N / \Omega_{LUM} \geq 6.7 h_{50}^{-1} \geq 3.3 . \tag{25}$$

If, instead of assuming that $\Omega_{TOT} = 1$, we turn to the data[23], we have $\Omega_{TOT} \gtrsim 0.2 \pm 0.1$. Since $\Omega_N \leq 0.12$, it is suggestive but, not yet mandatory, that the Universe is dominated by nonbaryonic dark matter.

Finally, we must remark on the excellent agreement between the cosmological bound to $N_\nu (\leq 3.4 - 3.8)$ and the very recent results from the SLAC and LEP $e^\pm$ colliders[24]. A weighted average of the ALEPH, DELPHI, L3 and OPAL collaborations at LEP and Mark II at SLAC yields,

$$N_\nu(\text{accel.}) = 3.2 \pm 0.2 , \tag{26}$$

confirming that a fourth family of neutrinos is excluded. This is the first time a cosmological prediction has been confirmed in a laboratory experiment. It gives us some confidence in our enterprise of using the early Universe to explore not only cosmology but, high energy physics as well.

VI. Acknowledgements

It is a great honor to have been invited to speak at this conference in honor of Felix Boehm. In his life as well as his science, Felix embodies the very highest standards. I sincerely hope my contribution is worthy of him and, I take great pleasure in wishing him well. My thanks go, too, to Peter Vogel for organizing this meeting and, for his patience.

Many of the results I presented here are from work with Keith Olive, Dave Schramm and Terry Walker and I am pleased to acknowledge their contributions. This work is supported at OSU by the DOE.

References

1. J. Yang, M.S. Turner, G. Steigman and K.A. Olive, *Ap. J.* 281, 493 (1984).
2. G. Steigman, D.N. Schramm and J.E. Gunn, *Phys. Lett.* B66, 202 (1977).
3. G. Steigman, K.A. Olive, D.N. Schramm and M.S. Turner, *Phys. Lett.* B276, 33 (1986).
4. A.M. Boesgaard and G. Steigman, *Ann. Rev. Astron. Astrophys.* 23, 319 (1985).
5. K.A. Olive, D.N. Schramm, G. Steigman and T.P. Walker, *Phys. Lett. B* (Submitted 11/89; OSU-TA-12/89).
6. T.P. Walker, G. Steigman, D.N. Schramm and G. Steigman, In Preparation (1989).
7. E. Palazzi etal., *Ap. J. Lett.*, In Press (1989).
8. W. Mampe etal., *Phys. Rev. Lett.* 62, 593 (1989).
9. G. Steigman, *Ann. Rev. Nucl. Part. Sci.* 29, 313 (1979).
10. H. Reeves, W.A. Fowler and F. Hoyle, *Nature* 226, 727 (1970).
11. T.P. Walker, G.J. Mathews and V.E. Viola, *Ap. J.* 229, 745 (1985).
12. N. Grevesse and E. Anders, AIP Conf. Proc. 183, 1 (1989).
13. D.S.P. Dearborn, D.N. Schramm, G. Steigman and J.W. Truran, *Ap. J.*, In Press (1989).
14. G. Steigman, "Cosmic Lithium: Going Up Or Coming Down?", In Preparation (1989).
15. D.S.P. Dearborn, D.N. Schramm and G. Steigman, *Ap. J.* 302, 35 (1986).
16. I. Iben, *Ap. J.* 147, 624, 650 (1967).
17. B. Pagel, A Unified View of the Macro- and Micro-Cosmos (eds., A. DeRujula, D.V. Nanopoulos and P.A. Shaver; *World Scientific*) p. 399 (1988).
18. K. Davidson, T.D. Kinman and S.D. Friedman, *A.J.* 97, 1591 (1989).
19. M. Peimbert and S. Torres-Peimbert, *Ap. J.*, 193, 327 (1974).

20. G. Steigman, J.S. Gallagher and D.N. Schramm, *Comm. Astrophys.* 14, 97 (1989).

21. G. Steigman, Proc. 5th IAP Astrophys. Mtg. (eds, M. Casse & E. Vangioni-Flam; Editions Frontieres) In Press (1989).

22. S. Faber and J.S. Gallagher, *Ann. Rev. Astron. Astrophys.* 17, 135 (1979).

23. B.E.J. Pagel, To appear in Baryonic Dark Matter (eds., G. Gilmore & D. Lynden-Bell; 1989).

24. ALEPH, DELPHI, L3, OPAL & Mark II Collaboration preprints (1989).

PARITY VIOLATION IN ATOMS

V. L. Telegdi
ETH Zürich

I am dedicating this short survey to my good friend Felix, whom I've had the pleasure of knowing since our graduate student days in Zürich, i.e. for over forty years. He stood out then, as he does today, not only by his deep and sincere interest in physics, but by his cultured interests and his refined manners. As opposed to others, he accepted me with all my shortcomings long before I could qualify as a successful physicist. For this, and for helping to arrange my regular visits to Caltech, I owe him a great debt.

This short survey does not aim to be a complete review of the field; its purpose is primarily pedagogical. I shall try to derive the necessary formulae more simply than in the literature and to focus on the essential features of the often very complex experimental arrangements.

Let me remark at the outset that atomic parity experiments have not yielded great surprises about the nature of the weak neutral current. They just supply another confirmation of the "standard" theory of electroweak interactions. Their interest lies in providing determinations of the key parameter ($\sin^2\theta_W$) of the theory in the limit of zero momentum transfer, and perhaps in providing a tool for the experimental discovery of nuclear magnetic anapole moments (which in turn originate from parity violating effects within the atomic **nucleus**).

1. THE PARITY VIOLATING POTENTIAL, V^{PV}

While the ordinary photon γ has only vector couplings to charges, the heavy photon Z° has **both** vector and axial vector couplings to all fermions. Parity violation in atoms arises from the existence of the two diagrams in Fig. 1, which we shall consider in the non-relativistic, $q^2 \to 0$ limit. Essentially, only the time component of the current, the "charge" matters, and the entire nucleus interacts **coherently** with the electron. It is thus useful to introduce, next to the nuclear charge Ze, a weak charge Q_W; this will be the sole quantity characterizing the nucleus. Nuclear spin effects, both electromagnetic and weak, are of order $1/A$, and can safely be neglected (like Z° diagrams with V and A vertices interchanged).

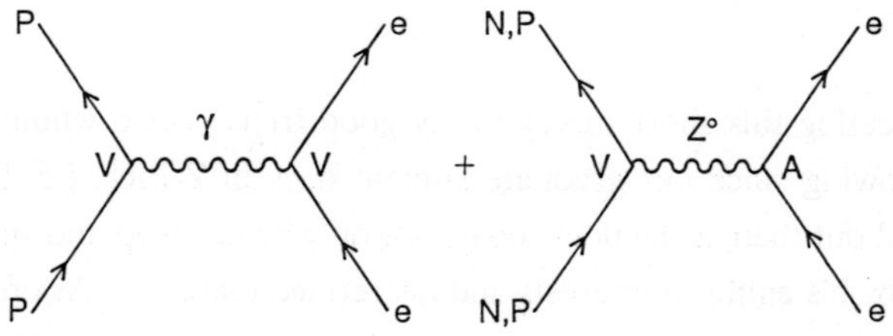

Figure 1. Interfering e.m. and Z° diagrams.

In the limit $q^2/M_Z^2 \to 0$ one recovers, of course, the familiar Fermi **contact** interaction, of strength $G_F/\sqrt{2}$. With these remarks, it is easy to guess the parity violating potential[1] (units $\hbar = c = 1$)

$$V^{PV} = \frac{G_F}{\sqrt{2}} \cdot \frac{1}{2m} \left[\vec{\sigma}\cdot\vec{p}\,\delta(0) + \delta(0)\vec{\sigma}\cdot\vec{p}\right] Q_W(N,Z) , \qquad (1)$$

where m, $\vec{\sigma}$ and $\vec{p}$ refer to the electron. The p.v. nature of this potential is evidenced by the fact that $\vec{\sigma}\cdot\vec{p}$ is a **pseudoscalar**. To obtain $Q_W(N,Z)$, one needs the dynamics of the electroweak theory; all we can anticipate is that it must be linear in N,Z and the parameter $\sin^2\theta_W$. The standard theory gives[1]

$$Q_W = - \left[N + Z\,(1 - 4\sin^2\theta_W)\right] . \qquad (2)$$

Since we know from the earliest high-energy neutral current experiments that $\sin^2\theta_W \approx 1/4$, we can readily see from Eq. (2) that parity experiments with atomic hydrogen look rather hopeless. Equation (1) suggests, furthermore, a fast increase of V^{PV} with Z.

V^{PV} mixes states of opposite parity; since in ordinary atoms (as opposed to muonic atoms) only s-states do not vanish at the nucleus, it suffices to consider s-p mixing. Thus, $|nS>$ becomes $\sim(|nS> + \eta\sum_{n'}|n'P>)$. Note that the mixing parameter η is pure imaginary because of $\vec{p} = i\nabla$ in Eq. (1). The quantitative usefulness of atomic parity experiments is limited by the accuracy with which η can be derived calculating $<nS|V^{PV}|n'P>$. It is generally believed that such calculations are most reliable for hydrogenic atoms. This makes the heaviest alkali, Cs, the most popular object of p.v. experiments. Its relevant levels are shown in Fig. 2.

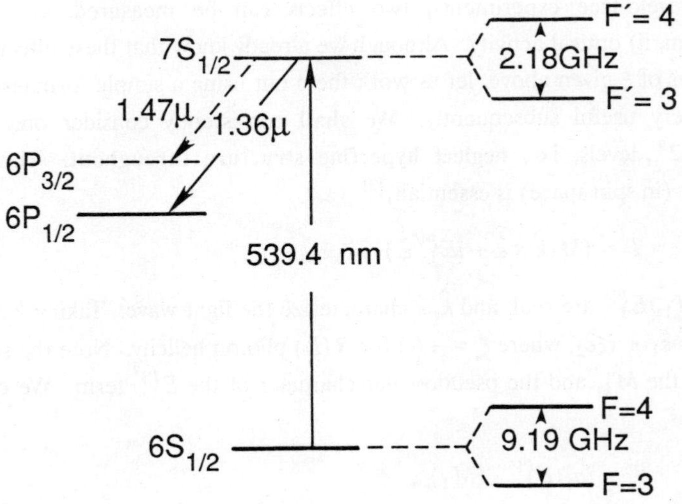

Figure 2. Cs levels.

2. THE VARIOUS TYPES OF P.V. EXPERIMENTS

The two levels with $\Delta J = 0,1$ and identical parity, such as the $6S_{1/2}$ and $7S_{1/2}$ levels of Cs (see Fig. 2), are connected by an $M1$ e.m. transition. The "parity contamination", η, leads to an additional electric dipole amplitude, E_1^{PV}. It is such an

E_1^{PV} that is determined in all the experiments. These can be divided into two classes, viz.

a) experiments without external fields,

b) experiments with $\vec{E}, \vec{B}$ fields.

2a. Experiments Without Fields

It is well known, e.g. from p.v. experiments in nuclear transitions (on which F. Boehm has worked extensively), that an M_1 - E_1^{PV} interference will lead to a circular polarization ξ of $\xi \simeq 2E_1^{PV}/M_1$. It is often argued that one should enhance this effect by choosing a **highly forbidden** M_1 transition. This is not a compelling argument, since the rate of **absorption** goes as M_1^2 and hence the statistical error as M_1. While the M_1 in Cs is indeed highly forbidden ($4 \cdot 10^{-5} \mu_B$), it is allowed ($\sim \mu_B$) in Bi, where the p.v. optical activity has been successfully measured.

In field-free experiments, two effects can be measured, viz. i) circular dichroism, ii) optical activity. Although we already know that these effects will be of the order of ξ given above, let us work them out using a simple formalism that will prove very useful subsequently. **We shall** consistently consider only a pair of $1/2^+, 1/2^+$ levels, i.e., **neglect hyperfine structure throughout.** The transition operator (in spin space) is essentially[3]

$$T = (M_1 \hat{k} \times \hat{\varepsilon} + iE_1^{PV} \hat{\varepsilon}) \cdot \vec{\sigma} , \tag{3}$$

where M_1, E_1^{PV} are real, and $\hat{k}$, $\hat{\varepsilon}$ characterize the light wave. Taking $\hat{k}$ along z, we have $\hat{\varepsilon} = \hat{\varepsilon}_1 + i\xi\hat{\varepsilon}_2$, where $\xi = +(-)$ for $R\,(L)$ photon helicity. Note the scalar character of the M_1, and the pseudoscalar character of the E_1^{PV} term. We can rewrite Eq. (3) as

$$T = i\sigma_\xi (E_1^{PV} - \xi M_1) , \tag{3'}$$

and obtain immediately the absorption amplitudes in Fig. 3.

Circular dichroism δ is defined as

$$\delta \equiv (\sigma_R - \sigma_L)/(\sigma_R + \sigma_L) ,$$

where the $\sigma_{R,L}$ are the absorption cross sections for R and L light. Squaring the amplitudes in Fig. 3,

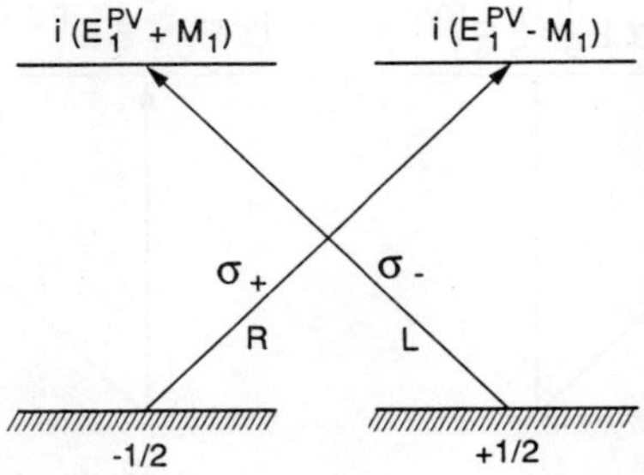

Figure 3. Transition amplitudes for circular dichroism.

$$\delta = 2E_1^{PV} M_1 / [\,(E_1^{PV})^2 + M_1^2] \simeq 2E_1^{PV}/M_1 \; .$$

Since $\delta \sim \mathrm{Im}\,(n_R - n_L)$, where the n's are indices, the **optical activity** is proportional to the same quantity.

A measurement of the optical activity in Bi, performed in Novosibirsk,[2] was the first successful p.v. atomic experiment. On the other hand, no circular dichroism p.v. measurement has been completed so far; the ETHZ experiment[3a] is still in progress.

2b. Experiments with external fields

The basic idea of such experiments, based primarily on detailed proposals made very early by the Bouchiats (a husband and wife team!) in Paris, is to **enhance** the forbidden M_1 transition by a static electric field $\vec{E}_0$: the Stark admixture of p-states yields an "allowed" p.c. E_1 amplitude. In radio ham terms, we have a "heterodyne" approach, with the Stark amplitude as the carrier wave, and the E_1^{PV} as the signal. It turns out that there arises a **p.v. electronic polarization,** $\vec{P}_{PV}$, perpendicular to the $(\vec{E}_0, \hat{k})$ plane. This we shall now compute.

The transition operator is (safely neglecting M_1) now

$$T = (\underbrace{\alpha \vec{E}_0 1}_{\text{p.c.}} + \underbrace{i E_1^{PV} \vec{\sigma}}_{\text{p.v.}}) \cdot \hat{\varepsilon} \; , \qquad (4)$$

where α = scalar electric polarizability. We take $\hat{k}$ and $\vec{E}_0$ perpendicular, and conveniently orient the axes as in Fig. 4. The light polarization is then

$$\hat{\varepsilon} = \hat{\varepsilon}_1 + i\xi\varepsilon_3 \; . \qquad (5)$$

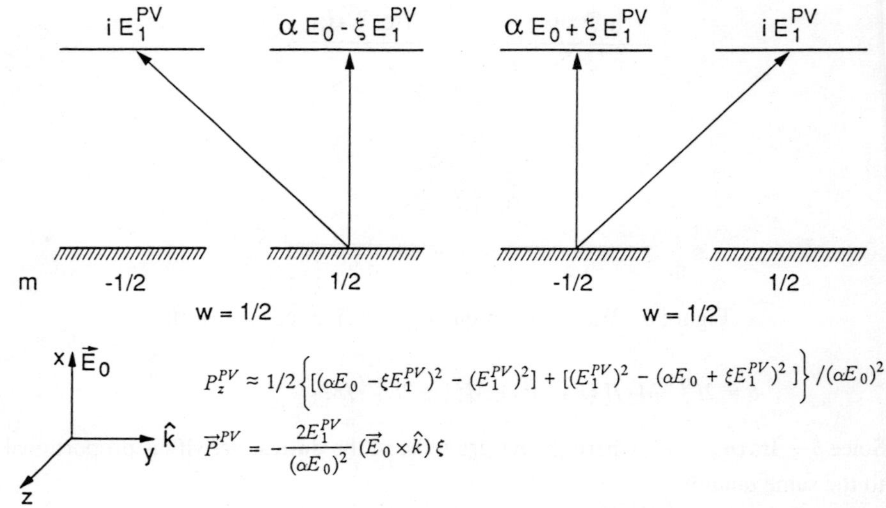

$$P_z^{PV} \approx 1/2\left\{[(\alpha E_0 - \xi E_1^{PV})^2 - (E_1^{PV})^2] + [(E_1^{PV})^2 - (\alpha E_0 + \xi E_1^{PV})^2]\right\} / (\alpha E_0)^2$$

$$\vec{P}^{PV} = -\frac{2E_1^{PV}}{(\alpha E_0)^2}(\vec{E}_0 \times \hat{k})\,\xi$$

Figure 4. Stark-E_1^{PV} interference. Levels and amplitudes.

Applying T as given by Eqs. (4) and (5) to the $m = \pm 1/2$ sublevels of the ground state, one readily obtains the amplitudes indicated in Fig. 4. These in turn yield immediately the polarization along y:

$$\vec{P}^{PV} = -\frac{2E_1^{PV}E_0}{(\alpha E_0)^2}(\hat{E}_0 \times \hat{k})\,\xi \,. \tag{6}$$

Note that formally this polarization is a **polar** vector (!); actually, it is an **axial** vector with a pseudoscalar coefficient, the helicity ξ.

A remarkable property of $\vec{P}^{PV}$ is that it is unchanged if one reverses $\hat{k}$ and ξ **simultaneously**. This is precisely what happens if a circularly polarized wave is inverted by reflection in a mirror. Reflection does not change the photon's angular momentum $\vec{J}_\gamma = \hbar\hat{k}\xi$. The Bouchiats have exploited this property to use a **multipass** cell, thereby increasing the effective cell length by over two orders of magnitude! On the other hand, $\vec{P}^{PV}$ changes sign when reversing $\vec{E}_0$ or ξ.

Things would be moderately easy **if** $\vec{P}_{PV}$ was the **only** polarization arising in the geometry idealized in Fig. 4. A polarization, $\vec{P}_1$, arises from the interference of the two parity conserving (p.c.) Stark and M_1 amplitudes (the latter was neglected so far). The relevant transition operator is

$$T = M_1(\hat{k} \times \hat{\varepsilon}) \cdot \vec{\sigma} + \alpha E_0 \hat{\varepsilon} \ . \tag{7}$$

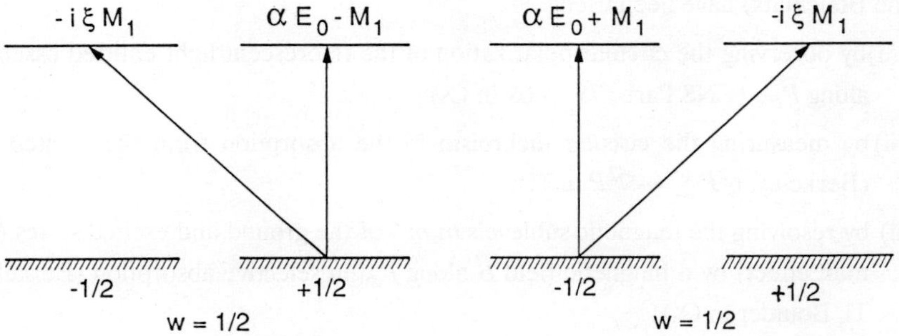

$$P = 1/2\left\{[(\alpha E_0 - M_1)^2 - \xi^2 M_1^2] + [\xi^2 M_1^2 - (M_1 + \alpha E_0)^2]\right\} / (\alpha E_0)^2$$

$$\vec{P}_1 \approx \frac{2\alpha \vec{E}_0 M_1}{(\alpha E_0)^2} \cdot \hat{\varepsilon}\, \mathrm{Re}(\hat{\varepsilon} \times \hat{k}) \ ; \ \vec{E}_0 \cdot \vec{\varepsilon}_x = \mathrm{Re}(\vec{E}_0 \cdot \hat{\varepsilon})$$

Figure 5. Stark-M_1 interference. Levels and amplitudes.

Figure 5 shows the resultant amplitudes. One readily obtains

$$\vec{P}_1 \equiv \frac{2M_1 \alpha E_0}{(\alpha E_0)^2} \varepsilon_3 = \frac{2M_1 \alpha \vec{E}_0 \cdot \hat{\varepsilon}}{(\alpha E_0)^2} \mathrm{Re}\, (\hat{\varepsilon} \times \hat{k}) \ . \tag{8}$$

The relevant things are (a) that $\vec{P}_1$ is independent of ξ, i.e. it depends only on the **linear** polarization, (b) that it reverses with $\hat{k}$, i.e., averages to zero in an **ideal** multipass system.

The pure M_1 amplitude will contribute a polarization along $\hat{k}$, namely

$$\vec{P}_0 \sim M_1^2 \xi \hat{k} \ . \tag{9}$$

Finally, a spin-dependent Stark effect term, the so-called "tensor polarizability" β_1 can also create a polarization, $\vec{P}_2$, along $\hat{k}$:

$$T = (\alpha \vec{E}_0 + i\beta \vec{\sigma} \times \vec{E}_0) \cdot \hat{\varepsilon} + \cdots \tag{10}$$

$$\vec{P}_2 \sim \frac{2\alpha\beta E_0^2}{(\alpha^2 + \beta^2)E_0^2} \xi \hat{k} \ . \tag{11}$$

$\vec{P}_0$ and $\vec{P}_2$ do not average to zero in the multipass interaction and do not reverse with $\vec{E}_0$.

Determination of $\vec{P}_{PV}$: Three different approaches (basically all suggested by the Bouchiats) have been used:

i) by observing the circular polarization of the fluorescent light emitted essentially along $\vec{P}_{PV}$ (ENS Paris, $7S \to 6S$ in Cs);

ii) by measuring the circular dichroism in the absorption from the excited state (Berkeley, $6^2P_{1/2} \to 7^2P$ in Tl);

iii) by resolving the magnetic sublevels m, m' of the ground and excited states (Zeeman effect) by a magnetic field $\vec{B}$ along $\vec{P}_{av}$ in selective absorption (Berkeley in Tl, Boulder in Cs).

2bi. Fluorescence analysis

The relevant signal detected in the ENS Cs experiment[4] is

$$S_{PV} \sim \vec{P}_{PV} \cdot \hat{k}_f \, \xi_f \ ,$$

i.e., a pseudoscalar quantity. The signals due to the other polarizations (which are axial vectors) are scalars and vanish under appropriate geometric inversions.

The apparatus used at the ENS is shown schematically in Fig. 6. The fluorescent light (1.3 μm) is analyzed with a rotating $\lambda/4$ plate, and detected with a small diode. The incident light comes from an ingenious modulator which produces a continuous sequence of elliptic polarizations, including R and L circular polarizations. The various signals are detected at specific frequencies by a lock-in amplifier.

In a first experiment, they used a transition where we can roughly apply Eq. (7), in a field $E_0 = 100$ V/cm. Numerically, $\alpha = 2.65 a_0^3 \cdots$, and $E_1^{PV} = 10^{-11} e a_0$ (see Appendix), and $e^2/a_0 \approx 5 \cdot 10^9$ V/cm; thus

$$| P_1^{PV} | = \frac{2 \cdot 10^{-11}}{2.65 \times 2 \cdot 10^{-6}} = \underline{3.8 \cdot 10^{-6}} \ .$$

In the second experiment, they used a $\Delta F \neq 0$ transition, where the tensor polarizability β plays the role of α, with a field $E_0 = 600$ V/cm. Since $\beta \simeq \alpha/10$, one has a factor 10/6 with respect to the first signal, i.e.,

$$| P_1^{PV} | = \frac{38}{6} \cdot 10^{6} \simeq \underline{6.3 \cdot 10^{-6}} \ .$$

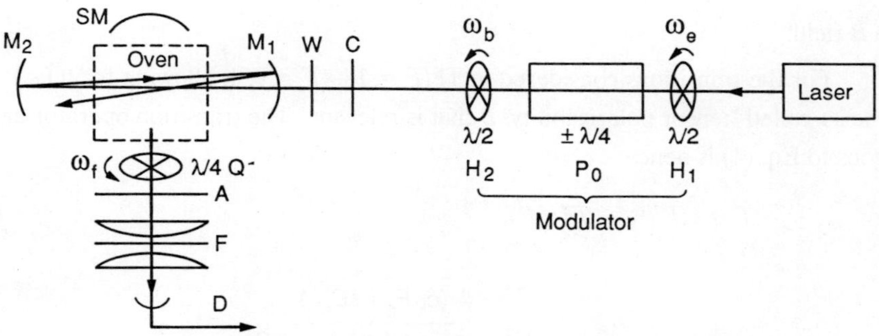

Figure 6. Apparatus used in the Paris Cs experiment.
H_1, H_2: rotating half-wave plates. Po: Pockels cell. C: compensator. W: entrance window. M_1, M_2: multipass mirrors. SM: spherical mirror. Q: rotating quarter-wave plate. A: Polaroid analyzer. F: interference filter. D: detector.

Note that the other p.v. polarizations are much larger, so the geometry has to be close to ideal. In this sense a field-free experiment sounds less prone to systematics.

Finally, we stress that the actual **measured** result, E_1^{PV}/α or E_1^{PV}/β (both [V/cm]), needs independent knowledge of α (or β) to yield E_1^{PV}.

2bii. Circular dichroism of the excited state

This was the first method used by Commins and his collaborators[5] to study parity violation in Tl, an atom in which 10 times larger effects are expected than in Cs, because of the Z^2Q_W factor in E_1^{PV} (see Appendix). The forbidden M_1 is here $6^2P_{1/2} \rightarrow 7^2P_{1/2}$, and the "scanning transition" is $7P_{1/2} \rightarrow 8S_{1/2}$.

We shall not discuss this method, which ultimately gave 27% accuracy, because it has been superceded by subsequent work using both $\vec{E}$ and $\vec{B}$ fields.

2biii. Simultaneous use of $\vec{E}$ and $\vec{B}$ fields

The key idea, advanced independently[6,7] by the Paris and Berkeley groups, is to perform the determination of P^{PV} (due to E_1^{PV}-Stark interference) by a measurement of the transition rates between resolved Zeeman levels, i.e., in the presence of a $\vec{B}$ field.

For the transitions considered in Tl ($F = 1 \rightarrow F = 0, 1$, allowing for h.f.s.) it is the so-called "tensor polarizability" β that is relevant. The transition operator analogous to Eq. (4) is hence

$$T = i\,(E_1^{PV} + \beta\vec{\sigma}\times\vec{E}_0)\cdot\hat{\varepsilon}\ . \tag{12}$$

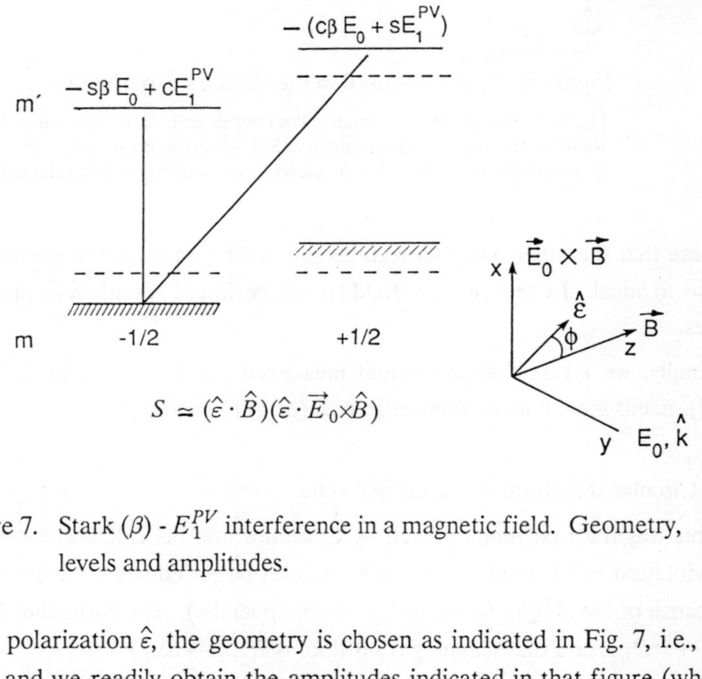

Figure 7. Stark (β) - E_1^{PV} interference in a magnetic field. Geometry, levels and amplitudes.

For linear polarization $\hat{\varepsilon}$, the geometry is chosen as indicated in Fig. 7, i.e., with $\hat{k}$ along $\vec{E}_0$, and we readily obtain the amplitudes indicated in that figure (where we neglect again h.f.s.). The intensities are

$$I(\Delta m = 0) \cong \beta^2 s^2 E_0^2 - 2cs\beta E_0 E_1^{PV} \tag{13}$$

and

$$I(\Delta m = +1) \cong \beta^2 c^2 E_0^2 + 2cs\beta E_0 E_1^{PV}\ ,$$

where $c = \cos\theta = \hat{\varepsilon}\cdot\hat{B}$, $s = \sin\theta$. The relevant pseudoscalar signal S_{PV} is proportional to $(\hat{\varepsilon}\cdot\hat{B})(\hat{\varepsilon}\cdot\vec{E}_0\times\hat{B})$; note that $\hat{B} = \vec{B}/B$. These simply derived expressions remain valid in a complete treatment that allows for h.f.s.

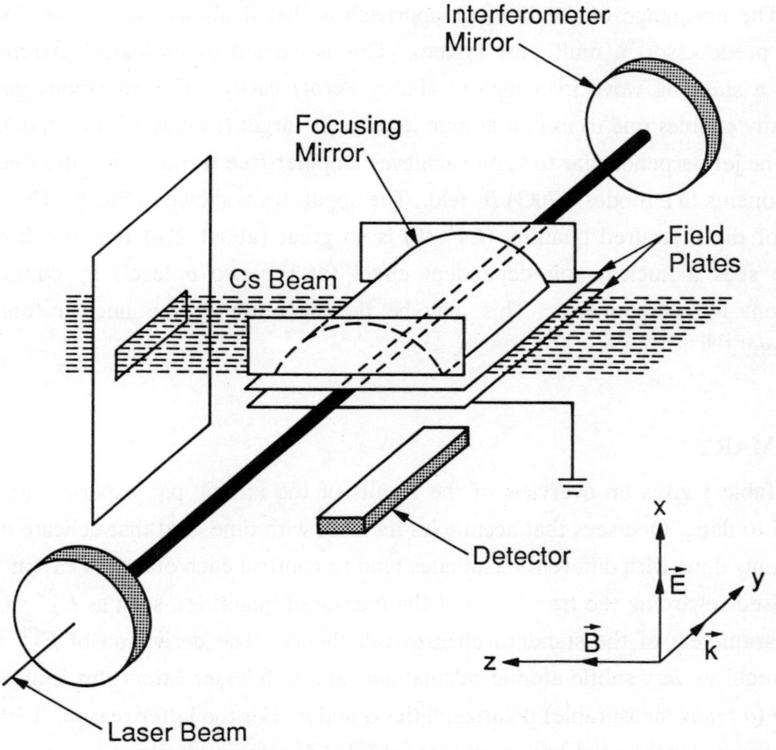

Figure 8. Geometry and apparatus of the Boulder Cs experiment.

The Tl experiment[8] is carried out (single-pass) in a high (3.2 kG) field in order to resolve the Doppler-broadened Zeeman components. $I(\Delta m)$ is measured at various angles θ through the fluorescent decay of the $7P_{1/2}$ state.

For circular polarization one keeps the geometry (E_0, B, $\hat{k}$ mutually perpendicular) that we discussed in relation to Cs earlier. In fact, a high-precision $6S \rightarrow 7S$ experiment was performed by the Boulder group[9] in this way.

Again neglecting h.f.s., we can readily write down the relevant intensity

$$I(\Delta m = \xi = \pm 1) \cong (2\beta^2 E_0^2 -+ 2\beta E_0 E_1^{PV}) \; . \qquad (14)$$

The characteristic pseudoscalar is $\cong (\vec{E}_0 \times \vec{B}) \cdot (\hat{k}\xi)$.

The advantage of the Boulder approach is that it allows one to use, like its Paris predecessor, a multi-pass system. This is carried to its logical extreme by using a standing wave in a high-Q (Fabry-Perot) cavity. The enormous gain in intensity enables one to use an atomic jet at a Cs target (instead of the vapor), and with the jet perpendicular to $\hat{k}$, one achieves Doppler-free resolution of the Zeeman components in a modest (70G) B field. The apparatus is shown in Fig. 8. The accuracy of the measured quantity (E_1^{PV}/β) is so great (about 2%) that the Boulder group sees a nuclear spin-dependent effect (at the two σ level) by comparing different h.f. components. This may be the signature of the nuclear "anapole moment".[10]

SUMMARY

Table I gives an overview of the results of the atomic p.v. experiments published to date. One sees that accuracies increase with time, and that delicate measurements done with different techniques tend to confirm each other. A caveat must be raised regarding the translation of the measured quantities, such as E_1^{PV}/β, into the parameters of the standard electroweak theory. The derivation of E_1^{PV} from V^{PV} requires very subtle atomic calculations, and to a lesser extent this is also true of the (*a priori* measurable) polarizabilities α and β. For the latter reason, field-free experiments (such as the one in progress at ETHZ) remain attractive.

A good review of experiments up to 1984 has been presented by Fortson and Lewis[11].

TABLE I (from G. L. Fogli[12])

Experimental data and theoretical estimates on which the analysis of the parity violation in atoms is based. For each experiment statistical and systematic errors are added quadratically. The number in the second column near Bi specifies that the 648-nm transition $6p^{3\,4}S_{3/2} \to 6p^{3\,2}D_{5/2}$ has been studied by the Novosibirsk, Oxford and Moscow groups, whereas in Seattle the 876-nm transition $6p^3\,{}^4S_{3/2} \to 6p^3\,{}^2D_{3/2}$ has been considered.

Experiment	Element	Experimental Estimate	Theoretical Estimate	Q_W
Novosibirsk[b)]	Bi 648	E_1^{PV}/M_1 = -(20.2±2.7) 10^{-8}	-(0.88±0.13)×$10^{-9}Q_W$[n)]	-(229.5±30.7±34.4)
Oxford[c)]	Bi 648	E_1^{PV}/M_1 = -(9.3±1.5) 10^{-8}	-(0.88±0.13)×$10^{-9}Q_W$[n)]	-(105.7±17.0±15.9)
Moscow[d)]	Bi 648	E_1^{PV}/M_1 = -(7.8±1.8) 10^{-8}	-(0.88±0.13)×$10^{-9}Q_W$[n)]	-(88.6±20.5±13.3)
Seattle '81[e)]	Bi 876	E_1^{PV}/M_1 = -(10.4±1.7) 10^{-8}	-(0.66±0.10)×$10^{-9}Q_W$[n)]	-(157.6±25.8±23.6)
Seattle '83[f)]	Pb	E_1^{PV}/M_1 = -(9.9±2.5) 10^{-8}	-(1.04±0.31)×$10^{-9}Q_W$[n)]	-(95.2±24.0±28.6)
Berkeley '79[a,g)]	Tl	E_1^{PV}/β = -(3.33±1.54)mV/cm	-(1.61±0.12)×$10^{-2}Q_W$mV/cm[o)]	-(206.8±95.7±15.4)
Berkeley '81[a,n)]	Tl	E_1^{PV}/β = -(1.80±0.49)mV/cm	-(1.61±0.12)×$10^{-2}Q_W$mV/cm[o)]	-(111.8±30.4± 8.3)
Berkeley '84[i)]	Tl	E_1^{PV}/β = -(1.73±0.27)mV/cm	-(1.61±0.12)×$10^{-2}Q_W$mV/cm[o)]	-(107.5±16.8± 8.0)
Paris '82[j)]	Cs	E_1^{PV}/β = -(1.33±0.25)mV/cm	-(2.20±0.12)×$10^{-2}Q_W$mV/cm[o)]	-(60.5±11.2± 3.2)
Paris '84[k)]	Cs	E_1^{PV}/β = -(1.75±0.27)mV/cm	-(2.20±0.12)×$10^{-2}Q_W$mV/cm[o)]	-(79.6±12.1± 4.3)
Boulder '85[l)]	Cs	E_1^{PV}/β = -(1.65±0.13)mV/cm	-(2.20±0.12)×$10^{-2}Q_W$mV/cm[o)]	-(75.1± 5.9± 4.0)
Boulder[m)]	Cs	E_1^{PV}/β = -(1.58±0.03)mV/cm	-(2.20±0.12)×$10^{-2}Q_W$mV/cm[o)]	-(71.7± 1.6± 3.8)

a) The measurement of E_1^{PV}/M_1 has been transformed into a measurement of E_1^{PV}/β making use of the experimental estimate of β[p)].

b) L. M. Barkov and M. S. Zolotorev, *Phys. Lett.* **85B**, 308 (1979); *Comm. Atom. Mol. Phys.* **8**, 79 (1979); *Sov. Phys. J. Exp. Theor. Phys.* **52**, 360 (1980).

c) J. D. Taylor, Oxford thesis. See also: P. E. G. Baird *et al., Phys. Rev. Lett.* **39**, 798 (1977); *Proc. Vavilov Conf.*, S. G. Rautian (ed.) (Novosibirsk, 1981) p. 22.

d) D. V. Saskyan, I. I. Sobel'man, and E. A. Yukov, *JETP Lett.* **29**, 258 (1979); Yu. V. Bogdanov *et al., JETP Lett.* **31**, 214 and 522 (1980).

e) J. H. Hollister *et al., Phys. Rev. Lett.* **46**, 643 (1981).

f) T. P. Emmons, J. M. Reeves, and E. N. Forston, *Phys. Rev. Lett.* **51**, 2089 (1983).

g) R. Conti *et al., Phys. Rev. Lett.* **42**, 343 (1979).

h) P. Bucksbaum, E. Commins, and L. Hunter, *Phys. Rev. Lett.* **46**, 640 (1981); *Phys. Rev.* **D24**, 1134 (1981).

i) P. S. Drell and E. D. Commins, *Phys. Rev. Lett.* **53**, 968 (1984).

j) M. A. Bouchiat *et al., Phys. Lett.* **117B**, 358 (1982); *Phys. Lett.* **134B**, 463 (1984).

k) M. A. Bouchiat *et al., J. Phys.* **47**, 1709 (1986).

l) S. L. Gilbert *et al., Phys. Rev. Lett.* **55**, 2680 (1985); S. L. Gilbert and C. E. Wieman, *Phys. Rev.* **A34**, 792 (1986).

m) M. C. Noecker, B. P. Masterson, and C. E. Wieman, *Phys. Rev. Lett.* **61**, 310 (1988).

n) A. M. Martensson, E. M. Henley, and L. Wilets, *Phys. Rev.* **A24**, 308 (1981).

o) M. J. Harris, C. E. Loving, and P. G. H. Sandars, *J. Phys.* **B11**, L749 (1978).

p) C. E. Tanner and E. D. Commins, *Phys. Rev. Lett.* **56**, 332 (1986).

APPENDIX

One quick glance at V^{PV} (Eq. (1)) shows that

$$<ns \mid V^{PV} \mid n'p> \cong \frac{\alpha}{\sqrt{2}} \left(\frac{G_F}{a_0^3} \right) Q_W f(Z; \nu, \nu') \ , \tag{A1}$$

where ν, ν' are the effective quantum numbers of the shielded valence electron ($n \equiv \nu - \sigma$, $\sigma =$ quantum defect). It is relatively easy to see that $f = Z^2 g(\nu, \nu')$. Hence we have, in atomic units,

$$<V^{PV}> \cong \frac{\alpha}{\sqrt{2}} G_F Q_W Z^2 g(\nu, \nu') \quad \text{a.u.} \tag{A2}$$

with

$$G_F = 10^{-5} (m / M_P)^2 \alpha = 3 \cdot 10^{-12} \alpha \quad \text{a.u.} \tag{A3}$$

Thus

$$E_1^{PV} = \frac{<V>}{\Delta E(\nu, \nu')} \varepsilon_1 \cong \frac{3 \cdot 10^{12}}{\sqrt{2}} \frac{\alpha^2 Z^2}{\Delta E} Q_W g \varepsilon_1 \ ,$$

where $\varepsilon_1 =$ electric dipole matrix element between the relevant s and p states. Putting brutally $g\varepsilon_1 / \Delta E = 1$, we get

$$E_1^{PV} \cong 2 \cdot 10^{-16} \, Q_W Z^2 \ .$$

For Cs, $Q_W Z^2 \approx 2 \cdot 10^5$, so that

$$E_1^{PV}(\text{Cs}) \simeq 4 \cdot 10^{-11} (e a_0) \ .$$

Careful evaluations, with allowance for relativistic corrections, yield

$$E_1^{PV}(\text{Cs}) \simeq 1 \cdot 10^{-11} e a_0 \ . \tag{A4}$$

With the **experimental value**

$$M_1 \simeq 4 \cdot 10^{-5} \, \mu_B = 2 \cdot 10^{-5} \alpha \quad \text{a.u.} \ ,$$

one predicts

$$\frac{2E_1^{PV}}{M_1} (\text{Cs}) \simeq 10^{-4} \ . \tag{A5}$$

REFERENCES

[1] M. A. Bouchiat and C. Bouchiat, *Phys. Lett.* **48B**, 111 (1974).

[2] L. M. Barkov and M. S. Zolotarev, *Pis'ma Zh. Eksp. Teor. Fiz.* **26**, 379 (1978).

[3] M. A. Bouchiat and C. Bouchiat, *J. Phys.* (Paris) **36**, 493 (1975).

[3a] P. P. Herrmann *et al., J. Phys. B* **19**, 1607 (1986).

[4] (a) M. A. Bouchiat *et al., Phys. Lett.* **117B** (1982); (b) *ibid.* **134B**, 463 (1984); (c) *J. Phys.* (Paris) **36**, 493 (1986).

[5] P. H. Bucksbaum, E. D. Commins, and R. L. Hunter, *Phys. Rev.* **D24**, 1134 (1981).

[6] M. A. Bouchiat, M. Poirier, and C. Bouchiat, *J. Phys.* (Paris) **40**, 1127 (1979).

[7] E. D. Commins and P. H. Bucksbaum, *Proc. Workshop on Parity Violation in Atoms*, Cargèse, Corsica, 1979 (unpublished).

[8] P. S. Drell and E. D. Commins, *Phys. Rev. Lett.* **53**, 988 (1984).

[9] C. E. Wieman *et al.* (a) *Phys. Rev. Lett.* **55**, 2680 (1985); (b) *Phys. Rev.* **A34**, 792 (1986); (c) *Phys. Rev. Lett.* **61**, 31P (1988).

[10] V. Flambaum, I. Kriplovitch, and O. Shushkov, *Phys. Lett.* **146B**, 367 (1984); P. A. Frantsuzov and I. B.l Kriplovitch, *Z. Phys.* **D7**, 297 (1988).

[11] E. N. Fortson and L. L. Lewis, *Phys. Rep.* **113**, 289 (1984).

[12] G. L. Fogli, *Z. Phys.* **C43**, 229 (1989).

A DETECTION OF HER-X 1 AT PeV ENERGIES, WITH ANOMALOUS MUON NUMBER: CONSTANCY OF THE PERIOD IN 1986

D.E. Nagle
Los Alamos National Laboratory

The CYGNUS collaboration[1] is operating a ground-based array of scintillation detectors to detect extensive air showers. The objective is to detect compact sources of Ultra High Energy (PeV) gamma radiation. At the present time the array consists of 106 scintillation detectors of $\sim 1m^2$ sensitive area deployed over $\sim 10^4 m^2$. It is supplemented by a tracking detector of muons, shielded by about 6 feet of steel: its purpose is to compare the number of muons from purported gamma-ray primaries with the number from ordinary cosmic-ray primaries.

Last year we reported[2] an unusual episode of emission from the direction of the binary source Hercules X-1, with the following features: (a) on 24 July 1986 an excess event rate in the direction of Her X-1 was detected. (b) the pulsar period observed, namely T=(1.23568s +/- 0.0003s), is blue-shifted by about 0.16% from the contemporary X-ray period. (c) the average number of muons detected in these events is not less than the number for ordinary cosmic ray events as seen in our detector. At Tev energies, two stations operating air-Cherenkov telescopes, namely Mt. Hopkins and Haleakala, had reported periods nearly identical to ours: Lamb et al.[3], T=(1.23579 +/- 0.0002), and Resvanis et al.[4], T=(1.23593 +/- 0.0002). Recently the Haleakala group[5] have added another detection, on May 23, 1987: T=23598s. Another report comes from the Tata Institute, Bombay, which operates the air-shower array at Ooty[6]. They found four episodes of enhanced emission in 1986, and decided to search for periodicity in a narrow interval about the above period. Their results are 1.23577s for all four episodes. They had no muon detectors operating. The several period measurements mentioned here are in agreement to about one part in 10^4. This is curious, since the Whipple observations, seven in number, between April 1984 and June 1985 showed deviations from -6 to +9 parts in 10^4, more or less centered on the pulsar frequency as seen in X-rays[7]. The energy distribution of our events is nothing like E^{-2} as is sometimes advocated for the "Spectrum" of a gamma-ray star, or $E^{-2.7}$, which is appropriate for cosmic rays; instead the events are bunched around a value a bit less than 1 PeV, higher than the average for our array.

The blue-shift from the pulsar period could be because the accelerator producing the UHE protons is extended in space and may have a moving target which produces neutral pions (and photons). The X-rays are made close to the neutron-star surface and are modulated with the true rotation period of the neutron star.

Our conclusion that we were seeing the source is reinforced, and the muon anomaly, previously reported for Cygnus X-3 by Kiel[8], here is seen in Her X-1, and remains as a challenge to physicists.

References:

1. The CYGNUS collaboration in 1988 included the following: B.L. Dingus, D.E. Alexandreas, R.C. Allen, R.L. Burman, K.B. Butterfield, R. Cady, C.Y. Chang, R.W. Ellsworth, J.A. Goodman, S.K. Gupta, T.J. Haines, D.A. Krakauer, J. Lloyd-Evans, D.E. Nagle, M. Potter, V.D. Sandberg, R.L. Talaga, C.A. Wilkinson, and G.B. Yodh.

2. B.L. Dingus et al., Phys. Rev. Lett. 61, 1906 (1988)

3. Ap. J. 328, L13, (1988)

4. Ap. J. 328, L9, (1988)

5. L. Kelley, Abstract, 21st International Conference on Cosmic Rays, Adelaide, Australia, 1990

6. S. Tonwar and S. Gupta, to be published

7. P. Gorham et al., in Very High Energy Gamma Ray Astronomy, K. Turver, ed., Reidel, Dordrecht, (1987)

8. M. Samorski and W. Stamm, Ap. J. 268, L170121, (1983)

A SEARCH FOR DOUBLE BETA DECAY IN ^{136}XE WITH A TIME PROJECTION CHAMBER

A Caltech-Neuchâtel-PSI Collaboartion

H.T. Wong [1,2], *F. Boehm* [1], *P. Fisher* [1,3], *K. Gabathuler* [3], *H.E. Henrikson* [1], *M.Z. Iqbal* [1], *L.W. Mitchell* [2], *B.M. O'Callaghan-Hay* [1], *J. Thomas* [1], *M. Treichel* [2], *J.-C. Vuilleumier* [2], *J.-L. Vuilleumier* [2].

(presented by Henry T. Wong)

The Caltech-Neuchâtel-PSI Collaboration has been involved in double beta decay experiments at the Gotthard Tunnel Underground Laboratory. A Time Projection Chamber (TPC) has been built to investigate this decay mode in ^{136}Xe (transition energy 2.48 MeV), in both the 0ν and 2ν channels[1)].

The Monte Carlo studies[2)], design[3)], electronics[4)], data acquisition system[5)] and performance[6)] of the TPC, as well as the background considerations of the experiment[6)], have been discussed elsewhere. The TPC has a fiducial volume of 208 liters. There are 168 readout channels in each of X and Y. The operating pressure is 5 atm, with an admixture of 4% methane. The chamber is built with OFHC copper, and will be shielded by 20 cm of lead and 5 cm of copper.

The track reconstruction capability of the TPC provides a powerful means of background rejection. A double beta decay event is identified as a continuous trajectory with the characteristic "end features": large angle multiple scatterings (that is, high second derivatives) and high charge depositions, at *both* ends. In contrast, single electron background exhibits such features at one end only. This forms the basis for background rejection.

Figure 1 shows a typical single electron trajectory. The XZ and YZ projections as well as the time evolution of the anode signals are displayed. The energy of the event is obtained by integrating the anode signal. It can be seen that the event does show the expected characteristics: minimum ionizing and relatively straight in the beginning, but staggered trajectory and charge "blob" at the end.

The best energy resolution achieved to date is 11% at 1 MeV, at 2.5 atm of xenon. A new anode system made of low background materials and with improved geometrical precision is currently under construction. Our goal is to achieve an

[1] Norman Bridge Laboratory of Physics, California Institute of Technology, Pasadena, California 91125, U.S.A.
[2] Institut de Physique, A.-L. Breguet 1, 2000 Neuchâtel, Switzerland.
[3] Paul Scherrer Institute (formerly SIN), 5234 Villigen, Switzerland.

energy resolution of 5% at 2.5 MeV.

The experiment will soon be installed at the Gotthard Laboratory. Data will be taken with 60% enriched ^{136}Xe, as well as xenon depleted in ^{136}Xe, at 5 atm. We anticipate to achieve a lifetime sensitivity of $10^{23} - 10^{24}$ yr for the 0ν mode in one year of runtime, depending on actual background.

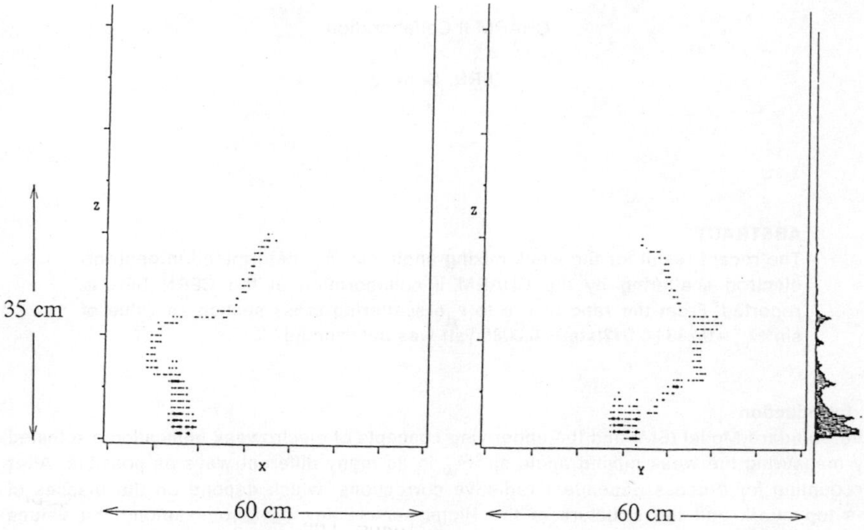

Figure 1. A typical single electron trajectory.

References

1. F. Boehm and M.Z. Iqbal, in *FestiVal - Festschrift for Val Telegdi*, ed. K. Winter, Amsterdam, North Holland (1988).
2. M.Z. Iqbal, B.M. O'Callaghan and H.T. Wong, Nucl. Instrum. Methods **A253**, 278 (1987).
3. M.Z. Iqbal et al., Nucl. Instrum. Methods **A259**, 459 (1987).
4. M.Z. Iqbal, B.M. O'Callaghan and H.T. Wong, Nucl. Instrum. Methods **A263**, 387 (1988).
5. J. Thomas et al., IEEE Trans. Nucl. Sci. **NS-34**, 845 (1987).
6. H.T. Wong et al., in *Tests of Fundamental Laws in Physics, Proc. of the XXIV Rencontre de Moriond*, eds. O. Fackler and J. Tran Thanh Van, Editions Frontières, Paris (1989).

A NEW DETERMINATION OF THE ELECTROWEAK MIXING ANGLE FROM MUON−NEUTRINO ELECTRON SCATTERING

Gabriele Zacek

CHARM II Collaboration

CERN, Geneva

ABSTRACT

The recent result for the weak mixing angle $\sin^2\Theta_W$, determined in neutrino electron scattering by the CHARM II collaboration at the CERN SPS is reported. From the ratio of $\nu_\mu e$ to $\bar{\nu}_\mu e$ scattering cross sections a value of $\sin^2\Theta_W = 0.233 \pm 0.012(\mathrm{stat}) \pm 0.008(\mathrm{syst})$ was determined.

1. Introduction

The Standard Model (SM) and the underlying concepts of electroweak unification are tested by measuring the weak mixing angle $\sin^2\Theta_W$ in as many different ways as possible. After accounting for process dependent radiative corrections, which depend on the masses of the top quark and the structure of the Higgs sector, any deviations among the values observed in individual experiments would signal new physics beyond the minimal version of the SM[1]. A very simple and well understood process to measure the weak mixing angle is the purely leptonic scattering of muon neutrinos on electrons, which proceeds via neutral currents and involves only pointlike fermions. In this context it is especially promising to exploit the functional relation between $\sin^2\Theta_W$ and the ratio of $\nu_\mu e$ to $\bar{\nu}_\mu e$ cross sections:

$$R = \frac{\sigma(\nu_\mu e)/E_\nu}{\sigma(\bar{\nu}_\mu e)/E_{\bar{\nu}}} = 3\,\frac{1-4\sin^2\Theta_W + 16/3\sin^4\Theta_W}{1-4\sin^2\Theta_W + 16\sin^4\Theta_W}$$

This method has the advantages that systematic errors largely cancel and that no absolute neutrino flux measurement is required. The weak mixing angle can also be derived from the masses of the intermediate vector bosons. Since the dependence of the radiative corrections on the top quark mass is similar for the two methods a comparison of both values allows a test of the SM, largely free of top mass effects. Alternatively $\sin^2\Theta_W$ obtained from deep inelastic νN scattering is essentially independent of the top quark mass and could thus in a comparison limit the range of possible top mass values[2]. Signature for neutrino electron scattering is an isolated electromagnetic shower initiated by a single electron, which is kinematically constrained to a strongly forward peaked cone of half opening angle $\Phi \leq \sqrt{2m_e c^2/E_e} \approx 32\mathrm{mrad}/\sqrt{E_e}$, where the electron energy E_e is given in units of GeV.

2. The CHARM II Detector

What requirements have to be imposed on a detector measuring electromagnetic showers induced by high energy neutrino reactions? Though the neutrino electron cross section rises linearly with energy it still is extremely small for GeV neutrinos ($\sigma/E_\nu \approx 2\times10^{-42}cm^2/GeV$) and consequently a large target mass is necessary. A good angular resolution is required to allow the precise determination of the direction of the electromagnetic shower. This is achieved by densely spacing active detector components within the target volume. In addition a high granular structure of the detector elements permits to separate electromagnetic and background induced hadronic showers due to the differences in their lateral shower profiles and their energy densities. This hadronic background originates from semileptonic νN reactions, which outnumber the purely leptonic νe scattering by three orders of magnitude. Following these design considerations the CHARM II collaboration has built at the CERN SPS wide band neutrino beam line a new neutrino detector working as electromagnetic calorimeter[3]. Glass plates, 4.8cm thick and with a surface area of 3.7m×3.7m are used as target material. Glass with the low atomic mass value $Z \approx 11$ is favoured since the angular spread due to multiple scattering is reduced. These glass plates alternate with planes of active detector elements consisting of 352 plastic streamer tubes with 1cm wire spacing. The sampling along the shower direction is thereby made in steps of 1/2 radiation length and in steps of 1cm in the transverse direction. To improve this lateral spatial resolution 176 aluminium strips (50μm thick, width 18mm) are glued to the back of each streamer tube plane orthogonal to the wires. A streamer generated by the ionization of a charged particle in the tubes, induces on the strips a charge profile extending over several pick-up strips. The centroid of this distribution allows to determine the crossing point of tracks with a precision of $\pm$3.5mm. The entire CHARM II detector is composed of 420 of these glass/streamer tube planes and has a total length of 35.7m and a total mass of 625t.

3. Background

After rejection of semileptonic neutrino nucleon reactions the remaining background is entirely of electromagnetic origin. Contibutions arise from coherent and incoherent production of π^0 on nuclei in neutral current neutrino reactions ($\nu(\bar\nu)A \to \nu(\bar\nu)\pi^0 A$) and from quasielastic scattering of electron neutrinos on nucleons ($\nu_e n \to p e^-$; $\bar\nu_e p \to n e^+$). Electron neutrinos are present in the neutrino beam due to a small (1%) beam contamination. Since the kinematic of these background reactions is determined by the nucleon mass, they lie to a major part outside the signal region ($E_e\Theta_e^2 \leq 2m_e c^2$) of νe-scattering events. Measuring the background in a larger $E\Theta^2$-window and using the theoretical prediction of their $E\Theta^2$-distributions allows the extrapolation of the background to small $E\Theta^2$ values and hence the subtraction from the observed νe signal. Figure 1 shows the resulting $E\Theta^2$-spectra after background subtraction recorded in ν- and $\bar\nu$-beams. The prominent peak for low $E\Theta^2$-values is due to neutrino electron scattering.

4. Neutrino Flux Normalization

Four independent monitor reactions are available to obtain the relative normalization of the event rates observed in different neutrino beam exposures. To make use of quasielastic ν_μN reactions ($\nu_\mu n \to p\mu^-$; $\bar\nu_\mu p \to n\mu^+$) the CHARM II detector is directly followed by a magnetic spectrometer, recording the charge and momentum spectra of the generated muons, and thereby providing an on-line monitoring of the fluxes and spectra of the incident neutrinos. In addition π^0's produced in ν_μN reactions, inclusive neutral and charged current neutrino reactions and the μ-flux from the decay of the parent π and K particles are used for normalization purposes. The relative neutrino flux normalization was evaluated for each of these reactions and the individual results agreed well within their systematic errors.

5. Result

For the present analysis data taken during the run periods in 1987 and 1988 were available[4]. This sample contained 792±43 $\nu_\mu e$ and 1017±51 $\bar{\nu}_\mu e$ events and constitutes the largest number of neutrino electron events collected so far. From the ratio of event numbers and including the neutrino flux normalization the following value for the weak mixing angle (without accounting for radiative corrections) is obtained: $\sin^2\Theta_W = 0.233 \pm 0.012(\text{stat}) \pm 0.008(\text{syst})$.

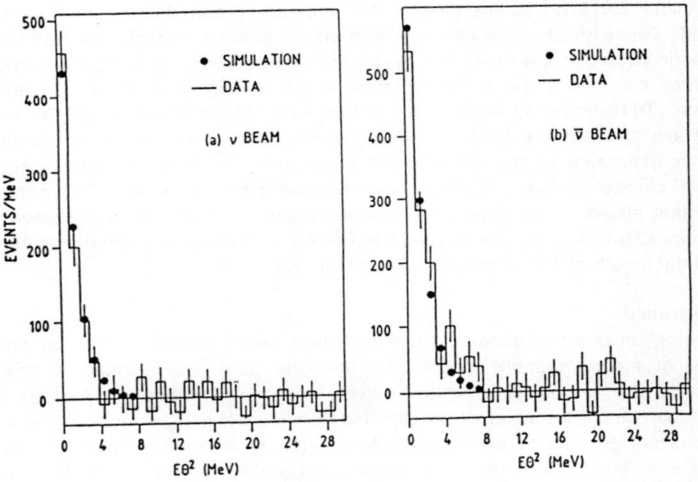

Figure 1: $E\Theta^2$ − spectra. Distribution of νe candidates as a function of $E\Theta^2$, after background subtraction; (a) ν − beam, (b) $\bar{\nu}$ − beam. The points represent the expected distributions.

References

[1] Marciano W.J. and Sirlin A., Phys.Rev.D29,945 (1984)

[2] Ellis J. and Fogli G.L., Phys. Lett. 213B, 526 (1988)

[3] De Winter K. et al., CHARM II Collaboration, Nucl.Inst.Meth, A278,670 (1989)

[4] Geiregat D. et al., CHARM II Collaboration, A New Determination of the Electroweak Mixing Angle from Muon−Neutrino Electron Scattering, to be submitted to Phys. Lett.

LOW TEMPERATURE DETECTORS

A. ZEHNDER
Paul Scherrer Institute
CH-5234 Villigen PSI, Switzerland

ABSTRACT
A review of the development of cryogenic detectors and their applications in astrophysics, dark matter search, and $\beta\beta$ decay, with emphasis on calorimeters and superconducting tunneling junction detectors.

1. INTRODUCTION

The discovery of violations of symmetries and of fundamental laws is often linked to new detectors. Improved energy, time, and spacial resolution allow deeper insight in physical processes and may reveal new structures, as illustrated by the following examples.

A new suite of both ground- and space-based instrumentation should be available for **astronomy** in the middle of the 1990s. This includes a wide variety of new telescopes including new detectors emphasizing improved sensitivity, spectral resolution, and spatial resolution. For **X-ray astronomy** to remain a science at the forefront of research, new telescope and detector developments, including cryogenic detectors, are actively pursued.

Presently the development of new large-scale detector systems, used in **high energy physics** experiments, is very active. Tracking of particles needs detectors with high spatial resolution. At the same time the detectors must be radiation hard. Cryogenic **vertex detectors** could fulfil both requirements.

One of the fundamental questions in cosmology is the unknown amount and nature of the **dark matter** that could pervade the universe. A direct observation of a class of dark matter candidates seems to be possible in the laboratory by using **phonon focusing detectors** at low temperatures.

Particles absorbed or scattered in a detector will lead to an energy deposit, creating a nonequilibrium state consisting mainly of excess charge carriers and phonons. An ideal detector measures both of these quantities. The energy resolution is determined by the statistical variations of the measured as well as the unmeasured excess quantities.

A more detailed analysis has to be given for each type of detector individually. Three types of cryogenic detectors exist:

- Calorimeters/Bolometers
  They are sensitive to the produced excess phonons and measure the deposited energy by detecting the heat pulses. Excess charge carriers should recombine to phonons.

- Tunneling junction detectors
  They are sensitive to excess charge produced by the Cooper pair breakup. Excess phonons should break up Cooper pairs.

- Superheated superconducting granules (SSG)
  A SSG detector consists of granules, the metastability of which is disturbed by radiation. The Meissner effect causes a change in the field distribution of the applied external field, which can be detected.

The present paper discusses the basic principle of calorimetric and tunneling junction detectors and some of their applications.

2. DETECTOR DEVELOPMENT

2.1 Bolometers

2.1.1 Principle

Simon [1] was the first to suggest the use of small low-temperature calorimeters, so-called bolometers. Niinikoski [2] pointed out in 1975 that single particles could be detected in this way. Experimental work started about 1983 by various groups, [3,4,5,6]. An overview is given by D. McCammon et al. [7] and in references therein.

In principle the detector works as follows: after absorbing an energy E_o in a material with the specific heat C_v the temperature rise is

$$\Delta T = E_0/C_v(T). \tag{1}$$

In an insulating crystal, the lattice vibrations (phonons) represent the dominant component of the specific heat C_v at low temperature. It decreases like T^3. Therefore to obtain large $\Delta T/T$, C_v has to be small, which limits the operation temperature to below 300 mK. The temperature rise can either be detected by the voltage drop across a thermistor in contact with the detector, a transition edge strip, a tunneling junction (see capt. 2) or a magnetic readout system (see below).

The equilibrium temperature is reached with a time constant $\tau = C_v/G$ in a few milliseconds via the thermal links with the conductivity G to the cooling system.

The theoretical estimates of the resolution for a **thermistor temperature sensor** is treated by Moseley et al. [8] and is summarized as follows:

In an ideal detector, the uncertainty of the measured energy depends on the quadrature sum of the magnitude of the following three noise sources:

- thermal fluctuations ("phonon noise")
- Johnson noise of the thermistor
- Johnson noise due to the load resistor of the amplifier

The contributions of the Johnson noise can be minimized by a careful detector design. Therefore the prevailing noise is caused by the thermal fluctuation. It can be estimated to be proportional to the square root of the number of phonon modes $\sqrt{C_vT/kT}$ times the energy per phonon mode kT, yielding

$$\Delta E_{rms} = \eta\sqrt{kT^2C_v} \propto T^{5/2} \qquad (2)$$

where η contains all the information on thermistors and thermal links ($1.5 < \eta < 2$). It is tabulated in [8]. The ideal energy fluctuation ΔE_{rms} for a $500 \times 500 \times 25\ \mu m^3$ Si crystal at a bath temperature of 100 mK ($C_v = 4\cdot10^{-15}$ J/K) should be $\Delta E_{rms} =$ 0.2 eV. This resolution has not yet been obtained, but it shows the potential of this type of detector.

In a realistic thermal detector not all the energy of the radiation is transformed into heat in a short time. The GSFC/Wisconsin group [3] estimates that only about 70% of the energy is promptly converted into phonons. The remaining 30% of the energy produces electron-hole pairs via ionization. They recombine only partly; some will be trapped by impurities and defect sites in the crystal. As long as the electron is trapped on one of these sites, its energy is lost, degrading the resolution. The GSFC/Wisconsin group solved this problem by absorbing their energy in a semimetal like Bi or in a semiconductor with a small band gap like HgCdTe or HgTe.

An interesting alternative to the thermistor temperature readout is the **magnetic temperature sensor** system described by [9]. In principle it consists of a two-level spin system in a magnetic field B. Absorbing the energy ΔE gives rise to $\Delta N = \Delta E/2\mu B$ spin flips. The associated change in the magnetic moment $\Delta M = \Delta N\ 2\mu$ is sensed by a SQUID. This type of readout showed the highest sensitivity, but is inherently slow, the decay constant is several seconds.

2.1.2 Experiments

2.1.2.1 X-Ray and α spectroscopy

Experimental works were done by various groups [3,4,5,6,7]. They either observed 5 Mev α particles or the 6 keV Mn X-ray. The GSFC/Wisconsin group reported an excellent resolution of about 12 eV FWHM [10] at 6 keV, including a base line noise of 7 eV. This value has to be compared to the 150 eV that is measured with a LN_2- cooled Si detector. The X-rays were absorbed in an element of a 1 x 12 array of 10 μg calorimeters, equipped with a HgTe semiconductor. The experiments were performed at 80 mK. An array of 12 detectors were fabricated. Fig. 1 shows the K_α line including the Lorentz broadening. The asymmetry due to the $K_{\alpha 1}$ - $K_{\alpha 2}$ splitting is clearly visible.

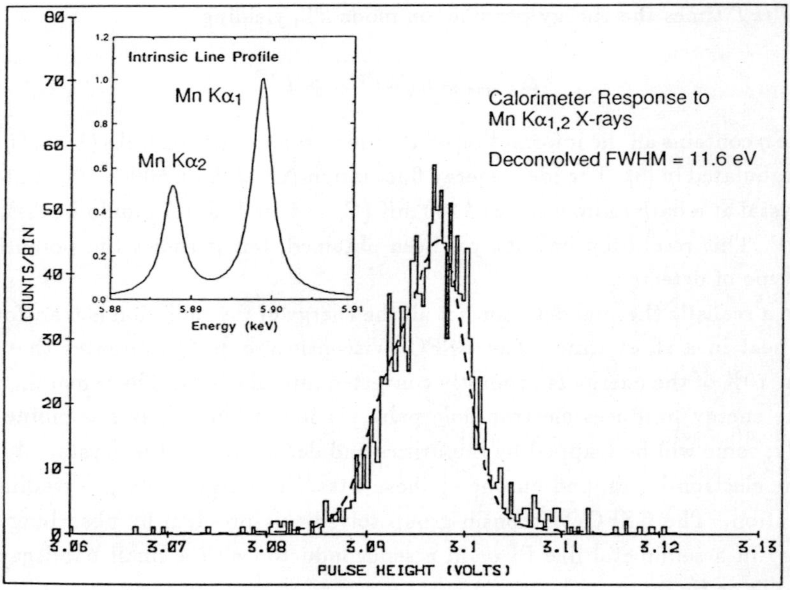

Figure 1: Energy spectrum from the Mn K_α lines. Correcting for the intrinsic line profile (see insert), a resolution of 11.6 eV (FWHM) is obtained. Fig. from Ref. [10].

Work with large Ge crystals were pursued by the Milano group [11]. They operated a 0.7 g detector at 71 mK and obtained a resolution of 1% for 5 MeV α particles.

Results from an even larger magnetic bolometer was reported by a group of the Meissner Institute in Garching [9] . They attached a Y(Er)Al garnet to a saphire crystal with a total weight of 7.5 g. The Y(Er)Al garnet served as magnetic readout system sketched above. The bath temperature was 400 mK (!) and the B-field 300 Gauss. The energy resolution for 5 MeV α particle is surprisingly good, namely 1.3%. The authors discuss the problems of observing a 100 eV energy deposition (coherent ν nuclei scattering) in a 100 g detector.

2.1.2.2 Ballistic phonon detection

In large insulating crystals (up to several cm^3) the initially generated phonons are rapidly converted down to about 1 THz [12]. At low temperature and at this frequency the phonon are traveling ballistically. For a single crystal the elasticity depends upon stress direction, and invariably the sound velocity is anisotropic [13]. Therefore phonon propagation is strongly anisotrope in Si crystals.

There exist ingenious experiments to demonstrate the effect of phonon focusing [14]. In these experiments the phonons are mainly generated by heaters or by laser light absorption. In contrast a high power density is generated in the crystal, if highly ionizing particles are absorbed. As an example, an α- particle deposits about $10^{11} W/cm^3$. It is therefore important that the phonon focusing effects are as well demonstrated under these conditions.

The TU-Munich group [15] absorbed 5 MeV alpha particles on the surface of a 3 mm thick Si crystal. Using a coded mask technique to select specific phonon directions, they showed, for the first time, that phonons generated by α particle absorption are indeed focused.

The PSI group in Villigen [16] observed strongly correlated signals between two STJ placed 100 μm apart on the surface of a $< 100 >$ oriented, 300 μm thick Si wafer. The front or back side was irradiated by α particles. Fig. 2 displays the correlated energy spectrum of the STJ 1 with STJ 2. Up to 4 strips are clearly visible, showing a strong focusing effect. The ongoing analysis includes phonon reflection at the surfaces as well as surface-phonon effects.

The Stanford group [17] reported the detection of 60 and 25 keV gamma radiation in a 1 mm thick Si crystal. The temperature was sensed by measuring the resistivity change of a meandering Ti transition edge strip with an active area of $1.5 \cdot 4.5$ mm^2. The estimated energy resolution was 2.5 keV.

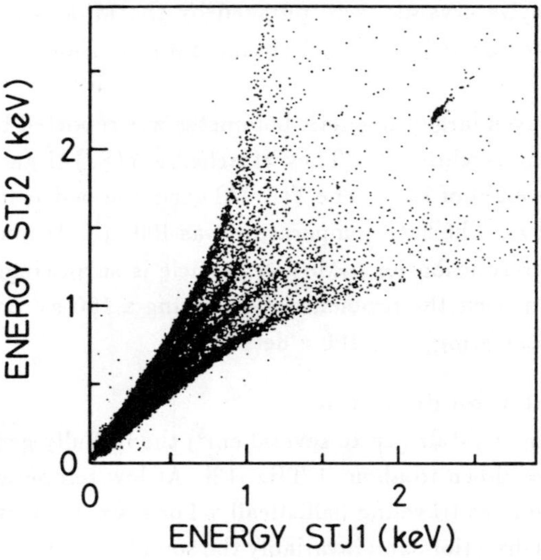

Figure 2: Typical energy correlation plot at T = 0.5 K. The four distinct strips are due to phonon focusing in the Si substrate, on top of which two superconducting tunneling junctions (STJ) are evaporated.

2.2 Superconducting Tunneling Junctions (STJ)

2.2.1 Principle

The possibility of using STJ's as detectors for ionizing radiation has been demonstrated for the first time in 1969 by Wood and White [18] and reviewed in various papers around 1981 by Kurakado and Mazaki [19]. Experimental results and theoretical studies are given in Ref. [20,21,22].

In this type of detector, the deposited energy is converted into excess quasi-particles (excited electronic states) resulting from the breakup of Cooper pairs. This charge is then collected via a tunneling barrier, similar to a semiconductor diode detector. The potentially high energy resolution arises from the fact that the Cooper pair binding energy Δ is in the order of 1 meV, which is about 1000 times smaller than the gap of a Si/Ge semiconductor detector. Assuming Poisson statistics, the energy resolution is given by

$$\Delta E_{rms} = \sqrt{f \cdot E_0 \epsilon} \tag{3}$$

where $\epsilon \geq \Delta$ is the minimal ionizing energy of the detector. This clearly favors small energy gaps. The Fanofactor f ≤ 1 describes the anticorrelation of the produced charge.

As an example, a semiconductor Si diode f=0.31 $\epsilon = 3.61$ eV yields

$$\frac{\Delta E_{FWHM}}{E}(@\,6\,keV) \sim 3\% \tag{4}$$

whereas a superconducting Sn detector f $\simeq 1$ $\Delta = 0.585$ meV should yield

$$\frac{\Delta E_{FWHM}}{E}(@\,6\,keV) \approx 0.1\%. \tag{5}$$

In contrast to semiconductor detectors, the energy deposit in phonons is not *a priori* lost in superconducting detectors, owing to the fact that the maximum phonon energy of $k_B\,\Theta_{Debye} \approx 10$ - 20 meV is larger than the gap, but much smaller than the excitation energy of a semiconductor. In fact, the Cooper pair breakup by phonons plays an essential role in the understanding of superconducting tunneling junctions.

The superconducting tunneling junction detector consists of two metallic films (at least one being superconducting) separated by a thin insulating barrier, which allows quantum mechanical tunneling of the quasiparticles. Such devices were first used by Giaever [23] to study the physics of quasiparticles and to determine the gap energy.

Fig. 3 shows the I-V characteristic of a STJ. If it is biased above a potential $\geq \Delta_1/e + \Delta_2/e$ (Δ_1, Δ_2 are the gaps of film 1 and 2 respectively), it displays ohmic behavior $\delta V = R_N \cdot \delta I$ and if biased below, a strong temperature-dependent, nonlinear behaviour is observed.

The tunneling time τ_{tun} can be expressed by

$$\tau_{tun} = e^2 R_N \cdot N_o \cdot At \tag{6}$$

where e is the electronic charge, N_o the single spin density at the Fermi energy ($\sim 1.4\ 10^{22}/eV/cm^3$ for Sn), and A and t the surface and thickness of the film considered. R_N is an important quantity, because it allows an easy calculation of the tunneling rate τ_{tun}^{-1}, which is typically about 10^6 sec^{-1}. In order to increase the signal current, τ_{tun} should be minimal. For a high quality junction $R_N \cdot A$ is in the best case as small as $10^{-7}\Omega$ cm^2. The thickness t is the only free parameter and should be minimal; this requirement limits the useful detector volumina. Increasing the surface area increases the junction capacitance and therefore the preamplifier noise. A possible solution for this problem is quasiparticle trapping as discussed below.

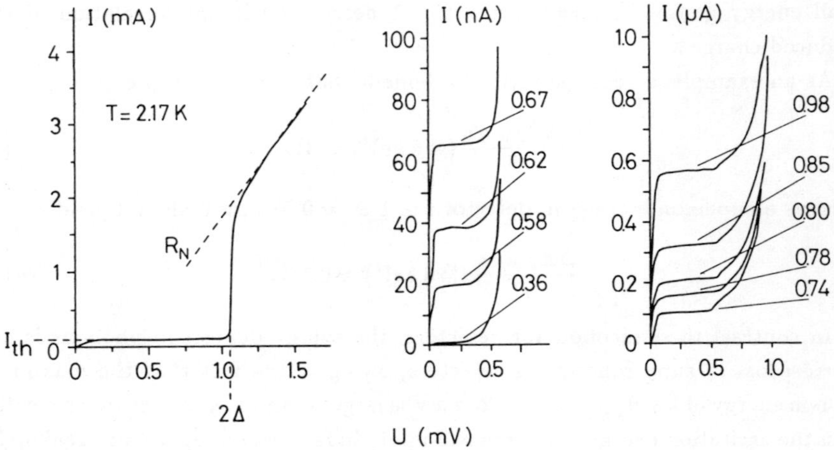

Figure 3: I-V characteristics of a high quality Sn/SnO$_x$/Sn ($g_n \approx 10^5 \Omega^{-1}\text{cm}^{-2}$) tunneling junction. It displays a strongly temperature-dependent subgap current. From Ref. [21].

Incident radiation will mostly produce a photoelectron that, in a very short time ($\sim 10^{-11}$ sec), breaks up Cooper pair, generating quasiparticles, or emitts phonons. Owing to the short phonon pair-breaking time of $\approx 10^{-10}$ sec, phonons with an energy above 2Δ, will very efficiently break up Cooper pairs. Once the quasiparticles relaxed to the minimal energy Δ, they can recombine emitting thereby 2Δ phonons. These 2Δ phonons are very likely absorbed, leading to the so-called phonon trapping. Ref. [24] gives the lifetimes for phonons and quasiparticles for the various relaxation processes to thermal equilibrium.

A set of coupled differential rate equations for phonons and quasiparticles in both films, the so-called Rothwarf-Taylor equations [25], describes the time evolution of phonons and quasiparticles. Details are found in [20]. Fig. 4 shows the time evolution for phonons and quasiparticles for a typical STJ at a working temperature of $(T/T_c) = 0.1$ and an absorbed energy of 6 keV. The phonons disappear in a short time ($\sim$ 1 nsec) leaving quasiparticles, which either tunnel, recombine, or diffuse out of the junction area.

The number of tunneling quasiparticles can be estimated

$$N_{tun} = N_0 \frac{\tau_{tun}^{-1}}{\tau_{tun}^{-1} + \tau_{rec}^{-1} + \tau_{loss}^{-1}} \tag{7}$$

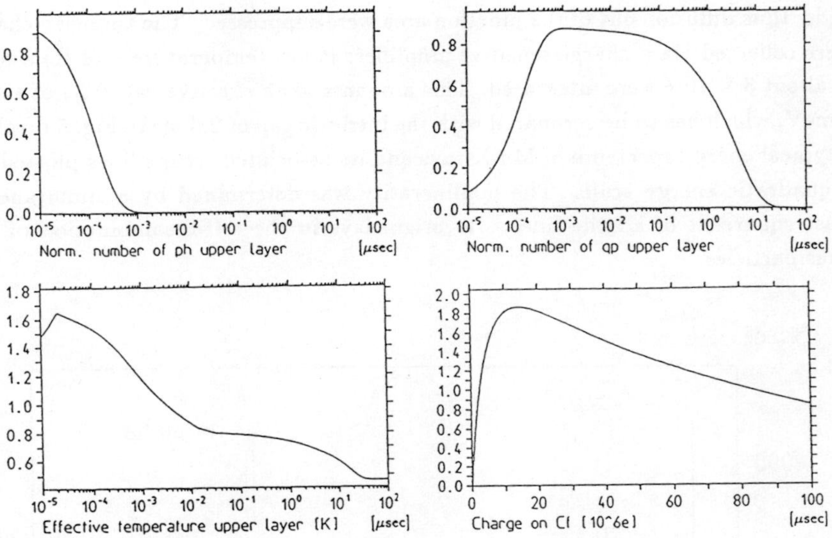

Figure 4: upper left and right: Total number of excess 2Δ phonons, and quasiparticles, respectively.
down left and right: Temperature of the quasiparticle system as a function of time, for a bath temperature of 0.4 K. Calculated preamplifier output signal.

where $N_0 = E_0/\Delta$ is the total of initial quasiparticles, τ_{tun}^{-1} is given by Eq. 6, and the recombination rate [24)] is

$$\tau_{rec}^{-1} \simeq (\pi)^{1/2}(2\Delta/kT)^{5/2}(T/T_c)^{1/2}\, exp(-\Delta/kT) \cdot \tau_0^{-1} \tag{8}$$

τ_0 is 2.30 nsec for Sn, other values can be found in [14]. The recombination time τ_{rec} is about 100 μsec at $T/T_c \approx 0.1$, which is much longer than τ_{tun}. However, owing to its strong temperature dependence, $\tau_{rec} \approx \tau_{tun} \approx 1$ $m$sec at $T/T_c \approx 0.2$.

Various processes are included in an effective loss time τ_{loss} and are discussed in 2.2.2.1.

2.2.2 Experimental work

2.2.2.1 X-ray detector

The best resolution obtained so far was by the PSI group with $Sn/SnO_x/Sn$ junctions with surfaces of about 50×50 μm^2. These junctions were fabricated by the photolithographic stencil lift-off technique. The current leads were about 5 μm wide, thus diffusion out of the junction area were suppressed. The tunneled charge were collected via a charge sensitive amplifier; at low temperature (0.4 K) a total of about 3×10^6e were integrated. This amounts to an effective gap Δ_{eff} of about 2 meV, which has to be compared with the intrinsic gap of 0.6 meV. Fig. 5 displays a typical energy spectrum of Mn X-rays and its associated escape lines plotted vs. a quadratic energy scale. The nonlineratity was determined by a simultaneous least square fit of all the lines. Its origin lays in the self-recombination of the quasiparticles.

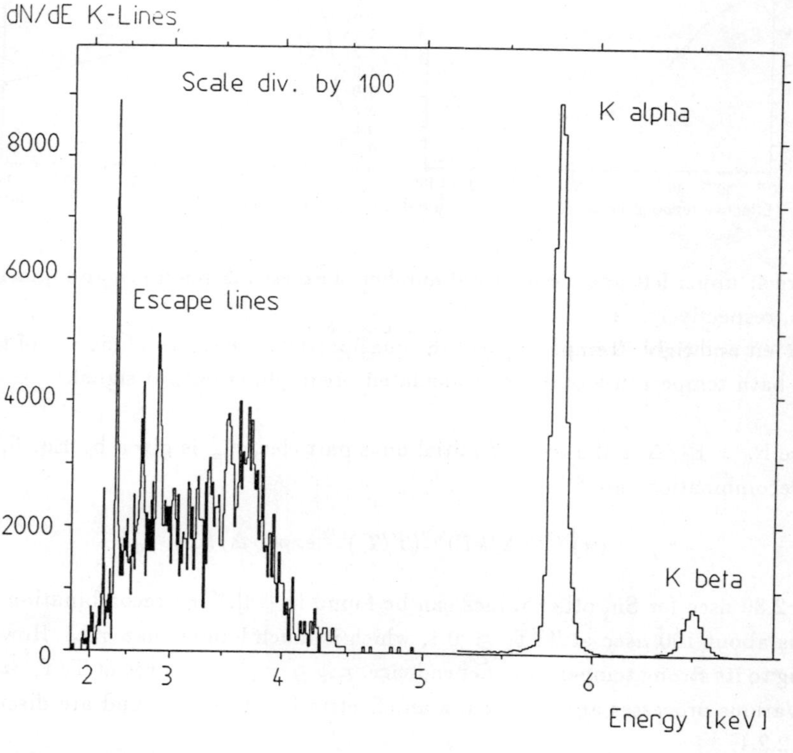

Figure 5: Mn X-ray spectrum in a Sn STJ detector. The lines at around 2 keV are resulting from the escape lines in Sn. Mark the pronounced nonlinearity.

The best resolution of a STJ is about 50 eV FWHM [29], including a simultaneously measured electronic noise contribution of about 24 eV. It amounts to an intrinsic resolution of about 42 eV. This resolution is not as good as expected (5 eV).

The origin of this discrepancy was investigated by additional measurements of the diffusion property [26] of quasiparticles. It was shown that an additional loss mechanism has to be included to understand the results. Its loss time is about 1 – 30 μsec, depending on various, not well controlled parameters. Therefore at low temperatures, next to tunneling time, this loss time determines the effective lifetime of quasiparticles. It could be attributed to the quasiparticle trapping and a later recombination in subgap structures of individual grains of the Sn film. It is not *a priori* granted that these effects are constant over the STJ area and could therefore explain the observed limited resolution. Further work is needed to clarify this point.

Fig. 6 shows a plot of energy vs risetime: one observes a repetition of the K_α, K_β line, which is due to absorption of the X-rays in the upper and the lower film respectively.

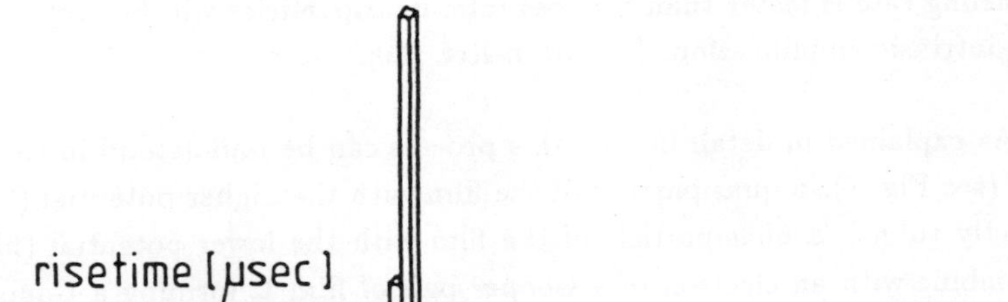

Figure 6: Two dimensional spectra from Sn STJ on a fused silica substrate. The K_α, K_β lines resulting from Mn X-ray absorption in the lower and upper film are clearly resolved.

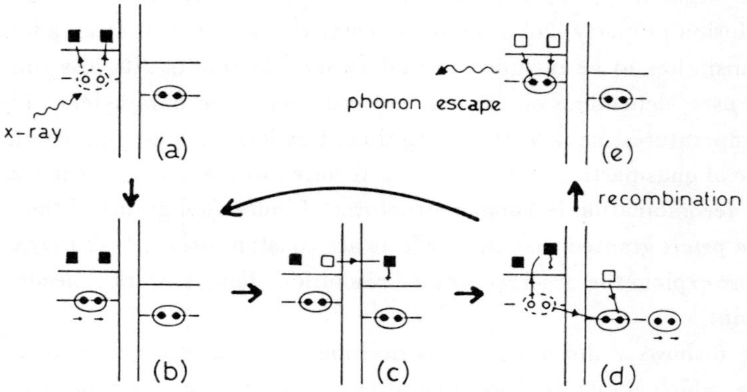

Figure 7: Tunneling processes in a junction with both films superconducting. If the tunneling rate is faster than the loss rate, quasiparticles will be recycled, leading to a intrinsic amplification. Fig. from Ref. [20].

As explained in detail in [20], this process can be understood in the following way (see Fig. 7): a quasiparticle of the film with the higher potential (film 1) can directly tunnel; a quasiparticle of the film with the lower potential (film 2) can recombine with an electron of a Cooper pair of film 1, forming a Cooper pair in film 2 and, by energy conservation, generating a quasiparticle in film 1. In both cases, a current of equal sign results. One sees that, in a single step, a symmetric situation (a quasiparticle on each side) is formed. If there are no loss mechanisms, the above described processes could be repeated leading to a current amplification [27].

An interesting extension of STJ's was proposed by N.E. Booth [28] by trapping quasiparticles in a two stage process. Fig. 8 (from [28]) shows the principle of the mechanism: Radiation is absorbed in a superconductor with a gap Δ_1 and generates quasiparticles. They diffuse fast ($v_F \approx 1$ μm/picosec) to an adjacent superconductor with a gap Δ_2 and are therein deexcited rapidly (less than 1 nanosec).

If the difference between the gaps Δ_1 and Δ_2 is larger than $2\Delta_2$, the phonons emitted during deexcitation can break up Cooper pairs in the superconductor 2 resulting in the multiplication of quasiparticles. In addition, the tunnel rate of the trapped quasiparticles is strongly enhanced, owing to their smaller diffusion volume (see Eq. 6).

At PSI a Pb alloy/Sn/SnO/Sn junction was fabricated [29]. The thickness was about 0.5 μm for Pb and about 0.1 μm for each Sn layer. Mn X-rays were absorbed

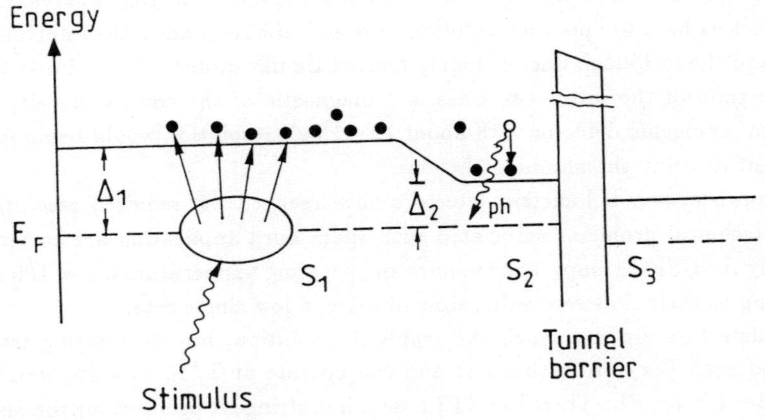

Figure 8: Energy diagram showing trapping of excess quasiparticles produced in superconductor S_1 by an adjacent superconductor S_2 of lower gap. Fig. from Ref. [28].

in the Pb alloy film. The tunnel time decreased to about 0.2 μsec compared to about 5 μsec in a typical Sn/SnO/Sn junction. The decrease was larger than calculated from the known thicknesses (see Eq. 6). The proximity effect could be a possible explanation: the Sn gap is widened towards the Pb alloy film and has its minima at the tunnel barrier where the quasiparticles are located. The tunneling current is therefore strongly enhanced.

An interesting experiment was recently carried out at the TU-Munich [30]. Two Al STJ were fabricated 470 μm apart on a Sn strip. Mn X-rays were used to generate quasiparticles in the Sn strip. They diffused to the Al STJ and were subsequently trapped. From a detailed analysis of the correlated signal of the two Al STJ, an energy resolution of about 60eV (FWHM) at 5.89 keV and a position resolution of better than 5 μm over the sensitive length was obtained.

In conclusion, STJ have interesting properties, that are fairly well understood. Possible applications are discussed below.

3. APPLICATION

3.1. High Resolution X-Ray

With the scheduled launch of large imaging X-ray astronomy satellites in the USA (AXAF), and in Europe (XMM), there exists a need for detectors with high throughput and high resolution in the 0.1 – 10 keV range. This region is rich in (highly excited) X-ray lines of cosmically abundant elements. Grating instruments have the required resolution, but are too inefficient at higher energies, whereas solid state detectors have too poor a resolution to resolve the resonance, the intercombination, and the forbidden lines of highly ionized He-like atoms. This is important since the ratio of the latter two lines is a diagnostic of the coronal density. A 5×5 mm^2 cryogenic detector with about 10 – 15 eV resolution would be an ideal instrument to study the abundant Fe ions.

As shown above, bolometric detectors have reached the required resolution, and the technical problems associated with space born application are currently studied by the GSFC group. They require an operating temperature below 100 mK and, owing to their slow recovering time of msec, a low single rate.

STJ detectors did not reach the required resolution, but no limiting factor was found yet. They are rather fast and can operate at $T/T_c \approx 0.15$, which is about 0.5 – 1 K for Pb. Therefore STJ pose a less stringent problem for the space cryogenics than bolometers. Trapping offers the possibility of a large detector area (1 mm^2) with a limited number of STJ's for the readout. Recent advances in GaAs preamplifier design [31] allow to operate the readout electronics near the STJ at cryogenic temperatures. Therefore the design would be facilitated and the electronic noise minimalized.

3.2. β Decay in True Calorimeters

Experiments that require accurate measurements of energy differences between an initial and a final state are often hampered by uncertainties about energy losses in the source, in the detector, and in the initial and the final state interactions. A true calorimeter that measures the Q-value, could solve these problems. A typical example is the measurement of the endpoint energy of the tritium β decay spectrum. This sets a limit to the neutrino rest mass. Tritium could be implanted in a detector (STJ or bolometer detector). The required resolution has been obtained with present bolometric detectors.

Since all β decays are measured, the count rate limitations are quite severe, but a calculation indicates that it would be practical to establish a mass limit of about 10 eV for the ν_e that would be relatively free of systematic errors [32].

3.3. Neutrino and Dark Matter Detection

Coherent neutrino scattering off nuclei has so far never been observed in laboratories, but it plays an important role in supernova dynamics [33]. Dark galactic halos may be needed to close the universe. This missing mass could make up as much as 90% of the total mass [34]. Both of these fundamental questions could be tackled using cryogenic detectors.

The detection of weakly interacting particles (WIP's) should be possible by observing the recoil of elastic scattering of WIP's off nuclei in a large insulating crystal [35,36].

The recoil energy and interaction rate for particles with coherent weak coupling to the target nucleus can be estimated as follows [37]: If a nonrelativistic halo particle with mass m and velocity v scatters off a target nucleus M, the recoil energy is $T_{max} = (2mv)^2/2M$ at the most. For v = 200 km/sec (virial velocity) and for a threshold energy of 10 eV per phonon-sensing detector with an overall efficiency of 2% per strip, one should be able to set a limit for a WIP's of $p \geq 2.6$ MeV/c, corresponding to a mass of halo particles of about 3.9 GeV/c^2.

The interaction rate can be calculated in the standard model for massive WIP's with vector coupling to Z_0 bosons:

$$R = 4.2 events/kg\, day[\frac{\sigma/A}{10^{-38}cm^2}] \cdot [\frac{\rho}{0.4GeV/cm^3}] \cdot [\frac{v}{200km/sec}] \quad (9)$$

where ρ is the mass density of the halo particle. ρ is estimated to be $\rho = 0.4$ GeV/cm^3. The reduced cross section σ/A is given by:

$$\sigma/A = (2.10^{-39}cm^2)\,[\frac{4mM}{(m\,+\,M)^2}]\,Y\,(N - (1 - 4sin^2\theta)Z)^2 \quad (10)$$

where N and Z are the numbers of neutrons and protons respectively, and Y is the weak hypercharge.

The event rate for coherent scattering of WIP's off Si nuclei is given in Table 1. The model-dependent hypercharge Y is assumed to be 1.

Mx (GeV)	Recoil (keV)	Rate (spin independent) kg/day	Rate (spin dependent) kg/day
1	0.03	23	4
5	0.82	86	15
10	3.3	130	22
100	330	110	19

Table 1: Recoil energy and interaction rate for coherent scattering off nuclei of WIP's

It is possible that dark matter consists of particles that interact with nuclei only via spin-dependent forces. In Table 1, we list as an example the rates of interactions between massive photinos and nuclei via the exchange of massive scaler quarks. The values in Tab. 1 were calculated assuming that $M_{\tilde{Q}} \simeq 100$ GeV and that the spin-dependent nuclear matrix elements are of order unity. Details are given in [37].

The cross section for neutral vector current scattering of neutrinos off nuclei is given by

$$\sigma_{tot} = G_F^2 E^2 (N - Z(1 - 4si - n^2\theta_w)^2/4\pi \tag{11}$$

and, in contrast to ν-e scattering, it is independent of the neutrino type. The event rate for ^{8}B solar neutrinos was calculated in [36], yielding a small recoil energy and a rate of 0.16 events/ton/day. This is most probably too small to be detected amidst the background.

Nuclear reactors are a powerful source of antineutrinos. For a detector located at a distance of 10 m from the reactor core with a flux of $2 \cdot 10^{13}/cm^2$/sec, one would get events with a recoil energy in the range of 0.5 keV $<$ T $<$ 6 keV at a rate of $\sim$ 40 kg/day. An interesting possibility arises by simultaneously detecting the recoil electrons from νe scattering in the same detector. They deposit a much greater energy (up to $\leq$ 10 MeV) at a rate of 4 events/kg/day. In contrast to the coherent ν-N scattering, the electron scattering is sensitive to ν oscillation; a movable recoil detector would therefore be an interesting tool to study ν oscillation.

In conclusion, a cryogenic ballistic phonon detector of a few kilograms may be a promising instrument to study the existence of certain dark matter candidates and the properties of reactor neutrinos. However, more studies must be made, particularly with regard to the background and to the detection of the ballistic phonons. First results are promising, therefore a major effort seems justified.

3.4. $\beta\beta$ Decays

It was pointed out by [38] that today's experiments with underground detectors set limits to $\beta\beta$ decays that are hard to improve. The neutrinoless $\beta\beta$ decay experiments are of particular interest to the testing of the lepton number non-conservation. The transition (A,Z) $\rightarrow$ (A,Z+2) + 2e would manifest itself in a monoenergetic line at the appropriate Q-value; the best efficiency is clearly given, if the detector is identical with the source. Cryogenic detectors should yield a better energy resolution than semiconductor detectors. It is also advantageous to have a larger selection of target materials. So far $\beta\beta$ decay search was restricted to Ge(Li) detectors and to Se and Xe TPC chambers. Calorimetric detectors must operate in a temperature range, where the material possesses a low specific heat (Eq. 1).

- In crystal insulators the specific heat is given by the dominant lattice specific heat $C_L = 1944$ (J/mol K)$\cdot$(T/ϕ_D)^3, where ϕ_D is the Debye temperature.

In this kind of detector, thermalization of phonons is mainly achieved by isotope scattering off crystal faults and off surfaces.

- Normal conducting metals are not well suited, because the specific heat is large and determined by conduction electrons $C_e \propto T$.

- Superconductors at low temperature have low specific heat, but the energy is not stored in phonons but in excess quasiparticles as explained above. If the surface is covered with tunneling junctions, a readout is possible in principle. The tunnel rate is extremely slow, owing to the large volume (see Eq. 6). An interesting possibility arises by coating the surface of the superconductor with tunneling junctions of a smaller gap, thus using the effect of trapping quasiparticles on the surface.

We consider the ^{100}Mo $\beta\beta$ decay to be an example: ^{100}Mo, having a $\beta\beta$ decay transition energy of 3033 keV, is a superconductor with a high T_c in a binary compound. The sensitivity is estimated in [38] to be $5 \cdot 10^{24}$ years for a 450 kg Mo cube. This limit is well below the present limits for the $\beta\beta$ decay lifetime; further studies are therefore justified.

4. CONCLUSION

The potential of cryogenic detectors is demonstrated in experimental work. A few examples show possible applications. However, further studies are needed and are pursued in various laboratories.

The work at PSI, Villigen has been performed by C. Hagen, P. Jokinen, W. Rothmund, and A. Zehnder

References

[1] S. Simon, Nature 135, 763 (1935).

[2] T.O. Niinikoski, in "Liquid and Solid Helium" (Halsted, New York), 145 (1975).

[3] D. McCammon et al., J. Appl. Phys. 56, 1263 (1984).

[4] N. Coron et al., Nature 314, 75 (1985).

[5] C. Borango et al., Cryogenics 24, 681 (1984).

[6] B. Sadoulet, Invited talk IEEE Nuclear Symposium, San Francisco, CA (1987).

[7] D. McCammon et al., Proc. 18th Conference on Low Temp. Phys., Kyoto, (1987).

[8] S.H. Moseley, J.C. Mather, and D. McCammon, J. Appl. Phys. 56, 1257 (1984).

[9] M. Bühler, and E. Umlauf, Europhys. Lett. 5, 297 (1988).

[10] D. McCammon, et al. in Low Temperature Detectors for Neutrinos and Dark Matter III (edition Frontières: Gif-sur-Yvette-CEDEX), (1989).

[11] A. Alessandrello et al., Phys. Lett. B202, 611 (1988).

[12] Y.B. Levison, Invited talk at PHONONS-89 Conference Heidelberg (Proceedings to be published) (1989).

[13] G.A. Northrop, and J.P. Wolfe, Phys. Rev. B22, 6196 (1980).

[14] W. Dietsche, G.A: Northrop, and J.P. Wolfe, Phys. Rev. Lett. 47, 660 (1981).

[15] T.H. Peterreins, et al., Phys. Lett. B 202, 161 (1988).

[16] C. Hagen, W. Rothmund and A. Zehnder, Proc. III Int. Workshop on Low Temperature Detectors for Neutrinos and Dark Matter, Gran Sasso, Italy, (1989)

[17] B. Cabrera, Invited talk at PHONONS-89 Conference Heidelberg (Proceedings to be published) (1989).

[18] G.H. Wood, and B.L. White, Appl. Phys. Lett. 15, 237 (1969).
G.H. Wood, and B.L. White, Can. Journ. Phys. 51, 2031 (1973).

[19] M. Kurakado, and H. Mazaki, Phys. Rev. B22, 168 (1980).
M. Kurakado, and H. Mazaki, Nucl. Instr. and Meth. 185, 141 (1981).
M. Kurakado, and H. Mazaki, Nucl. Instr. and Meth. 185, 149 (1981).
M. Kurakado, Nucl. Instr. and Meth. 196, 275 (1982).
M. Kurakado, J. Appl. Phys. 55, 3185 (1984).

[20] D. Twerenbold, Phys. Rev. B34, 7748 (1986).

[21] D. Twerenbold, and A. Zehnder, J. Appl. Phys. 61, 1 (1987).

[22] H. Kraus et al., Europhys. Lett. 1, 161 (1986).

[23] I. Giaever, Phys. Rev. Lett. 5, 147 (1960).
I. Giaever, Phys. Rev. Lett. 5, 464 (1960).
I. Giaever, and K. Megerle, Phys. Rev. 122, 1101 (1961)

[24] J.J. Chang, and D.J. Scalapino, Phys. Rev. B15, 2651 (1977).
S.B. Kaplan et al., Phys. Rev. B14, 4854 (1976).

[25] A. Rothwarf, and B.N. Taylor, Phys. Rev. Lett. 29, 27 (1967).

[26] W. Rothmund, and A. Zehnder, Proc. of the European Workshop on Low Temperature Devices, Annecy, 1988.
P. Jokinen, W. Rothmund, and A. Zehnder, PSI Annual Report, Annex III, 59 (1988).

[27] K. E. Gray, Appl. Phys. Lett. 8, 392 (1978).

[28] N.E. Booth, Appl. Phys. Lett. 50,2 (1987).

[29] A. Zehnder, Festi-Val Festschrift for V. Telegdi, K. Winter, editor (Elsevier Science Pub. BV) (1988)

[30] H. Kraus et al., to be published in Phys. Lett. B (1989).

[31] D. V. Camin et al., Alta Frequenza 56, 347 (1987).

[32] D. McCammon et al., Neutrino Mass and Low Energy Weak Interactions, Telemark 1984, eds. V. Barger and D. Cline (World Scientific Singapore), 329 (1985).

[33] E. Commins, and P. Bucksbaum, Weak Interactions of Leptons and Quarks (Cambridge Univ. Press, New York, 1983) and references therein.

[34] G.R. Blumenthal et al., Nature 311, 517 (1984).

[35] B. Cabrera, J. Martoff, and B. Neuhauser, Nucl. Instr. and Meth. A275, 97 (1989).

[36] B. Cabrera, L.M. Krauss, and F. Wilczek, Phys. Rev. Lett. 55, 25 (1985).

[37] M.W. Goodmann and, E. Witten, Phys. Rev. D31, 3059 (1985).

[38] E. Fiorini, and T.O. Niinikoski, Nucl. Instr. and Meth. in Phys. Research 224, 83 (1984).

A New Look at an old Problem: Testing the Equivalence Principle in the Field of the Earth

E.G. Adelberger
Physics Department
University of Washington, Seattle WA 98195

According to Einstein's weak equivalence principle it is impossible, in a sufficiently small laboratory, to distinguish an external gravitational field from an acceleration of the laboratory. Because of the fundamental role this principle plays in theories of gravity, it has been tested with exquisite precision. Traditionally the tests are viewed as probing whether gravitation is described by a metric theory (such as Einstein's general relativity) in which gravity is ascribed to a curvature of space. However, one can also view the experiments as probing whether any long range fields other than gravity act between electrically neutral and shielded bodies.

Because of the general consensus that the current standard model of physics is not complete, there are many speculations about new particles and interactions, based on a wide variety of theoretical ideas ranging from string theories to attempts to quantize gravity. One normally accounts for the fact that these new particles have not been seen by assuming that their masses are too large to have been produced in any existing accelerator. However, there is another possibility: perhaps some of these predicted particles have so little mass that they produce macroscopic forces, and the forces have not been detected simply because they are so weak. On general grounds any new macroscopic interactions are expected to produce forces that are not exactly proportional to a test body's *mass* but rather to a *charge*, q_5, that is carried by its constituent particles. Such interactions would produce a potential between two macroscopic bodies which can be written as

$$V_{12}(r) = \alpha_5 \, (q_5/\mu)_1 \, (q_5/\mu)_2 \, G \, \frac{m_1 m_2}{r} \, e^{-r/\lambda} , \tag{1}$$

where G is the gravitational constant, α_5 may be thought of as the strength of the new interaction relative to gravity, μ represents a test body mass in AMU, and $\lambda = \hbar/m_b c$ is the Yukawa range of an interaction mediated by bosons of mass m_b. Some possible candidates for q_5 are baryon number, B, lepton number, L, or the quantity $B - L$ which is conserved in most Grand Unified Theories. In this scenario, equivalence principle tests provide a new window on particle physics.

The equivalence principle is tested by comparing the acceleration of two different materials placed in the gravitational field of a massive, third body. The equivalence principle states that the acceleration difference should vanish. The most precise tests employ torsion balances containing two different materials mounted on a frame suspended from the finest possible fiber. If the "gravitational" force on a test body is not exactly proportional to the inertial (centrifugal) force, the balance will twist slightly. This twist is detected by changing the orientation between the torsion balance "detector" and the external "source".

Such tests of the equivalence principle were pioneered by von Eötvös[1]. His best results were obtained by using the earth as a source and detecting the fiber twist by rotating his apparatus by 180 degrees about a vertical axis. However, as recognized originally by Eötvös, one can avoid rotating the apparatus (which inevitably disturbs the sensitive instrument) if one uses the sun as a source. Then the apparatus is rotated very gently "for free" once every twenty four hours. Although a given difference in the ratio of inertial to gravitational mass produces a smaller effect in a solar source experiment, the cost is only about a factor of three. As a result, the best modern tests of the equivalence principle (by Dicke *et al.*[2] and by Panov and Braginsky[3]) used a solar source; the experimenters decided that the benefit of the smooth rotation more than compensated for the smaller signal.

However, if one views these experiments as probes of possible macroscopic interactions arising from the exchange of heretofore undetected particles, then (as was emphasized recently by Fischbach *et al.*[4]), the solar experiments have very poor sensitivity for interactions with ranges less than the distance to the sun (*i.e.* for exchanged particles with a mass greater than about 10^{-17}eV). Fischbach *et al.* even claimed that evidence for a new interaction (a "fifth force" that coupled to baryon number) was already at hand; they reanalysed of the Eötvös data and uncovered a striking correlation between the acceleration of a test body in the field of the earth and the number of baryons per gram it contains.

A group of us at the University of Washington, calling ourselves Eöt–Wash, has been developing a continuously rotating torsion balance designed to search for new, finite-range macroscopic interactions by testing the equivalence principle in the field of the earth. Because our signal would come from a force that is fixed in the laboratory, problems from residual magnetism and gravity gradients were more severe than those encountered in experiments using a solar source. The problems were overcome by designing a balance that is so symmetrical that there is almost no way for the known interactions to produce a significant spurious signal.

In a recent issue of Physical Review Letters[5)], we presented results for the differential horizontal acceleration of beryllium and copper and of beryllium and aluminum in the field of the earth. Because the results have just been published, I will not repeat the description given in my talk at the Felix Fest. I will merely summarize the conclusions. We see no effects ascribable to new macroscopic interactions. Our upper limits on infinite–range interactions whose charge is B and $B - L$, shown in Fig. 1, are significantly tighter than those obtained from the Princeton experiment[2)], and in addition we extend constraints down to interactions with a range of 1 meter. In fact, our limits on interactions coupled to B and $B - L$ are more stringent than those from the best astronomical tests of the $1/r^2$ law[6)]. In Fig. 2 we add our two data points to the reanalysed Eötvös data. It is clear to us that the remarkable correlation in the Eötvös data was not a result of "new physics", and that there is no credible evidence at this point for new macroscopic interactions.

We were not lucky; we did not find "new physics" even though we built an instrument that has very good sensitivity. (Our upper limits on Δa correspond to a kinetic energy in the fundamental mode of our 70 gram pendulum of less than 5 μeV. This is roughly 10^{-29} eV/atom. One can appreciate just how small 10^{-29} eV is by recognizing that it is the electrostatic energy of to electrons $\approx 10^4$ light years apart.) However we are still from the fundamental limits of the torsion balance. We are therefore continuing to improve our device, and also using it to ask new questions. In this regard, I would like to think that we are following in Felix's footsteps. He did lovely experiments to look for a finite neutrino mass. He probed farther into *terra incognita* than anyone had gone before, and showed that claims of positive effects from other laboratories were not credible. Although he was unlucky in not yet finding new physics, he advanced our understanding of Nature and severely constrained theoretical speculations. And I am delighted to see that Felix is continuing to advance the frontiers, designing more sensitive detectors to explore farther into unknown territory. It has been a pleasure to join this celebration of Felix and his physics.

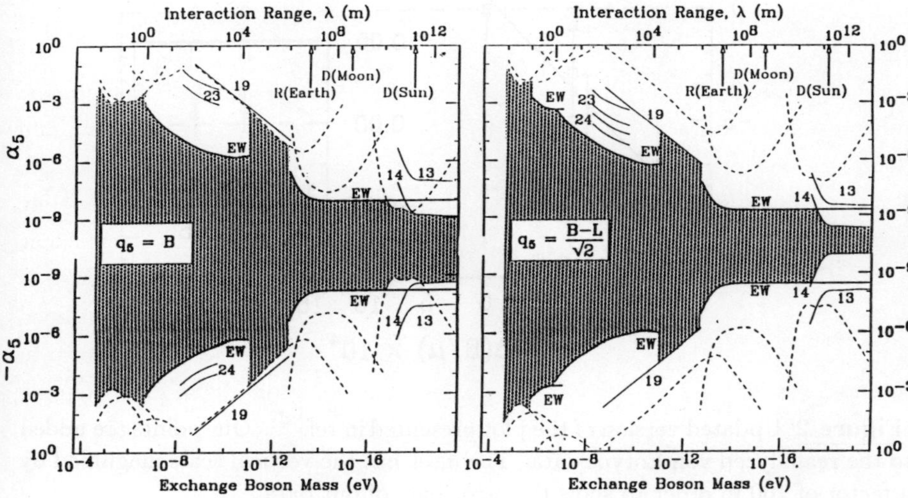

Figure 1 Constraints on interactions of ultra–low mass vector bosons from this and previous experiments. Upper limits on α_5 as a function of boson mass $m_b c^2$, or range λ. Although one expects α_5 to have a positive sign corresponding the the repulsive vector interaction between like charges, for simplicity in interpreting experimental results which can have either sign we consider the possibility of negative α_5 as well. Dashed curves are from tests of the $1/r^2$ law summarized in ref. [6]; solid curves are from composition dependence experiments. Left panel: $q_5 = B$, right panel: $q_5 \propto B - L$. The curves labelled EW are from this work. Numbered curves correspond to results from other groups (see ref. 5 for details).

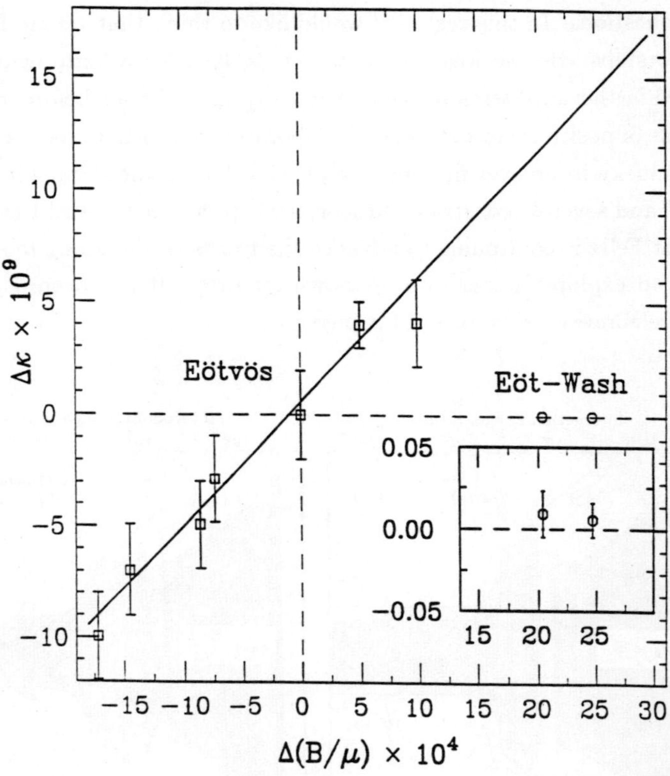

Figure 2 Updated version of the plot presented in ref. [4]. Our points are added to the reanalyzed von Eötvös data. The inset has the vertical scale magnified by a factor of 100 in order to show the error bars of our data.

References

1. Eötvös von R., Pekar D., and Fekete E., Ann. Phys. (Leipzig) **68**, 11 (1922).

2. Roll P.G., Krotkov R. and Dicke R.H., Ann. Phys. (NY), **26**, 442(1964).

3. Braginsky V.B. and Panov V.I., Zh. Eksp. Toer. Fiz. **61**, 873 (1971) Sov. Phys. JETP **34**, 463 (1972).

4. Fischbach E., Sudarsky D., Szafer A., Talmadge C. and Aronson S.H. , Phys. Rev. Lett. **56**, 3 (1986).

5. Heckel B.R., Adelberger E.G., Stubbs C.W., Su Y., Swanson H.E., Smith G. and Rogers W.F., Phys. Rev. Lett., to be published.

6. C. Talmadge, J.-P. Berthias, R.W. Hellings, and E.M. Standish, Phys. Rev. Lett. **61**, 1159 (1988).

DEEPLY BOUND PIONIC STATES

Toshimitsu Yamazaki

Institute for Nuclear Study, University of Tokyo
Tanashi, Tokyo 188, Japan

We discuss the structure and formation of deeply bound π^- states in heavy nuclei, which are expected to be narrow due to the repulsive π^--nucleus interaction. Possible experiments to produce those states are described.

1. INTRODUCTION

The 1s state of π^- bound by proton or deutron is the extreme outermost 1s orbital of pionic atom, whose Bohr radius is 200 fm. A high-precision pionic x-ray spectroscopy was carried out by Felix Boehm's group at LAMPF using a crystal diffraction spectrometer [1]. Their observation revealed an extremely small but sizable strong-interaction shift of the 1s binding energy [1]. The small strong-interaction shift for π^-d [2] is already indicative of the repulsive character of the S-wave optical potential. If we convert this very very small shift (5 eV!) to the nuclear potential depth, we see that V_0 is about 10 MeV. The strong-interaction shift increases rapidly as the nuclear charge increases, and the strong-absorption width becomes also appreciably large, and beyond Z=13 the 2p-1s transition is no more observable because of the too large nuclear absorption. One can, however, address the following question; namely, what happens on an extremely innermost 1s orbital, say, the 1s π^- state of ^{208}Pb.

In the present paper we discuss this problem following our recent work [3-6] and discuss experimental methods to study the structure of deeply bound π^- states.

2. COULOMB-ASSISTED HYBRID BOUND STATES

Let us discuss some general properties of a negatively-charged hadron in nuclear medium. For instance, a Σ^- hyperon forms an exotic atom, but only outer orbitals are known because of the strong absorption by the nucleus. On the other hand, a nucleus which involves a Σ^- as a constituent is called Σ^- hypernucleus. What is the distinction between Σ^- atoms and Σ^- hypernuclei?

No matter whether the strong interaction is attractive or repulsive, the long-range Coulomb attraction helps accommodate bound states. The low-lying states are of more nuclear character and can be called "Coulomb assisted hybrid bound states" [7]. The density of such a state distributes partly inside and partly outside the nucleus, and its width is expressed by

$$\Gamma_{n\ell m} = -2W_0 \int_{r<R_0} \rho_{n\ell m}(\mathrm{r})\mathrm{d}\tau. \tag{1}$$

Obviouslty, inner states have larger widths because of larger overlapping with the nucleus.

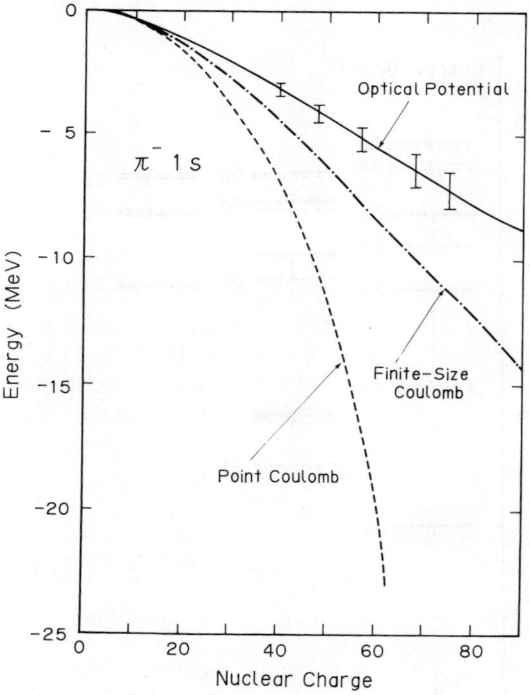

Fig.1 The binding energy of the 1s state of pionic atom versus Z, calculated with the pion-nucleus optical potential, with the finite-nuclear -size Coulomb potential only, and with the point Coulomb potential only. The widths of the 1s level are indicated by vertical bars for representative pionic atoms. Taken from Toki and Yamazaki [3].

The inner states (defined as those inside the last orbital where the atomic cascades ceases) can be populated by strangeness-exchange reactions such as (K^-,π^+). Due to practical difficulties the experimental studies of Σ hypernuclei are still in a premature stage [8,9]. On the other hand, the pionic atoms have been studied rather thoroughly and the pion-nucleus potential has well been established by now, but it is surprising to see that virtually no thought has been put on the inner structure of π^- bound states so far. Very recently, Toki and Yamazaki [3] studied this problem theoretically and clarified that the inner states are expected to exist with narrow widths and can be populated by charge-exchange nuclear reactions.

This new spectroscopy will provide a unique way to embed "real pions" inside nuclear medium and will shed a light on the intriguing question, namely, what is the pion and whether the pion keeps its free identity (mass and size) in nuclear medium.

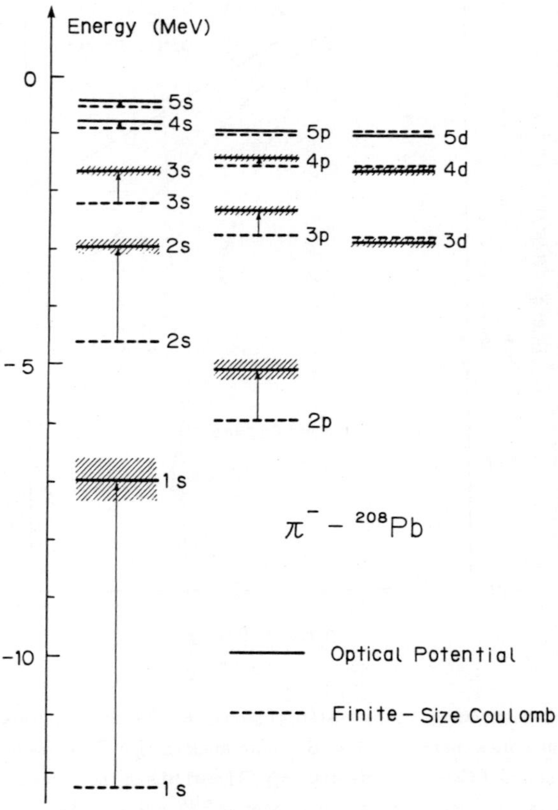

Fig.2 Energy levels of pionic atom of ^{208}Pb, calculated with the finite Coulomb potential (dashed bass) and with the optical potential (solid bars) with hatched area indicating the level widths. Taken from Toki *et al.* [4].

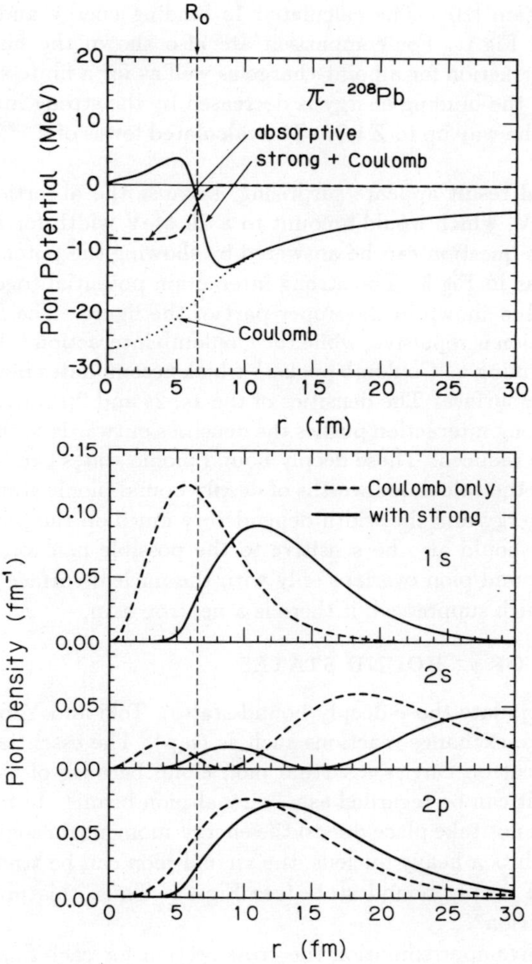

Fig.3 (Upper) The π^- optical potential (local part) for ^{208}Pb. (Lower) The pionic atom densities of the 1s, 2s and 2p states, calculated with the finite-size Coulomb potential (broken curve) with the optical potential (solid curve). From Toki *et al.* [4].

3. STRUCTURE OF π^- BOUND STATES

Toki and Yamazaki [3] and Toki *et al.* [4] have calculated the structure of deeply bound pionic atoms using the pion-nucleus optical potential with the parameters of Seki and Masutani [10]. The calculated 1s binding energy and width versus Z are depicted in Fig.1. For comparison are also shown the binding energies without strong interaction for a point-charge as well as for a finite-size nucleus. It is remarkable that the binding energy is decreased by the strong interaction. The width is small all the way up to Z=82. The calculated levels of $\pi^- {}^{208}$Pb are shown in Fig.2.

This theoretical result appears surprising, because the absorption potential is as much as 10 MeV, which would amount to a 20 MeV width for a deep enough nuclear state. This question can be answered by showing the potential shape and the π^- densities, as in Fig.3. The strong-interaction potential together with the Coulomb potential is shown in the upper part of the figure. The S-wave part of the strong-interaction is repulsive, while the Coulomb interaction is long range and attractive, thus forming a "Coulomb pocket" which accomodates bound states just outside the nuclear surface. The densities of the 1s, 2s and 2p states are shown in the figure. The strong interaction pushes the densities outwards so that the overlap with the nucleus is reduced. Those deeply bound pionic states can be called "pion halo". This is why the calculated widths of deeply bound pionic states are narrow.

The binding energy and the width depend very much on the potential parameters. The width should also be sensitive to the possible neutron skin of heavy nuclei; since the bound pion overlaps only with the nuclear surface region, its absorption is very much suppressed, if there is a neutron skin.

4. FORMATION OF π^- BOUND STATES

How can we populate those deeply bound states? Toki and Yamazaki [3] proposed to use charge-exchange reactions such as (n,p). The essential ingredient is as follows. The neutron carries a virtual pion cloud because of the dissociation n$\rightleftharpoons\pi^-$p, and thus it can be regarded as a "virtual-pion beam". In free space, such a dissociation does not take place due to the energy-momentum conservation, but, when the neutron hits a heavy nucleus, the virtual pion can be transferred to the nucleus, forming a deeply bound state (see Fig.4). Thus, this process is called "pion-transfer reaction".

In the plane-wave approximation the cross section for each final state (nl) is expressed as

$$\left(\frac{d\sigma}{d\Omega}\right)_{\rm PW} = \frac{M^2}{(2\pi)^2}\frac{p_f}{p_i}\left(\frac{f}{\mu}\right)^2 \frac{q_L^2}{\mu}\sum_m |\psi_{n\ell m}(\vec{q})|^2 \tag{2}$$

where M is the nucleon mass, μ is the pion mass, f is the coupling constant of the Yukawa vertex (taken to be 1) and the ψ_{nlm}(q) is the Fourier transform of the pion wavefunction at the momentum transfer $\vec{q} = \vec{p_i} - \vec{p_f}$ of the reaction. The $q_{\rm L}$ is the Lorentz-invariant momentum transfer. The momentum transfer is given by

$$q = \omega \frac{E_i}{p_i},$$

$$q_L = \omega \frac{M}{p_i} \tag{3}$$

at forward direction. Typically, for $T_i = 400$ MeV, $q = 200$ MeV/c and $q_L = 140$ MeV/c at $\omega = 140$ MeV. For deeply bound states the internal momentum is large enough so that the cross section is substantial (as much as 1 mb/sr in the plane-wave approximation).

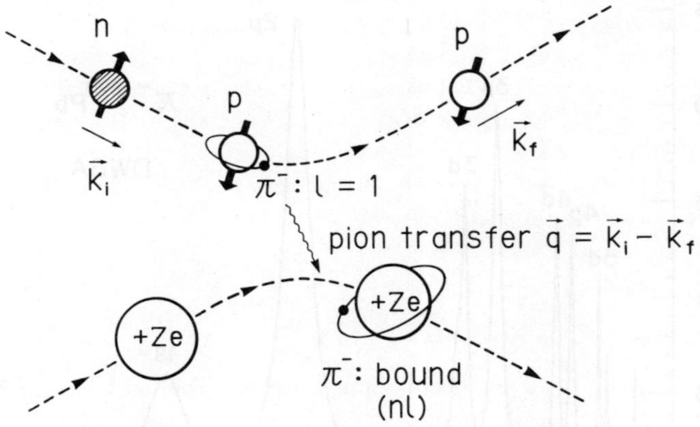

Fig.4 "Pion-transfer" reaction. The virtual pion in the neutron beam is transferred to the target nucleus to form a real bound state. The reaction is characterized by charge-exchange ($\Delta I = 1$) and spin-flip ($\Delta S = 1$).

Recently, Toki *et al.* [11] estimated the distortion effect of the incoming and outgoing nucleons. The cross section is generally reduced by a factor of 10. The calculated line profile of the ^{208}Pb(n,p) reaction is shown in Fig.5. The 2p peak is the largest of all.

The pionic bound states appear in a continuum in the range of 140 MeV excitation. They are characterized by $\Delta S = 1$ and $\Delta T = 1$ and thus constitute a new family of Gamow-Teller type resonances.

5. POSSIBLE EXPERIMENTS

An experimental search using the (n,p) reaction was attempted at TRIUMF.

The TRIUMF CHARGEX facility [12] permits (n.p) reaction spectroscopy with a medium resolution (the mass resolution is around 1 MeV). Iwasaki *et al.* [13] obtained a spectrum of ^{208}Pb(n,p), revealing no distinct peak within the statistics attained. It excludes the large cross section as estimated by the plane-wave approximation, but is consistent with the distorted-wave calculation. Since the (n,p) spectroscopy has an inherent difficulty in statistics and in background, it seems to be difficult to obtain a spectrum with an overwhelming statistics. One can make use of other charge-exchange reactions such as (d,^{2}He).

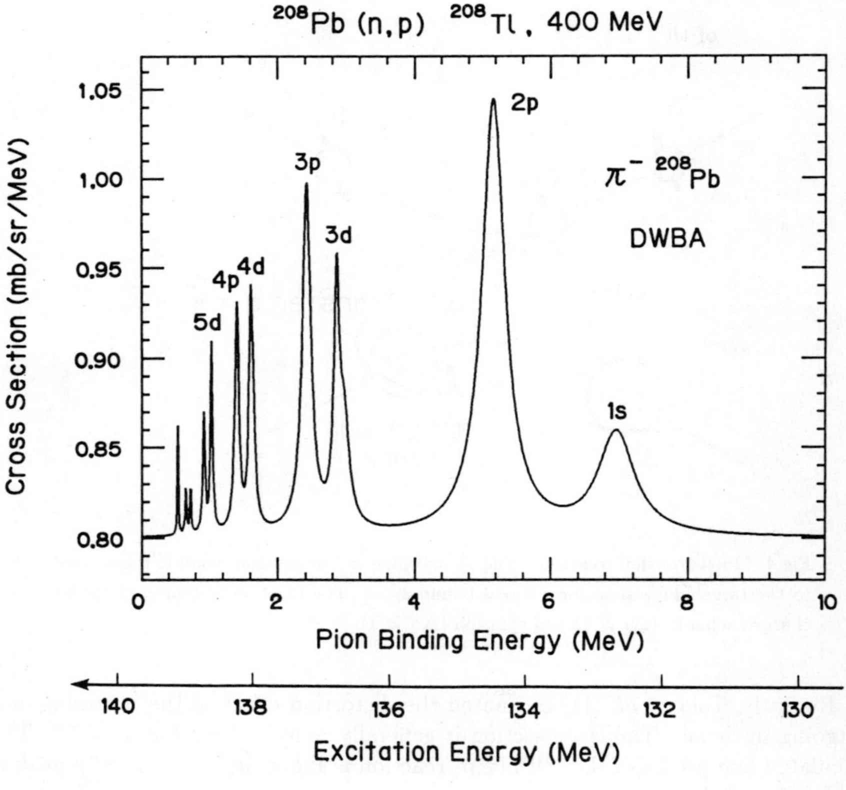

Fig.5 A DWBA calculation of the line profile of the ^{208}Pb(n,p) reaction in the region of pionic bound states at $T_n = 420$ MeV. Calculated by Toki *et al.* [11].

The (d,^{2}He) type charge-exchange reaction at intermediate energy was studied by Ellegaard *et al.* [14] at SATURNE at Saclay. The region we are interested in lies between the low-energy Gamow-Teller resonance and the high-energy Δ resonance.

Since the pion-transfer reaction has a spin-flip operator $\vec{q}\cdot\vec{\sigma}$, the forward amplitude is enhanced in the case of transverse polarization of deuteron; the cross section increases by a factor of 3. An experimental search for deeply bound π^- states with the (d,^{2}He) reaction is to be carried out [15].

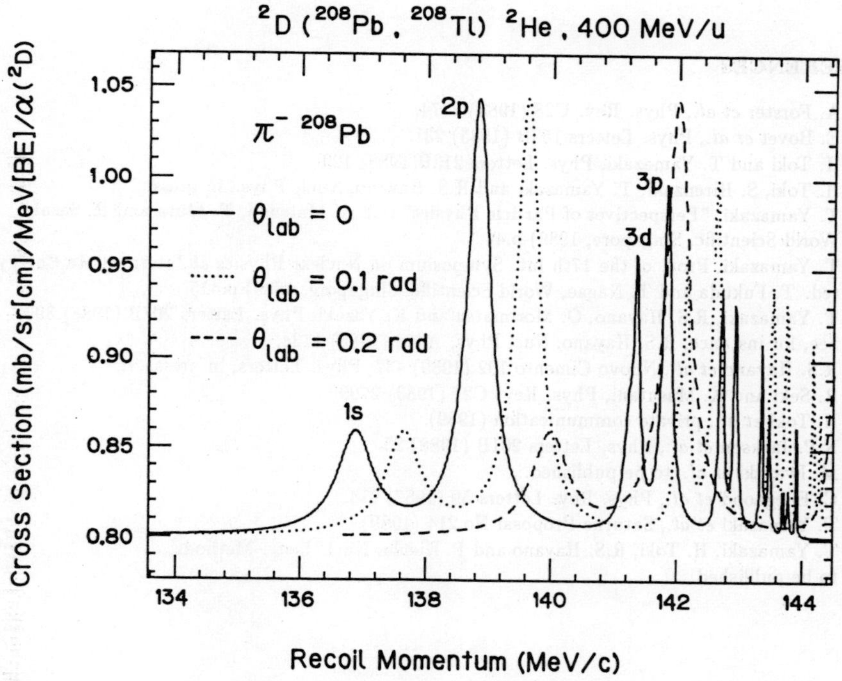

Fig.6 Recoil momentum spectrum of "^{2}He" (two correlated protons) in the inverse kinematics where a deuteron target is hit by a ^{208}Pb beam (400 MeV/u).

In the future one can make use of the charge-exchange reactions with inverse kinematics, namely, a medium energy ^{208}Pb beam hits a deuteron target and the two recoil protons (to form ^{2}He) are detected at forward angle. It is shown [16] that the spectrum of Fig.5 should be transformed into a recoil spectrum, as shown in Fig.6, since the recoil momentum is nearly equal to the Lorentz-invariant momentum transfer q_L as given by eq.(3). This method will offer a high resolution spectroscopy (the mass resolution as good as 100 keV) so that one can determine not only the binding energies but also the widths of the 1s and 2p states. It requires a cooled heavy ion beam, as will be obtained at the SIS-ESR ring at GSI. With this method one can in principle form pionic bound states in unstable nuclei, hitherto unconceived objects.

Acknowledgement

This paper is based on the collaborative work with Professors H. Toki, R.S. Hayano and P. Kienle, whom the author would like to thank. It is supported by the Grant-in-Aid of Special Project Research on Meson Science of the Japanese Ministry of Education, Science and Culture.

REFERENCES

[1] A. Forster *et al.*, Phys. Rev. C28 (1983) 2374.
[2] E. Bovet *et al.*, Phys. Letters 153B (1985) 231.
[3] H. Toki and T. Yamazaki, Phys. Letters 213B(1988) 129.
[4] H. Toki, S. Hirenzaki, T. Yamazaki and R.S. Hayano, Nucl. Phys., in press.
[5] T. Yamazaki, "Perspectives of Particle Physics" (ed. S. Matsuda, T. Muta and R. Sasaki, World Scientific, Singapore, 1989) p.44.
[6] T. Yamazaki, Proc. of the 17th Int. Symposium on Nuclear Physics at Intermediate Energy (ed. T. Fukuda and T. Nagae, World Scientific, Singapore, 1989) p.415.
[7] T. Yamazaki, R.S. Hayano, O. Morimatsu and K. Yazaki, Phys. Letters 207B (1988) 393.
[8] See, for instance, R.S. Hayano, Nucl Phys. A478 (1988) 113c.
[9] R.S. Hayano *et al.*, Nuovo Cimento 102 (1989) 437; Phys. Letters, in press.
[10] R. Seki and K. Masutani, Phys. Rev. C27 (1983) 2799.
[11] H. Toki *et al.*, private communication (1989).
[12] K.P. Jackson *et al.*, Phys. Letters 201B (1988) 25.
[13] M. Iwasaki *et al.*, to be published.
[14] C. Ellegaood *et al.*, Phys. Rev. Letters 59 (1987) 974.
[15] T. Yamazaki *et al.*, Saturne Proposal No.214 (1989).
[16] T. Yamazaki, H. Toki, R.S. Hayano and P. Kienle, Nucl. Instr. Methods., to be published

EARLY WORK ON DOUBLE BETA DECAY

Chien-Shiung Wu
Dept of Physics, Columbia University

ABSTRACT

In this brief review, no theoretical conceptions of the Double Beta Decay and the nature of the neutrinos will be introduced. Two major different types of experimental methods and their recent developments will be presented and discussed. The present status of Double Beta Decay events of ^{48}CA, (1,2,3.); ^{76}Ge,(4,5,6,7.); ^{82}Se,(8,9,10,11,12.); ^{100}Mo,(13); ^{136}Xe,(14,15); and ^{128}Te/^{130}Te(16, 17,18) will be described. The outlook is hopeful.

I INTRODUCTION

The Search of Double Beta Decay

In 1959, exactly 30 years ago, the difficulties of the search of double beta decay were summarized by Dubrokhotov et.al. (Dub 59) as an amazing example of a fantastic succession of periods of hope and disillusionment. He wrote: Two times in the course of a single decade 1949-59, the phenomenon has been "discovered" and both times the discovery has been found to be "erroneous". This is due to the fact that the measurements of these rare events were not only difficult to observe but, they were also treacherously confused by many kinds of "look-alike" double beta decay backgrounds. The results obtained during the period around the 1980's were most baffling and disappointing. Happily, the baffling situations were gradually cleared up after 1986 as the importance of radiation backgrounds in the direct detection method became understood and its reduction was greatly improved.[(Wu-83)]. In the mean time, the early history and impurity of the old rock ores in the geochemical methods also became a very critical and important selection factor. Thus, the recent experimental results from geochemical methods became consistent and reliable among different groups.[(Ki-83)]

I.I THE DIRECT DETECTION METHOD

Its sensitivity has been improved by several orders of magnitudes by introducing the following modifications:

(a) Mounting a large and thin film of double beta decay source inside of a "track-visualizing device" such as

A streamer chamber or

A cloud - chamber or

A time projection chamber (TPC) etc.

(b) Applying Anti-coincidence circuits, magnetic fields and low radiation environment.

(c) Application of energy-sum methods in scientillation detectors such as

CaF_2 ^{48}Ca in CaF (1966)

^{76}Ge in $GeLi_2$ (1967)

^{100}Mo Sandwich Method (1966 or 1980)

^{136}Xe in "Xe TPC" (1986)

II GEOCHEMICAL METHOD

In this method, the daughter nucleus of the ($\beta\beta$) decay must be a rare gas and occluded in a rock of known old age (> 10^9 yrs). The isotopic contents can be accurately determined by a highly sensitive mass-spectrometer. Although leptons were not detected in these experiments, the determined ratio of the double beta decay half lifes such as in ^{128}Te VS ^{130}Te

$$S \equiv T_{1/2}(^{128}Te)/\ T_{1/2}(^{130}Te)$$

Comparing to the theoretically predicted ratios based on either from "zero neutrino" 'DBD'

$$S_{0\nu} \equiv \{T_{1/2}(^{128}Te)]^{0\nu}/[T_{1/2}(^{130}Te)]^{0\nu} = 10^{2.2}$$

or from two neutrio 'DBD' decay:

$$S_{2\nu} \equiv [T_{1/2}(^{128}Te)]^{2\nu}\ /\ [T_{1/2}(^{130}Te)^{2\nu} = 10^{4.0}$$

Of course, it is also important to directly compare the half lives which were determined either from the direct detection method or that of by the geochemical method. In order to understand the apparent experimental discrepencies, a more detailed discussion will help.

III MORE DETAILED DISCUSSIONS

(a) Direct Detection Method:

At first glance, the coincidence technique of simultaneous detection of the two e^-(or $2e^+$) in double deta decay may be sensitive to observe this rare process. Many experiments had been carried out with coincidence technique prior to 1967 in the search for double beta ($\beta\beta$) decay, but the sensitivity of those experimental searches were limited by high radiative backgrounds counting rates which may even be much larger than the expected $(\beta\beta)_{2\nu}$ rates.

To realize the treacherous coincidence counting rates (Fig 1) the background results principally from local gamma rays are either compton scattering from counter to counter (1B) or producing recoil electrons in one counter which strike another counter (1C). The other mechanisms from (D) to (J) are more rare than (B) and (C) but all the events can be reduced or eliminated by surrounding the double beta decay source with some device showing the tracks of particles leaving the source and placing the apparatus in a magnetic field. Such a device, by the Columbia group (Ba-67), is shown schematically in Fig (2). A source, unstable to double beta decay, is mounted inside a Track Visualizing Device (such as a Streamer Chamber, a Cloud Chamber, or a "TPC".) Outside the chamber are mounted a pair of number '16' wedge shaped scintillation counters. A coincidence between any two opposite counters makes the chamber operate and produce a picture of all the particle tracks in it. A magnetic field is directed normal to the plane of the source. The radiation background environment of gamma spectra taken with a 12.7 and X 12.7 cm NaI crystal under various conditions on the earth's surface and in the salt mine is shown in Fig 3. The Intrinsic background in the mine was found by placing the NaI crystal in a freshly dug pit in the salt, well-shielded from all other man-made objects. The 1.46 MeV gamma ray from the decay of ^{40}K in the salt was the only significant radiation naturally present in the salt. The 2.62 MeV line of the ThC (^{208}Th) was visible throughout the experimental area and appeared to be associated with objects brought in from the surface of the earth.

A large component of the background came from airborne radon gas (^{222}Em, with a 3.8d half-life). Coming from the decay of ^{226}Ra, the level of this activity increased noticeably in rainy weather even in the deep mine.

The data film was scanned for possible $(\beta\beta)_{o\nu}$ events by firstly selecting all events with a total energy above 3 MeV containing a single pair of tracks meeting at the source. In 1,103 hrs. of life-time with the ^{48}Ca Source, 191 such events were found. Of the remaining 119 events, all but 23 were rejected for track curvature clearly of the wrong sign or particle momentum clearly inconsistant with recorded scintillation information. These were mostly high-energy compton recoils or cosmic-ray muons crossing the chamber. Only one of the surviving ^{48}Ca events was within the two standard deviations of 4.24 MeV, the expected $(\beta\beta)_{o\nu}$ sum energy. A half-life limit for the $(\beta\beta)_{o\nu}$ decay in ^{48}Ca was arrived at. In 1967 Columbia group (Ba-67)

$$^{48}\text{Ca: } T_{1/2}(o\nu) > 2\times10^{21} \text{ yr at (the 80\% C.L.)}$$

The scanning of the two-neutrino decay $(\beta\beta)_{2\nu}$ events obtained requires changing the criteria above. A lower limit of the two-neutrino decay half-life $T_{1/2}$ obtained at an 80% confidence level is the 1967 Columbia group:

$$^{48}\text{Ca: } T_{1/2}\ (2\nu) \geq 10^{20} \text{ Yr.}$$

However, because the radiation background rates are high, therefore the correction uncertainty is higher than the $T_{1/2}(2\nu)$ rate. In other words, the confidence level C.L. was very poor: #^{48}Ca investigation in 1988-89 by a Chinese Academic Sinica group: In July, 1989, a large group of more than ten scientists from Academia Sinica and World Laboratory in Beijing, China submitted a paper on "A Search for Neutrinoless Double Beta Decay of ^{48}Ca*". The search was carried out in a coal mine near Beijing, China, which is equivalent to about 1,300 M of water. Large scintillation crystal of natural CaF_2 were used as both the detector and the beta source. The energy resolution is determined to be $R=18.7\%\ E^{-1/2}$. After the first 1,700 hrs of data was taken, the lower limit of the half life of the neutrinoless double beta decay of ^{48}Ca obtained was $^{48}\text{Ca}[T_{1/2}(o\nu)]>3.6\times10^{21}$ yr. The investigation was still progressing in July 1989. From the description in the paper, the natural radiation background could be further reduced by using additional shielding or the experiment be carried out at a greater depth. The schematic drawing of the detector assembly is shown in Fig (4). The counts per channel around 4.27 Mev for 1,700 hours are shown in Fig (5). However, no peak is present near 4.27 Mev. The total count for 1,700 hours within an energy interval 4.27±. 194 Mev are 52. They adopted the "sensitive formula" introduced by Fiorinic (Fi67)

$$T_{\frac{1}{2}} \geq A(MT/\Delta E.B)^{\frac{1}{2}} \text{ years}$$

Using $A=(o.76xL\mu 2x6.023x10^{23}x^a/876xm)$

$A=(8.35x10^{14}$, $B=4.28x10^{-9}$ cts/hr/Kev/g

$M=18.400g$ $\Delta E=388$ Kev, $T=1700$ hrs;

Then $T_{\frac{1}{2}} \geq 3.6x10^{21}$Years at 68% C.L.

Table (1) shows the results of $^{48}Ca(0\nu)$ DBD experiments. The Chinese result (3) is compatible to that of Columbia's result (2) in regards to the quantities of ^{48}Ca and the time of data taken.

Table (1)

Authors	Mass of ^{48}Ca	Hrs Taken	$T_{\frac{1}{2}}\geq$
Goldhaber (1)	11.4 g	689 hrs	$\geq 2x10^{20}$Yr
Columbia (2)	10.6 g	1150 hrs	$\geq 2x10^{21}$Yr
Academia Sinica (3)	20.0 g	1700 hrs	$\geq 3.6x10^{21}$ Yr

The results from the Academia Sinica in Beijing, China from sum energy scintillation method is compatible to that of Columbia group with a track visualizing streamer chamber method and in regards to the quantities of ^{48}Ca (See Table 1) and the time of data taken. Further improvements on the lowering of radiation background and the statistics in the Beijing, China experiments which are still progressing should yield a better result.

#^{82}Se:

The apparatus and measurement technique similar to that used for the study of ^{48}Ca by Bardin et.al (Ba-67) were also applied to the study of (ββ) decay of ^{82}Se by Cleveland et.al. (C1-75) in 1975. The selenium source contained approximately 46g of metallic powder enriched to 56.5% of ^{82}Se formed the center place (Diam.=50cm.; t=58 mg/cm^2) of a helium filled double gap streamer chamber. The selenium was purified by precipitation of radium with separated ^{138}Ba and by multiple passes through ion-exchange columns. In order to reduce background, the experiment was also conducted in the morton salt mine (600m below ground level) at Cleveland. The total number of events recorded was 65,500. The vast majority of these were background induced, primarily Compton scattering of γ-rays in a scintillator, the recoil electron then passing through the chamber and hitting a second counter. Only 201 events are two-tracks events with the signature of ββ decay. Many of these were caused by double scattering within the source and it was difficult to distinguish them from true ββ events. By restricting the energy range to between 2.4 and 3.2 MeV and imposing appropriate acceptance criteria on the track curvature, the overall selection efficiency became 19% and no events were found in this energy region. At a confidence level of 68%, this null result implies the following lower limit on the half-life for no-neutrino (ββ) decay of ^{82}Se:

$$T_{\frac{1}{2}}^{0\nu} \geq 3.1\times10^{21} \text{ C.L. } 68\%$$

Moe and Lowenthal's New Approach

In the Direct Detection used by Bardin et.al. in 1967 and Cleveland et.al. in 1975 for the study of (ββ) decay of ^{48}Ca, and ^{82}Se respectively, among the "$2e^-$" backgrounds, they noticed unusually large number of double-beta-like events indistinguishable from the true double beta decays. After thorough investigations, it was concluded that the principal radioactive contaminant occuring uranium series; its multiple ($\underline{\beta\text{-}\gamma}$) cascades cause "two electron" events to emerge from the source resembling the ($\underline{\beta\beta}$) decay (See FigII and IJ). The second electron was produced by Møller or Compton scattering or by internal conversion. Although

^{214}Bi nucleus is short lived; its presence after a few days is maintained by its long-lived progenitor ^{225}Ra. It had been hoped that by chemical purification technique substantially lowers levels of radiation in isotopically enriched double beta sources could be achieved, however, success had not been adequately demonstrated. The level of ^{226}Ra contamination that would prove troublesome in this case only needs to be on the order of one part in 10^{15}. Moe and Lowenthal (MO87) have made special efforts to modify the direct detection method in order to identify and eliminate a large fraction of double-beta-like events from ^{214}Bi. The general principle is as follows: ^{214}Bi beta decay is followed in 164 μs by emission of a 7.7 MeV alpha particle. The ^{214}Bi induced "two-electron" events thus can be identified if the α-particle can be detected (see Fig. 6). For the 164 μs delayed α-particle from ^{214}Bi to be detected, a track visualization chamber of longer sensitive duration of post trigger period as compared to 164 μs must be used. A large cloud chamber was chosen by Moe and Lowenthal for this purpose and its duration of post trigger sensitivity was found to be easily to cover the delay of alphs-particles attending ^{214}Bi decay.

Observed Events of DBD in ^{82}Se in Cloud Chamber Method:

At the end of 37 live days, 36 negatron pairs from the ^{82}Se source (13.34 g) strips had been observed. Of these, 20 were clean events ($2e^-$) and 16 were accompanied by α-particles ($2e^- + \alpha$) track pictures.

Since the observed total number of $2e^-$ events were 20 and the calculated ($2e^-$) from ^{214}Bi with a α-particles trapped in the source 4.8± 1.2 Thus ($2e^-$) from ^{82}Se were only (15.2±4.6.).

Using the cal. overall detection efficiency as 0.22±0.007, the half-life of the two neutrinos (2ν) decays can be calculated to be:

$T_{1/2}^{2\nu} = 1.0 \pm 0.4 \times 10^{19}$ year with sum energy below the 3.3 MeV

Because of its extremely low source mass and trigger efficiency, it is not sufficient to detect for neutrinoless $(\beta\beta)_{0\nu}$ decays. No events at high energies were seen.

However, in replacing a velvet screen to reduce the light reflection in the photography of the cloud chambers, the authors noted that the background radiation suddenly jumped up a factor of two probably due to the gas oculation in the velvet screen. The oculation could be due to the "Rn" air borne Rn gas. The authors decided then to replace the cloud-chambers by a "Time-Projection" Chamber which had been thoroughly developed by Dr. Herb Chen for a "Xe TPC".

On August 31, 1987, Moe et.al. reported the direct evidence for two neutrino double beta decay in ^{82}Se: $T^{2}_{\frac{1}{2}} \geq (1.1 + 0.6\text{-}0.3) \times 10^{20}$Yrs. by the "TPC" method. This result from direct counting confirms the earlier geochemical measurements and helps provide a standard by which to test this double beta decay matrix elements of nuclear theory.

Moe	^{82}Se	$(TPC)^{2\nu}$	$T_{\frac{1}{2}}$	1.1×10^{20}	Yrs	1987	(9)
Kirsten	^{82}Se	$(Geo)^{2\nu}$	$T_{\frac{1}{2}}$	1.3×10^{20}	Yrs	1986}	(10) (11)
Manuel	^{82}Se	$(Geo)^{2\nu}$	$T_{\frac{1}{2}}$	1.0×10^{20}	Yrs	1986	(12)

The agreements between the results of direct detection method (TPC II, Moe's Group) with the geochemical methods (Kirsten's and Manuel's Group) and also between the Kirsten's and Manuel's group imply that the principles and assumptions used in these major methods were very solid and absolutely unquestionable. The reconciliation observed when the large discrepancy of one order of magnitude between the half-life of $^{82}Se(2_L)$ from the direct detection method by cloud chamber method by Moes' group of 10^{19}yr was eventually overturned by the half-life of $^{82}Se(2_L)$ determined by "TPC II" of 10^{20} years and in agreement with the half-life of ^{82}Se by geochemical method of (Scinivasan 2.76×10^{20} Yr) and others as shown in table list above.

Energy Sum Method in Ge-Detector (Fi-73)

Another ingenious direct detection of double beta decay was

applied to ^{76}Ge by Fiorini et.al. (Fi-73) to the high resolution Ge(li) detector in 1973. Similar principle was also applied to detection of (ββ) ov in ^{48}Ca by Goldhaber and des Mastesian (1) however, the background was much too high. Ge(li) detectors are fabricated out of high purity Ge metal with a natural abundance of ^{76}Ge of 7.67% the Ge(li) detector used by Fiorini et.al. had an active volume of 68.5 cc and an energy resolution of 6 Kev at 2,615 MeV. A diagram of the experimental apparatus is shown in (Fig 7). The experiment was located in the Mont Blanc tunnel connecting Italy to France and the crystal was shielded from local radioactivity by layers of paraffin, cadmium, low-activity lead, bi-distilled mercury, nylon and high purity electrolytic copper. After 4400 hours of running, no peak was found in the 2.045 MeV region of expected neutrinoless (ββ) decay (Fig-8). Tests indicated that the residual backgrounds observed in this energy region were likely due to ^{40}K, ^{235}U, ^{238}U, and ^{232}U contaminants of less than 10^{-5}ppm which originated inside the local shield, probably in the Ge(li) crystals crystat structure. Possible ^{222}Rn contaminants in the liquid nitrogen coolant were also suspected. Fig. 9 shows the spectrum obtained in the final run of the experiment lasting 2300 hours. The background counting rate in the 2.045 MeV region was $(2\pm0.2) \times 10^{-4}$ counts KeV $^{-1}$hr^{-1}, allowing a limit to set on the half-line for the neutrinoless (ββ) decay of ^{76}Ge, $T_{\frac{1}{2}}^{0\nu} \geq 5\times10^{21}$ year at a 68% confidence level. With improved background, the limit has improved to $T_{\frac{1}{2}} \geq 2\times10^{22}$ yr. in the summer of 1982.

The energy sum method in Ge detector (Fi-73) was developed by Fiorino extensively in the past twenty years and so have groups by Avignone et.al. at pnl-use (prl 54,2309) and Caldwell (AV-85) et.al at UCSBLBL (P.R. D33.2737). As a result, we now have an extremely high limit on the half-life of ^{76}Ge (ov) > 6×10^{23} years. (Ca-87) A group in the Soviet Union (SO-87) (Proceedings of Hamburg

Conference on Leptons photons interactions at H.E. July, 1987, C.P. No 27-31-07-87) is working with an 85% enriched source of ^{76}Ge and so one may anticipate even tighter limit or perhaps the observation of the decay in the future.

IV GEOCHEMICAL METHODS

Life-Time Ratio "S" $\equiv T_{\frac{1}{2}}$ (^{128}Te) / $T_{\frac{1}{2}}$ (^{130}Te).

In an earlier experiment, by Takaoda & Ogata (Ta-66) (1966), using different tellurium ores, an excess of ^{128}Xe was found; but the authors cautioned that it was difficult to assign the excess of ^{128}Xe entirely to ^{128}Te double beta decay, because a small, persistent background in the mass spectrometer disturbed exact measurements at ^{128}Xe. If one takes the total amount of excess ^{128}Xe observed as due to ^{128}Te double beta decay then the ^{128}Te half-life is $10^{22.5 \pm 0.5}$ yr. Furthermore, it is not unreasonable to assume that nuclear matric elements for ^{128}Te$\rightarrow$^{128}Xe and ^{130}Te $\rightarrow$ ^{130}Xe are approximately equal. Under this assumption, the ratio of double beta decay rates should be proportional to the ratio of the available phase spaces, which is the 8th through the 11th power of energy release for $(\beta\beta)_{2\nu}$ and the 4th through 5th power of energy release for $(\beta\beta)_{o\nu}$ decay. Since the energy release for ^{130}Te is three times that for ^{128}Te, one would expect S $\equiv T_{\frac{1}{2}}$ (^{128}Te)/$T_{\frac{1}{2}}$(^{130}Te) $\cong 3^{8.4}$ = $10^{4.0}$ for $(\beta\beta)_{2\nu}$

" " $\cong 3^{4.6}$ = $10^{2.2}$ for $(\beta\beta)_{o\nu}$

When the specific warnings given by the authors about a small persistent background on the ^{128}Xe excess was not only ignored, instead, used to calculate the ratio "S", then we have

$$\text{"S"} = T_{\frac{1}{2}}(^{128}\text{Te})/T_{\frac{1}{2}}(^{130}\text{Te}) = 10^{22.5 \pm 0.5}/10^{21.34 \pm 0.12}$$
$$= 10^{1.2 \pm 0.6}$$

This ratio seems to be in better agreement with that predicted for non-neutrino double β-decay.

Pontecorvo Proposal:

On the basis of this rather questionable* experimental indication in ^{128}Te, Pontecorvo proposed that the decays of ^{130}Te and ^{128}Te are, predominately, the first order effect of a new super-weak($\Delta Q = \pm 2$, $\Delta S = 0$) interaction which mediates non-neutrino double beta decay, rather than the second-order effect of the usual weak ($\Delta Q = \pm 1$, $\Delta S = 0$) interaction. This super-weak interaction also causes the observed slight CP violation in K^0 decay.

In 1975, Hennecke, Manual and Sabu carried out another investigation on the ratio "S" of $T_{1/2}$ (^{128}Te)/ $T_{1/2}$(^{130}Te) by extracting Xenon by stepwise heating of the sample, and its isotopic abundance were measured in the mass region A = 122 to A = 136 in a mass spectrometer. For ^{130}Xe the excess over atmospheric abundance (^{130}Xe excess)/^{130}Xe atmospheric) was = 712±1, which was more than an order of magnitude greater than in the previous experiments. Many sources of systematic errors in measuring the individual ββ decay half-lives, such as errors in ore age T, in the tellurium determination N(^{130}Te) and in the Xenon content N^{excess} (^{130}Xe), cancel out in the determination of the ratio "S" of the half-lives. The experimental results found by Hennecke, Maulel and Sabu were:

$$T_{1/2}\ (^{130}\mathrm{Te}) = 10^{21.34 \pm 0.10}\ \mathrm{yr.}$$

$$T_{1/2}\ (^{128}\mathrm{Te}) = 10^{24.54 \pm 0.12}\ \mathrm{yr.}$$

and $\text{"S"} = 10^{24.54 \pm 0.12}/10^{21.34 \pm 0.10} = 10^{3.20 \pm 0.01}$

Since "S" = $10^{3.20}$ which lies between the predictions for (2ν) and (oν) (ββ) processes, it was natural to interprete that the final state must be a combination of these two decays. In 1978, Bryman and Pissiotto used the idea of non-conserving parameter "n" introduced by Primakoff and Rosen (Pr-69) in the lepton current $\{{}_{L}^{lep} + n\ j_{R}^{lep}\}$ and obtained a value for $\eta = (4.3 \pm 0.1)^{-5}$ from the ratio "S" = $10^{3.20}$.

By early 1980, the theoretical and experimental interests in searching for the solutions of the mass of neutrino by using double

* Authors had warned that the mass-spectrometer had been used pre-

beta decay results were greatly aroused. Several theoretical attempts, using implicit helicity breaking by neutrino mass m_ν, to interpret the observed "S" value (S ≡ 3.20) were reported. The mass values obtained vary from ∿ 10ev by Haxton (Ha-82) to ∿34ev by Doi (DO-82). These values are supposed to be compared with the mass value $T_{1/2}(^{128}Te) \geq 8 \times 10^{24}$ yr

$T_{1/2}(^{130}Te = 2.60 \pm 0.28 \times 10^{21}$ yr.

$$\rho = \frac{128\lambda}{130\lambda} = \frac{1}{\text{"S"}} = (1.3)\,^{+1.13}_{-1.03} \times 10^{-4}; \text{ (where } \lambda = \frac{\ln 2}{T_{1/2}}$$

Actually, the half-lives of ^{130}Te determined from various attempts are rather in good accord such as $(2.27\pm0.60)\times10^{21}$yr. and $(2.72\pm0.27)\times10^{21}$yr. and the present one $(2.60\pm0.28)\times10^{21}$yr.. The suggested "best value" is $^{130}T_{1/2} = (2.55\pm0.20)\times10^{21}$yrs..

Continuous Reduction of Value of "ρ"

While the $^{130}T_{1/2}$ determined were in rather good accord; however, the value of "ρ" has varied from 295×10^{-4} from Takaoka's experiment ('66) to 194×10^{-4} by Srinivasai's study ('72). In 1972 Kirsten et.al obtained a much lower "ρ" value equal to "5", but it was kept from publishing for waiting to improve on it. In 1975 and 1978, Hennecke et.al. published their results on $\rho=(6.29\pm0.20)\times10^{-4}$ to the present value(2)

$$\rho = (1.03^{+1.13}_{-1.03}) \times 10^{-4}$$

The theoretically expected ratio for pure Dirac decay is ($\rho_{2\nu} \sim 2\times10^{-4}$). Since the measured value $\rho = (1.03\,^{+1.13}_{-1.03}) \times 10^{-4}$; it implies that there is very little room for an admixture of Majorana decay or non-neutrino decay in DBD.

The relationship between the value of "ρ" and "η" with various values of m_ν for ^{128}Xe and ^{130}Xe. It is quite obvious, from the Heidelberg results, limits on $m_{\nu e} \leq 5.6$ev (95% Confidence) and the lepton violation parameter "η" $\leq 2.4\times10^{-5}(2\sigma)$. It is also consistent

*viously in ^{128}I experiment and they had difficulties in removing the occluded gas on the glass wall. Probably, that was the villain, should have been disregarded.

with zero neutrino mass $m_\nu = 0$ and for the lepton conservation $\Sigma \ell_e \equiv 0$.

However, the final conclusion on mass of neutrino will have to depend on most precise and confident various experimental results. It's also interesting to observe the "neutrino oscillation conclusion in progress. Professor Boehm may have done something to report to us on his 70th birthday. We are most anxiously waiting for it!

† N. Takaoka and Hitoshi Sagawa (89)
the DBD of ^{128}Te and ^{130}Te by Takaoka and Sagawa was reported at the Yamada Conference on 06/12-15/89 and the proceeding was received after Boehm's conference on 09/7-9/89. The comparison shown in the following table:

Relative decay rate	Reference
P= < (3.1+/-1.3) x 10^{-4}	Takaoka and H.
P= (3.9+/-0.9) x 10^{-4}	Lin et.al. ('87)
P= (6.4+/-0.4) x 10^{-4}	Hennecke at.al. ('78)
P= (1.0+/-1.1) x 10^{-4}	

In general good agreement. I appreciate the speeding completion of the publication of the proceeding of the Yamada Conference. The puzzle was started by Takaoka and Segawa, now happily solved by them too. The Iodine content in the Te-Ore might have caused the trouble as I suggested.

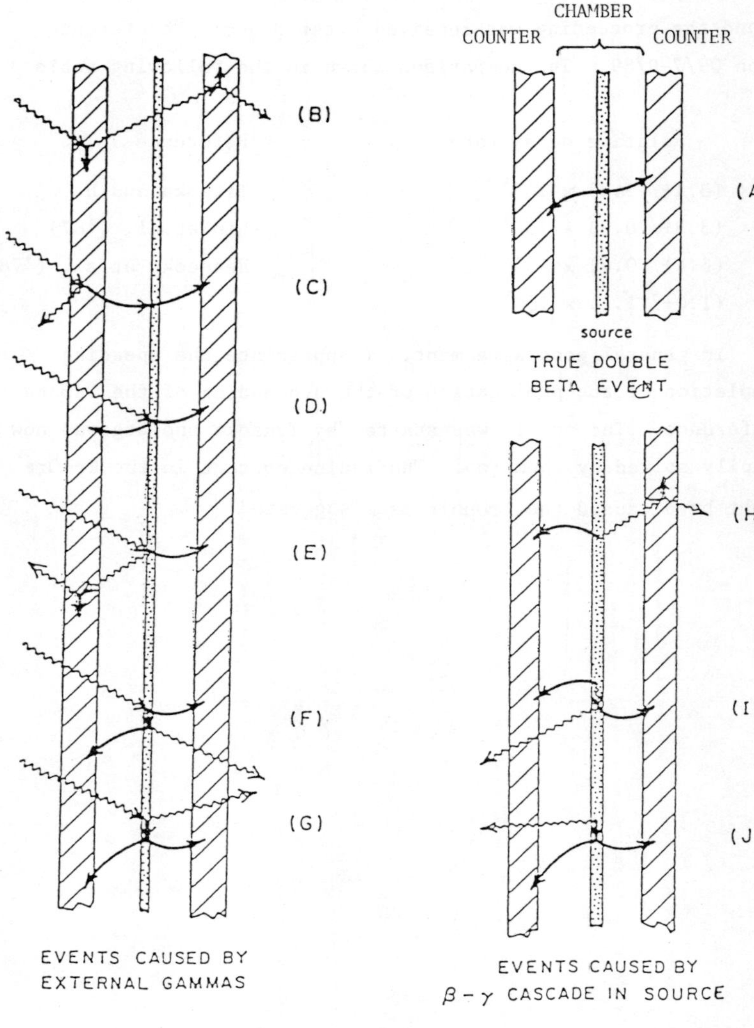

FIG 1 Appearance of various events in a track-visualizing chamber.

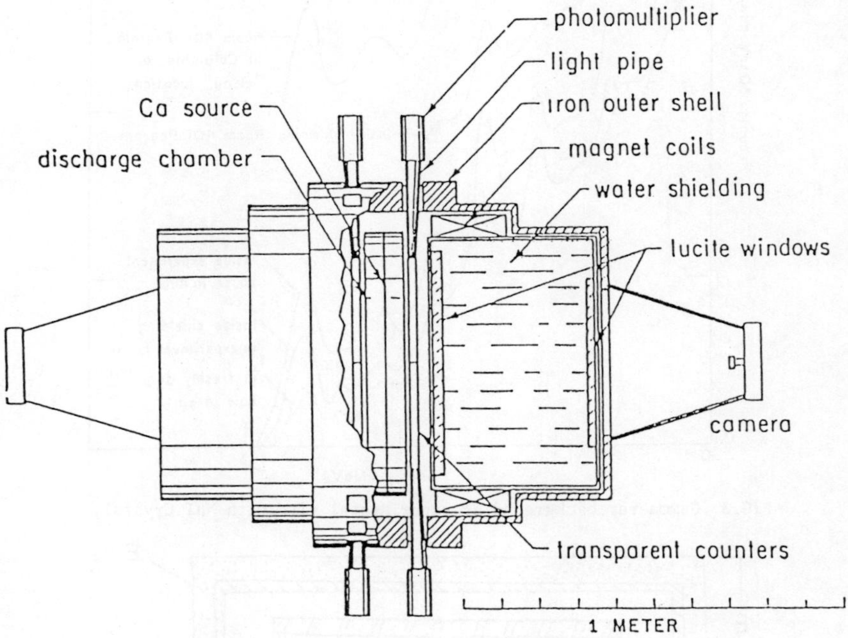

FIG 2 The cutaway drawing of DBD apparatus.

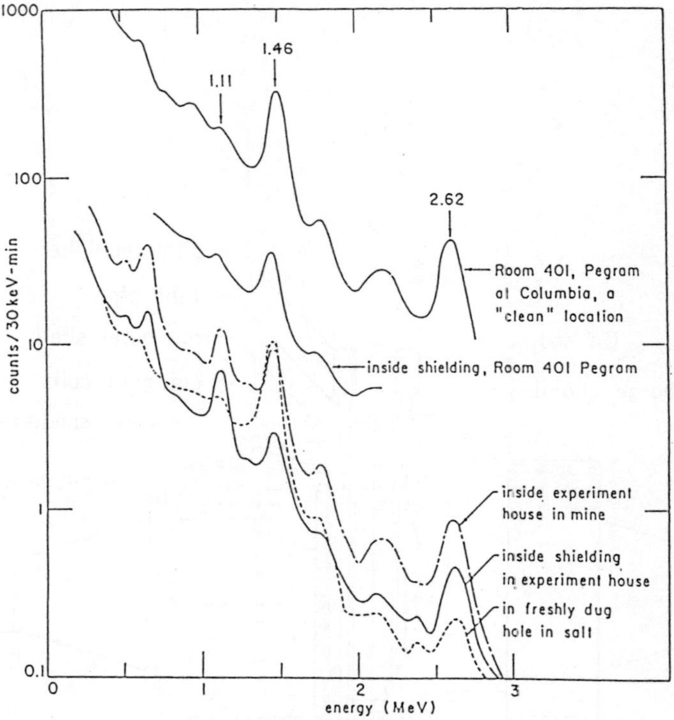

FIG.3 Gamma-ray background at experimental site with NaI Crystal.

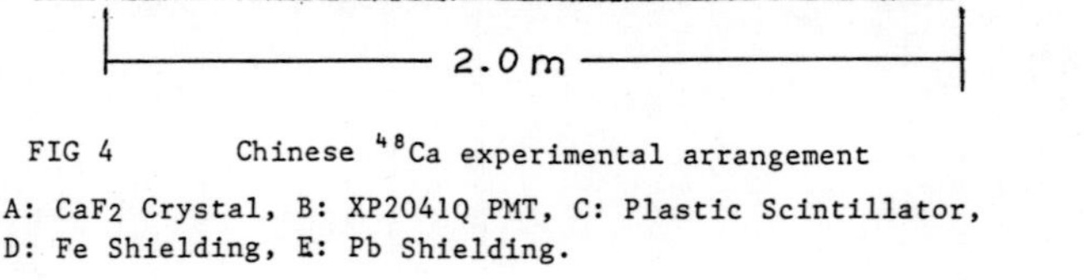

FIG 4 Chinese ^{48}Ca experimental arrangement

A: CaF_2 Crystal, B: XP2041Q PMT, C: Plastic Scintillator, D: Fe Shielding, E: Pb Shielding.

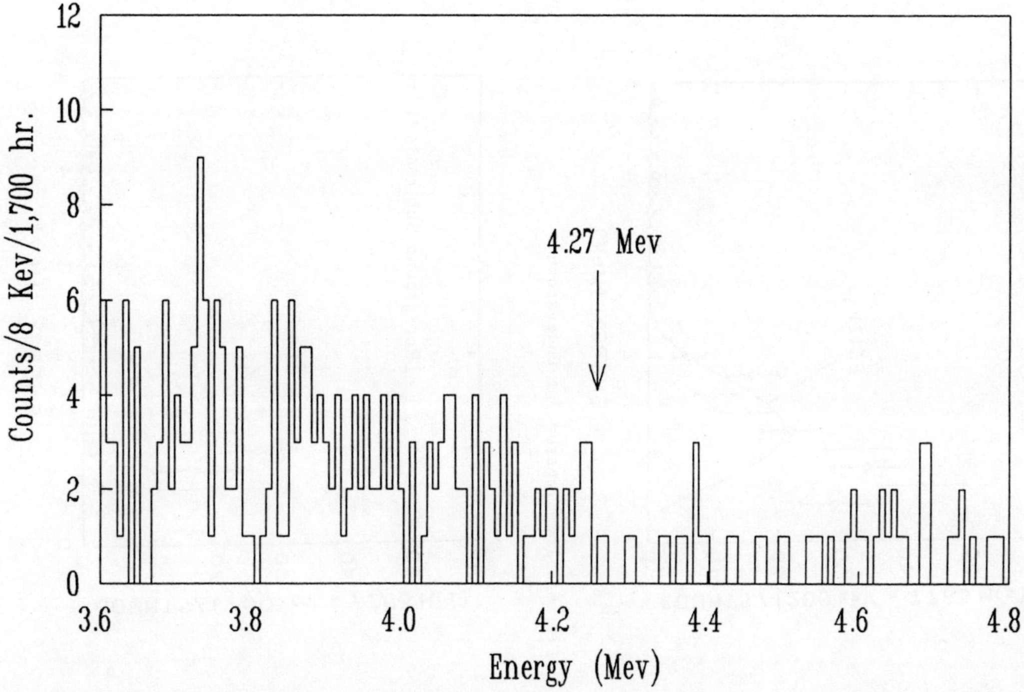

Fig. 5 The Energy Spectrum near 4.27 MeV.

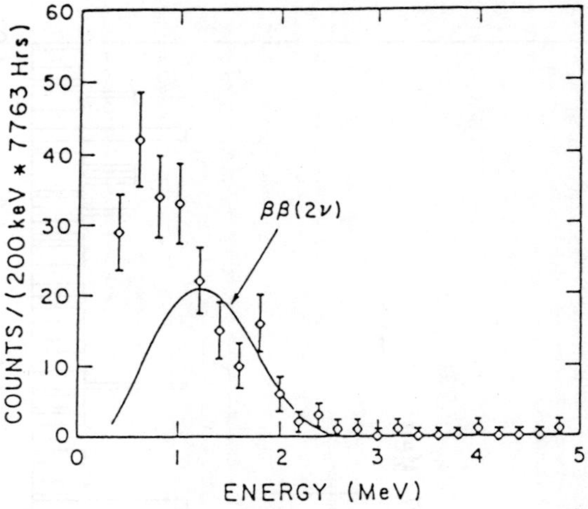

Fig. 6A The Electron Sum Underground.

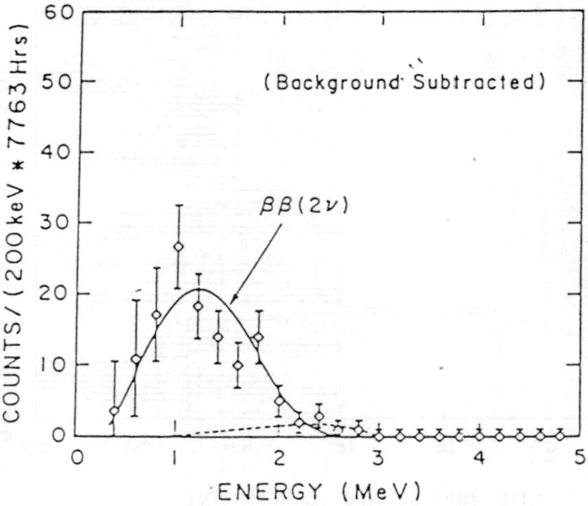

Fig. 6B Spectrum of DBD accumulated with TPC 2.

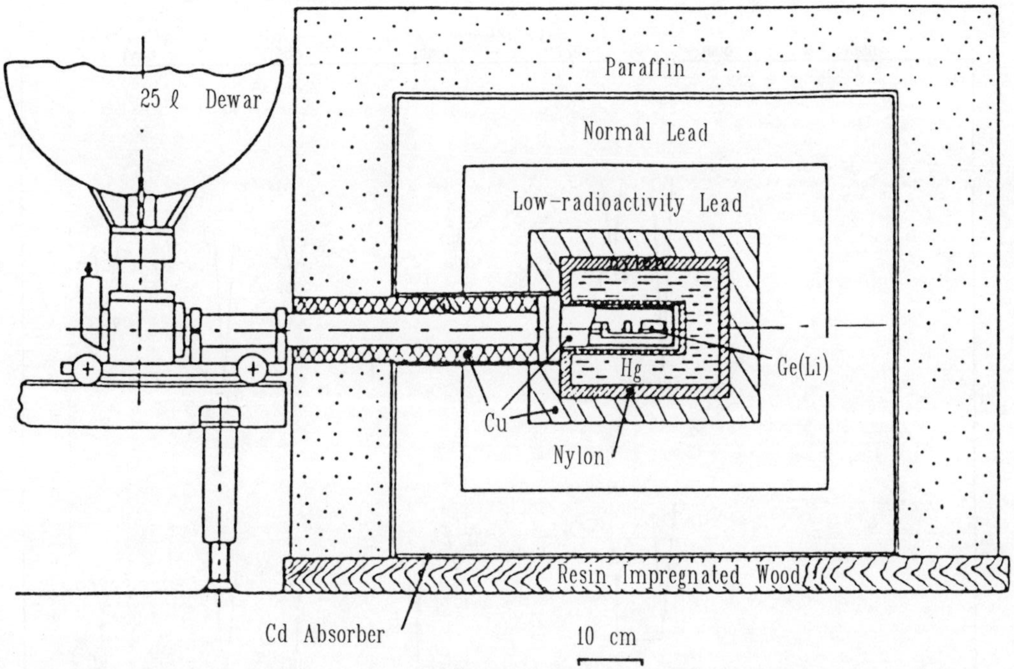

Fig. 7 Experimental Set-up for ^{76}Ge Sum-energy spectrum.

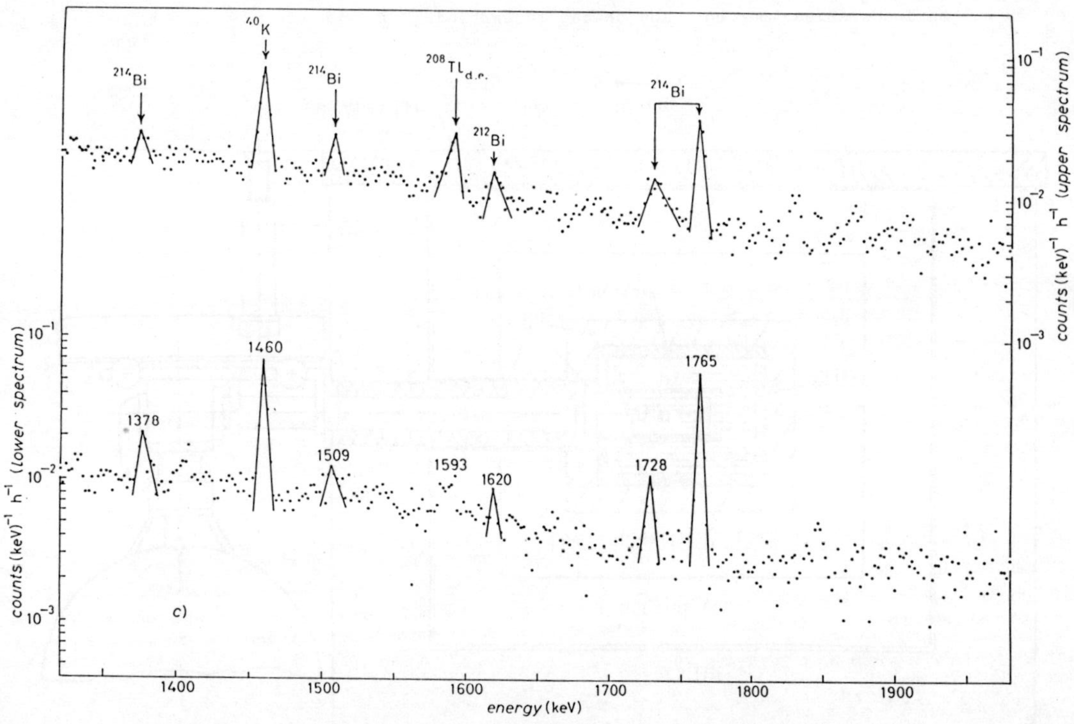

Fig. 8 Background Counts even with Low Radiation Environment

REFERENCES

Dub-59 E.I. Dubrokhotov et.al. J.Exptl. Theoret. Phys. (U.S.S.R.) 36(9)54(1959)

Wu-83 in "Science Underground" edited by M.M. Mieto et.al. (A.I.P.) New York, 1983. p. 374

Ki-83 T. Kirsten in "Science Underground", edited by M.M. Nieto et.al. (A.I.P., New York, 1983) p 396

Ba-67 R.K. Bardin et.al. Phys. Lett. 26B (1967) 112

Fi-67 E. Fiorini et.al. Phys Lett. 26B (1967) 602

Cl-75 B.T. Cleveland et.al. Rev. Lett. 35, 737 (1975)

(1) E. der Mateosian and M. Goldhaber, Phys. Rev. 146, 810 (1966)

(2) See above (Ba-67)

(3) Academia Sinica, Beijing, China (1988)

Mo-87 S.R. Elliott et.al Irvine Preprint UCI 87-34 and Phys. Rev. Lett. (to be published)

(9) " " " " "

(10) (Ki-83)

(11) (Ki-83)

(12) (He-75) E.W. Hennecke, O.K. Manual and D.D. Sabu Phys. Rev. C11, 1378, (1975)

Fi-73 E. Fiorini et.al. Nuovo Cimento A13, 747 (1973)

Av-85 F.T. Avignone et.al. Phys. Rev. Lett. 54, 2309 (1985)

Ca-87 D. Caldwell et.al. "In Telemark IV". edited by V. Barger and M. Marshak (1987)

So-87 Proceding of Hamburg Conf. on Leptons, Photons Interactions at H.E. July C.P. No-27-31-07-88)

Ta-66 N. Takaoka and K. Ogata, Z. Naturforsch 21A, 84 (1966)

Photograph Identification

1st row from left to right: Roberson Bowman Gould Vanderleeden Vogel Garcia Boehm Wolfenstein Krane Vuilleumier Stech Eichler Henley

2nd row: Lee Iqbal Lipson Barett G.Zacek Steigman Herczeg Rosenblum Winther Miller Alder Kankeleit Nagle

3rd row: Mak Brockmeier Imel Varnell Bonvin Weidenmüller unknown V. Zacek Rosen Khriplovich Hamilton de Boer Blumberg Temmer

4th row: Ramsay Henrikson Hatch Wong van Bladel Mampe Nykreim Adelberger Kayser Becker Robertson Wark Thomas Leslie

List of Participants

Professor Eric G. Adelberger
Nuclear Physics Lab G1-10
University of Washington
Seattle, WA 98195

Professor Kurt Alder
Universität Basel
CH-4056 Basel
Switzerland

Professor R. Barrett
Department of Physics, Bld. 81
University of Arizona
Tucson, AZ 85721

Dr. Hans Becker
Physics Department
Caltech
Pasadena, CA 91125

Dr. Herbert Blumberg
AKZO/ENKA AG
Enka Glanzstoff A.G.
Wuppertal, Germany

Professor Felix Boehm
Physics Department
Caltech
Pasadena, CA 91125

Dr. Emmanuel Bonvin
Physics Department
Caltech
Pasadena, CA 91125

Dr. David Bowman
Los Alamos National Laboratory
H-846
Los Alamos, NM 87545

Dr. Richard Brockmeier
Department of Physics
Hope College
Holland, MI 49423

Dr. Nim Kwan Cheung
Bellcore, Room 3Z-219
331 Newman Spring Rd.
Red Bank NJ 07701-5699

Professor Jorrit deBoer
Sektion Physik
Universität München
D-8046 Garching
West Germany

Professor Jorg Eichler
Hahn-Meitner-Institute
D-1000 Berlin 39
West Germany

Dr. Albert Forster
Inno Vision Medical Inc.
2361 McGaw
Irvine, CA 92714

Dr. James Gimlett
Bell Communications Research
Box 7020
Redbank, NJ 07701

Dr. Chris Gould
NC State University
Physics Department
Box 8202
Raleigh NC 27695

Dr. Dennis Hamilton
University of Sussex
College of Math., and Physics
Brighton, BN1 9QH
United Kingdom

Dr. Shahram Hatamian
Physics Department
Caltech
Pasadena, CA 91125

Dr.Eastman Nibley Hatch
Dept. of Physics
UMC 4415
Utah State University
Logan, UT 84321

Bridget Hay O'Callaghan
Physics Department
Caltech
Pasadena, CA 91125

Professor E. M. Henley
Physics Department
University of Washington
Seattle, WA 98195

Mr. Herb Henrikson
Physics Department
Caltech
Pasadena, CA 91125

Dr. Peter Herczeg
LANL Group T5
MS B283
Los Alamos, NM 87545

Mr. David Imel
Physics Department
Caltech
Pasadena, CA 91125

Dr. Muhammed Zafar Iqbal
Bell Communications Research
331 Newman Springs Road
Box 7020
Redbank, NJ 07701

Professor E. Kankeleit
Institut für Technische Kernphysik
der Technischen Hochschule Darmsta
61 Darmstadt
Schlossgartenstrasse 9
WEST GERMANY

Professor Boris Kayser
Division of Physics
NSF
Washington, DC 20550

Mr. Ken Karn
Physics Department
Caltech
Pasadena, CA 91125

Professor I. B. Khriplovich
Institute of Nuclear Physics
Academy of Sciences USSR
Novosibirsk 630090
USSR

Professor K. Krane
Dept. of Physics
Oregon State Univ.
Corvalis, OR 97331-6507

Dr. Heemin Kwon
Research Laboratories
Kodak Park
Eastman Kodak Company
Rochester, NY 14650

Dr. Paul Lung Sang Lee
Department of Physics and Astronomy
California State University, Northridge
18111 Nordhoff Street,
Northridge, CA 91330

Dr. J. Leslie
Queen's University
Physics Department
Kingston, ONT K7M 2S5
Canada

Dr. Edward D. Lipson
Department of Physics
Syracuse University
Syracuse, NY 13210

Dr. H. B. Mak
Queen's University
Physics Department
Kingston, ONT K7M 2S5
Canada

Dr. Walter Mampe
Inst. Laue-Langevin
B.P. 156 X
F-38042 Grenoble Cedex
Grenoble, France

Dr. James Miller
Boston University
590 Commonwealth Ave
Boston, MA 02215

Prof. Steven A. Moszkowski
Department of Physics
UCLA
405 Hilgard Ave
Los Angeles, CA 90024

Professor Darragh Nagle
MP-DO
MS H830
Los Alamos National Lab.
Los Alamos, NM 87545

Dr. W.D. Ramsay
TRIUMF
4004 Wesbrook Mall
Vancouver, B.C. V6T 2A3

Dr. N. Russell Roberson
Department of Physics
Duke University
Durham, NC 27706

Dr. Hamish Robertson
Los Alamos National Lab.
MS-D449
Los Alamos, NM 87545

Professor S. Peter Rosen
TOD
MS B288
Los Alamos National Lab.
Los Alamos, NM 87545

Dr. Stephen Rosenblum
Varian Associates
MS K307
611 Hansen Way
Palo Alto, CA 94303

Professor B. Stech
Institut für Theoretische Physik
Universität Heidelberg
D-6900 Heidelberg
West Germany

Professor Gary Steigman
Physics Department
Ohio State University
174 West 18th Av.
Columbus, OH 43210

Professor Valentine Telegdi
Eidg. Technische Hochschule
Hönggerberg
CH-8093 Zürich
Switzerland

Professor Georges M. Temmer
Rutgers University
Dept. of Physics, Box 849
Piscataway, NJ 08854

Dr. James Thomas
L-397
LLNL
P.O. Box 808
Livermore, CA 94550

Dr. J.C. Vanderleeden
4208-62 Street
Camrose, Alberta
Canada, T4V2W1

Dr. Larry Varnell
Jet Propulsion Laboratory
169-327
Pasadena, CA 91109

Dr. Petr Vogel
Physics Department
Caltech
Pasadena, CA 91125

Professor Jean-Luc Vuilleumier
Inst. de Physique
Universite de Neuchatel
1 Rue A.-L. Breguet
CH-2000 Neuchatel
Switzerland

Dr. Keh-Chung Wang
Rockwell International Corp.
1049 Camino Dos Rios, MS A-4
Thousand Oaks, CA 91360

Dr. David Wark
MS-D449
Los Alamos National Laboratory
Los Alamos, NM 87545

Professor Hans Weidenmüller
Max Planck Institute für Kernphysik
Postfach 10 39 80
D-6900 Heidelberg 1
West Germany

Professor A. Winther
Niels Bohr Institute
Blegdamsvej 17
DK-2100 Copenhagen
Denmark

Professor Lincoln Wolfenstein
Physics Department
Carnegie-Mellon University
Pittsburgh, PA 15213

Mr. Henry Wong
Institut de Physique
A.-L. Breguet 1
CH-2000 Neuchatel
Switzerland

Professor Chien-Shiu Wu
Department of Physics
Columbia University
538 W. 120 St.
New York, NY 10027

Professor T. Yamazaki
Institute of Nuclear Study
University of Tokyo
7-3-1 Mirodi-cho
Tanashi, Tokio 188
Japan

Dr. G. Zacek
Division EPI
CERN
1211 Geneva 23
Switzerland

Dr. Victor Zacek
Division EPI
CERN
1211 Geneva 23
Switzerland

Dr. Alex Zehnder
Paul Scherrer Institute
CH-5234 Villigen
Switzerland